Computational Learning and Probabilistic Reasoning

Computational Learning and Probabilistic Reasoning

Edited by

A. Gammerman
Royal Holloway, University of London

JOHN WILEY AND SONS
Chichester · New York · Brisbane · Toronto · Singapore

Copyright © 1996 by John Wiley & Sons Ltd,
                    Baffins Lane, Chichester,
                    West Sussex PO19 1UD, England

                    National   01243 779777
                    International (+44) 1243 779777

All rights reserved.

No part of this book may be reproduced by any means,
or transmitted, or translated into a machine language
without the written permission of the publisher.

Other Wiley Editorial Offices

John Wiley & Sons, Inc., 605 Third Avenue,
New York, NY 10158-0012, USA

Jacaranda Wiley Ltd, 33 Park Road, Milton,
Queensland 4064, Australia

John Wiley & Sons (Canada) Ltd, 22 Worcester Road,
Rexdale, Ontario M9W 1L1, Canada

John Wiley & Sons (Asia) Pte Ltd, 2 Clementi Loop #02-01,
Jin Xing Distripark, Singapore 0512

British Library Cataloguing in Publication Data

A catalogue record for this book is available from the British Library

ISBN   0 471 96279 1

Produced from camera-ready copy supplied by the authors
Printed and bound in Great Britain by Bookcraft (Bath) Ltd
This book is printed on acid-free paper responsibly manufactured from sustainable forestation,
for which at least two trees are planted for each one used for paper production.

Contents

Preface	xiii
List of Figures	xvii
List of Tables	xxi
List of Contributors	xxiii

I Generalisation Principles and Learning   1

1 Structure of Statistical Learning Theory
   V. Vapnik   3
- 1.1 Function Estimation Model 3
- 1.2 Problem of Risk Minimization 4
- 1.3 Three Main Learning Problems 4
  - 1.3.1 The Problem of Pattern Recognition 4
  - 1.3.2 The Problem of Regression Estimation 5
  - 1.3.3 The Problem of Density Estimation 5
  - 1.3.4 The General Setting of the Learning Problem ... 5
- 1.4 Empirical Risk Minimization Induction Principle 6
- 1.5 Four Parts of Learning Theory 6
- 1.6 Theory of Consistency of the Learning Processes 7
  - 1.6.1 The Key Assertion of the Learning Theory 7
  - 1.6.2 The Necessary and Sufficient Conditions for Uniform Convergence 8
  - 1.6.3 Three Milestones of Learning Theory 9
- 1.7 Bounds for the Rate or Convergence of the Learning Processes   11
  - 1.7.1 The Structure of the Growth-function 12
  - 1.7.2 VC-dimension of the Set of Functions 12

	1.7.3	Distribution Independent Bounds for the Rate of Convergence of the Learning Processes	13
	1.7.4	Problem of Constructing Rigorous (distribution dependent) Bounds	15
1.8	Theory for Controlling Generalization Ability of Learning Processes		16
	1.8.1	Structural Risk Minimization Induction Principle	16
	1.8.2	Examples of the Structure for Neural Nets	18
	1.8.3	The SRM Principle in the Problem of Polynomial Regression	19
	1.8.4	Problem of Local Function Estimation	19
1.9	Theory of Constructing Learning Algorithms for the Pattern Recognition Problem		20
	1.9.1	Sigmoid Approximation of Indicator Functions and Neural Nets	21
	1.9.2	Method of Optimal Separating Hyperplane and Support Vectors Networks	22
	1.9.3	Why Can Neural Networks and Support Vectors Networks generalize?	26
References			30

2 Stochastic Complexity — an Introduction
J. Rissanen   33

2.1	General	33
2.2	Stochastic Complexity	36
2.3	Predictive Coding	39
References		41

3 MML Inference of Predictive Trees, Graphs and Nets
C.S. Wallace   43

3.1	A Critique of Computational Learning Theory		43
	3.1.1	Computational Learning Theory	44
	3.1.2	Gold's Model	44
	3.1.3	Gold's Theory	44
	3.1.4	Valiant's Model	45
	3.1.5	The Deductive Fallacy	46
3.2	Learning from Data		47
	3.2.1	Learning is NOT Deduction	47
	3.2.2	Learning as a Competition	47
	3.2.3	Explanations	48
3.3	The Message Length Criterion		49
	3.3.1	Parameter Coding	50
	3.3.2	Bayesian Parallel	50
3.4	Theoretical Results		51

		3.4.1	Sufficiency	51
		3.4.2	Avoidance of Overfitting	51
		3.4.3	Consistency	52
		3.4.4	Oracular Imitation	52
		3.4.5	Invariance	53
		3.4.6	Kullback-Liebler Distance	53
	3.5	The Kolmogarov-Chaitin Formulation		54
		3.5.1	Theoretical Results from Kolmogorov-Chaitin Form	54
	3.6	Prediction vs. Induction		55
		3.6.1	Prediction in the Kolmogarov-Chaitin Formulation	55
	3.7	Applications		56
		3.7.1	Learning the Class Structure of Populations ("unsupervised learning")	56
		3.7.2	Learning the Shape Patterns of Megalithic Stone Circles	56
		3.7.3	Learning Simple Grammars	57
		3.7.4	Learning Rules in the Form of Decision Trees	57
		3.7.5	Learning Decision Graphs	57
		3.7.6	Inferring Relationships among DNA Strings	58
		3.7.7	Learning Existence and Effects of Hidden Variables in Linear Models	58
		3.7.8	Inference of Prolog Programs	58
	3.8	Coding Techniques		58
		3.8.1	Encoding Real-Valued Data	59
		3.8.2	Coding Real Parameters	61
		3.8.3	Coding Discrete Parameters	62
		3.8.4	Prior and Progressive Codes	63
	References			64

4 Learning and Reasoning as Information Compression by Multiple Alignment, Unification and Search
J.G. Wolff   67

	4.1	Introduction		67
	4.2	Multiple Alignment Problems		68
		4.2.1	What is an "Optimal" Alignment?	68
		4.2.2	Searching for Optimal Alignments	69
	4.3	Unifications and Tags		69
	4.4	Learning as Multiple Alignment with Unification and Search		70
		4.4.1	Learning Segmental Structure	71
		4.4.2	Learning Disjunctive Classes	72
	4.5	Reasoning as Multiple Alignment with Unification and Search		74
		4.5.1	Recognition and Reasoning	74
		4.5.2	Modus Ponens	74
		4.5.3	Modus Tollens	76

	4.5.4	Abduction	77
	4.5.5	Chains of Reasoning	78
	4.5.6	Inductive Reasoning	78
	4.5.7	Analogical Reasoning	81
4.6	Conclusion		81
References			83

5 Probabilistic Association and Denotation in Machine Learning of Natural Language
P. Suppes and L. Liang — 87
- 5.1 Our Approach to Machine Learning 88
- 5.2 Problem of Denotation 89
- 5.3 Background Cognitive and Perceptual Assumptions 90
- 5.4 The General Axioms of Association and Denotation 93
- 5.5 Denotational Learning Models 94
- 5.6 Some Empirical Results 99
- 5.7 Conclusion 100
- References 100

II Causation and Model Selection — 101

6 Causation, Action, and Counterfactuals
J. Pearl — 103
- 6.1 Introduction 103
- 6.2 Action as a Local Surgery 106
  - 6.2.1 Mechanisms and Surgeries 106
  - 6.2.2 Laws vs. Facts 107
  - 6.2.3 Mechanisms and Causal Relationships 108
  - 6.2.4 Causal Ordering 108
  - 6.2.5 Imaging vs. Conditioning 109
- 6.3 Formal Underpinning 111
  - 6.3.1 Causal Theories, Actions, Causal Effect, and Identifiability 111
  - 6.3.2 Action Calculus 115
  - 6.3.3 Historical Background 118
- 6.4 Counterfactuals 118
- References 120

7 Another Semantics for Pearl's Action Calculus
V.G. Vovk — 125
- 7.1 Introduction 125
- 7.2 Main Result 128
- 7.3 Formal Action Calculus 132
- 7.4 Proof of Theorem 7.1 137

Table of Contents ix

    7.5   Discussion 141
    References 143

8 Efficient Estimation and Model Selection in Large Graphical Models
    D. Wedelin   145
    8.1   Introduction 145
    8.2   An Orthogonal Interaction Model 146
          8.2.1   Independence Properties in the Orthogonal Model . . . 148
    8.3   Efficient Parameter Estimation in a Given Undirected Model . 149
          8.3.1   The Reverse Algorithm 150
          8.3.2   An Improved Reverse Algorithm 151
    8.4   Efficient Model Selection 152
    8.5   Directed Models 154
    8.6   Computational Results 155
    8.7   Conclusion 158
    References 158

9 T-Normal Distribution on the Bayesian Belief Networks
    Yu.N. Blagoveschensky   161
    9.1   Introduction 161
    9.2   Necessary Definitions and Denotations 162
    9.3   Results and some New Problems 163
    9.4   Conclusions 166
    References 166

III Bayesian Belief Networks and Hybrid Systems 167

10 Bayesian Belief Networks with an Application in Specific Case Analysis
    C.G.G. Aitken et al.   169
    10.1 Introduction 169
    10.2 The CATCHEM Project 170
          10.2.1 Offender Variables 170
          10.2.2 Victim Variables 171
          10.2.3 Combination of Offender and Victim Variables 171
    10.3 PRESS — a Bayesian Belief Network System 171
          10.3.1 Network Construction 173
          10.3.2 System Testing and Performance 173
    10.4 Examples of Networks 174
          10.4.1 Seven Node Network 174
          10.4.2 Ten Node Network 178
          10.4.3 Networks with a Large Number of Nodes 180
    10.5 Comments 180

10.6 Discussion 182
    10.6.1 Measures of Performance 182
    10.6.2 Models in Real Time 182
    10.6.3 Interpretation of Probabilities 182
    10.6.4 Incorporation of Detective Expertise 182
10.7 Conclusion 183
References 184

11 Baysian Belief Networks and Patient Treatment
L.D. Meshalkin and E.K. Tsybulkin   185
11.1 Introduction 185
11.2 Scale of Severity of a Patient's State and Mathematical Tools . 186
11.3 Mathematical Tools Adjustment 188
11.4 Notation, Time Structure and Leading Graph 189
11.5 The Choice Between Different Trajectories of Disease
    Development as a New Problem 191
11.6 Example 191
11.7 Conclusion 195
References 196

12 A Higher Order Bayesian Neural Network for Classification and Diagnosis
A. Holst and A. Lansner   199
12.1 The Classification Task 199
12.2 The Bayesian Confidence Propagation Neural Network 202
12.3 The Diagnosis Application 205
12.4 Discussion and Conclusions 207
References 208

13 Genetic Algorithms Applied to Bayesian Networks
P. Larrañaga et al.   211
13.1 Introduction 211
13.2 Genetic Algorithms 212
13.3 Genetic Operators in Relation to the TSP 213
    13.3.1 Crossover Operators 213
    13.3.2 Mutation Operators 216
13.4 Decomposition of Bayesian Networks 216
    13.4.1 Introduction 216
    13.4.2 Description of the Experiments 217
    13.4.3 Results and Conclusions 217
13.5 Structure Learning 219
    13.5.1 Introduction and Related Work 219
    13.5.2 Searching in the Space of Network Structures 219
    13.5.3 Searching for the Best Ordering 223

13.5.4 Remarks and Conclusions	224
13.6 Fusion	225
13.6.1 Introduction	225
13.6.2 Proposed Approach	225
13.6.3 An Example	225
13.6.4 Remarks and Conclusions	227
13.7 Conclusions	227
References	227

IV Decision-Making, Optimization and Classification   235

14 Rationality, Conditional Independence and Statistical Models of Competition
J.Q. Smith and C.T.J. Allard   237

14.1 Introduction	237
14.2 Making Deductions in Statistical and Decision Problems	240
14.3 Common-Knowledge and Rationality	244
14.4 Equilibral Rationality	248
14.5 Statistical Models Structured from Evoking Rationality	251
14.6 Conclusions	256
References	256

15 Axioms for Dynamic Programming
P.P. Shenoy   259

15.1 Introduction	259
15.2 Valuation Networks and Optimization	261
15.3 The Axioms	264
15.4 A Fusion Algorithm	265
15.5 Mitten's Axioms for Dynamic Programming	270
15.6 Conclusion	271
15.7 Proofs	271
References	273

16 Mixture-Model Cluster Analysis Using the Projection Pursuit Method
S. Aïvazian   277

16.1 Mixture-Model and Cluster Analysis Problem	277
16.2 Estimating the Unknown Parameters of the Mixture-Model	278
16.3 Problem of Estimating the Unknown Number of Classes	279
16.4 Detecting the Number of Clusters by Means of Projection Pursuit	281
16.4.1 Projection Pursuit Technique	281

16.4.2 Discriminant Subspace 282
16.4.3 Projection Index Suitable for Detecting the Number of Clusters 283
16.5 Mixture-Model Cluster Analysis and Bayesian Belief Networks 284
References 285

17 A Parallel k_n-Nearest Neighbour Classifier for Estimation of Non-linear Decision Regions
A. Kovalenko   **287**

17.1 Introduction 287
17.2 Parallel Classifier 288
17.3 Signal Recognition Against a Noise Background: A Numerical Example 291
References 293

18 Extreme Values of Functionals Characterizing Stability of Statistical Decisions
A.V. Nagaev   **295**

18.1 Introduction 295
18.2 Stability within a Belief Network 296
    18.2.1 Properties of Elliptically Contoured Distributions 296
    18.2.2 Extreme Problems for the Variation Distance 298
    18.2.3 On the Stability of Predictions 301
18.3 Extreme Problems for Quantiles 303
    18.3.1 Lower Extreme Values for Quantiles of the Greater Level 304
    18.3.2 Lower Extreme Values for Quantiles of the Smallest Level 305
    18.3.3 Upper Extreme Values 306
References 307

Index   **309**

Preface

Advances in Probabilistic Reasoning and Computational Learning

This book is devoted to two interrelated techniques for solving some important problems in machine intelligence and pattern recognition, namely **probabilistic reasoning** and **computational learning**. These techniques were discussed at a conference on Applied Decision Technologies, ADT '95, held in London in April 1995. The conference had four major streams, and I had the pleasure of leading the Computational Learning and Probabilistic Reasoning stream. The conference attracted a wide audience, and papers were published in the Proceedings of ADT '95. Given the considerable interest shown in this area of research, it has been decided to develop some papers and publish them in a separate volume, together with several extra chapters written by the leading experts in the field especially for this book.

The book is divided into four parts. The first part describes several new inductive principles and techniques used in computational learning.

Vladimir Vapnik in *Structure of Statistical Learning Theory*, describes the *empirical risk minimization induction* principle in brief, and sets up four major parts of a statistical learning theory. The chapter also considers the consistency of the learning process, bounds for the rate of convergence of the learning processes, control of the generalization ability and the *structural risk minimization induction* principle, and the theory of constructing the learning algorithms for pattern recognition.

Jorma Rissanen in *Stochastic Complexity — an Introduction* discusses a problem of model selection, and points out the difficulties associated with the use of Kolmogorov complexity as a basis for inductive inference. Then he introduces the *minimum description length* principle, and presents a formula for the stochastic complexity for the model classes. This allows one to calculate a predictive code length to approximate the stochastic complexity.

Chris Wallace in *MML Inference of Predictive Trees, Graphs and Nets*, reviews different models of learning. The chapter starts with a critique of computational learning theory, then considers the learning from data, the message length criterion, and presents some theoretical results with a number of applications.

Gerry Wolff in *Learning and Reasoning as Information Compression by Multiple Alignment, Unification and Search*, aims to develop further his "computing as compression" conjecture, *i.e.* the idea that different types of reasoning may be understood as information compression by pattern matching, unification and metrics-guided search.

Patrick Suppes and **Lin Liang** in *Probabilistic Association and Denotation in Machine Learning of Natural Language*, describe two denotational learning models. They formulate two general axioms of association and denotation, and present some empirical results using those models.

Part two contains chapters on Causal Probabilistic Models, *i.e.* the graphical probabilistic models that exploit the independence relationships presented in the graph. What is especially interesting is that these graphs can also be used as a formal language for the communicating and processing of **causal** information. It also has chapters on model selection and application of Bayesian networks to multivariate statistical analysis.

Judea Pearl in the chapter *Causation, Action, and Counterfactuals*, summarizes the recent advances in causal reasoning and causal graphs. It considers actions as a form of surgeries and the causal relation as an abbreviation for the surgery process. It also connects actions to Lewis' theory of counterfactuals using imaging to replace conditioning.

Vladimir Vovk, in a closely related chapter entitled *Another Semantics for Pearl's action calculus*, proposes a new definition of indentifiability of causal effects. It discards an assumption that the manipulated variables are governed by stable stochastic mechanisms.

Dag Wedelin in *Efficient estimation and model selection in large graphical models*, develops a method that determines the interaction structure in a multidimensional binary sample. To find the best model he uses a heuristic search where the structure is determined incrementally, and some causal directions could be established on the structure.

Yuri Blagoveschensky in *T-normal Distribution on the Bayesian Belief Networks*, considers how much help one can get out of the graphical (Bayesian) models to solve some problems in multivariate statistical analysis, especially when the distribution is not normal and the number of parameters is large.

The third part on Bayesian Belief Networks and hybrid systems has four chapters. It includes case studies and description of several hybrid systems.

Colin Aitken et al. in *Bayesian Belief Networks with an Application in Specific Case Analysis*, describe an interesting application of Bayesian Belief Networks to statistical modelling of offender profiling. It describes the problem, and four different statistical techniques that have been used in predicting the offender profiles. It has been shown that Bayesian networks perform on the level of the best available techniques, but have the advantage of being simple and flexible.

Lev Meshalkin and **Eduard Tsybulkin** in *Bayesian Belief Networks and Patient Treatment*, discuss the problem of the prediction of a disease development. They introduced a hierarchical sequence of graphs and a leading (in time) initial graph that makes it easier to use Bayesian networks for prediction.

The other two chapters in this part deal with the so-called "hybrid" approach using neural networks and genetic algorithms to enhance the performance of Bayesian networks.

Anders Holst and **Anders Lansner** in *A Higher Order Bayesian Neural Network for Classification and Diagnosis*, use data to establish a causal structure of the problem, and then to estimate conditional probabilities of the classes. The Bayesian confidence propagation neural network technique has been used to diagnosis problems in telephone exchange computers.

Pedro Larrañaga et al. in *Genetic Algorithms applied to Bayesian Networks*, show how genetic algorithms can be applied to the problem of optimal decomposition, and some other learning problems.

The final part, comprising five chapters, describes some related theoretical work in the field of probabilistic reasoning.

Jim Smith and **Crispin Allard** in *Rationality, Conditional Independence and Statistical Models of Competition*, demonstrate that conditional independence can be used on its own, *i.e.* without probabilities to define a reasoning system. The main ideas of game theory are expressed in terms of conditional independence that can determine models of rational players.

Prankash Shenoy in *Axioms for Dynamic Programming*, describes a formal framework, called a valuation network, for representing and solving discrete optimization problems. It can be shown that the local computational algorithms in Bayesian Belief Networks are just instances of the dynamic programming method.

Sergei Aïvazian in *Mixture-Model Cluster Analysis Using Projection Pursuit Method*, develops an estimation procedure for the standard multivariate normal mixture model using the EM-algorithm. It also connects mixture-model cluster analysis with Bayesian Belief Networks.

Andrei Kovalenko in *A Parallel k-Nearest Neighbour Classifier for Estimation of Non-Linear Decision Regions*, considers a nonparametic classifier. The algorithm has a linear and spatial complexity and has been implemented on a parallel computer system.

Alexander Nagaev in *Extreme Values of Functionals Characterizing Stability of Statistical Decisions*, discusses a number of problems in belief networks. They are mainly related to the stability of decision rules within the framework of a semiparametric model of elliptically contoured distributions.

Acknowledgements

First and foremost I would like to thank the authors whose chapters reflect so well the advances being made in computational learning and probabilistic reasoning and our referees who worked hard to meet rather tight deadlines.

I would also like to take this opportunity to thank Mark Salter of Unicom Seminars, for his effort in handling all of the organizational problems connected with this publication.

I am especially grateful to my parents and my brother Misha, for their support and encouragment in all my work.

Finally, my wife Sue has aided and helped me throughout and our children, Yasha and Anya, put up with me while I was engaged in this work. I warmly appreciate their understanding and support.

Alexander Gammerman
University of London

List of Figures

1.1	Three components of the model of learning processes.	4
1.2	Convergence to minimal possible risk $R(\alpha_0)$.	8
1.3	The optimal hyperplane is the one which has a maximal margin.	24
1.4	Examples of a pattern with labels from a US Postal Service database.	28
1.5	Labelled examples of training errors for the 2nd degree polynomial.	28
4.1	An alignment amongst five DNA sequences.	68
4.2	An alignment amongst some sentences.	72
4.3	Unified patterns from the alignment in Figure 4.2, with word class symbols as tags.	73
4.4	*Modus ponens* reasoning by alignment and unification.	75
4.5	Abductive reasoning by alignment and unification.	77
4.6	Multiple alignment in a chain of inference.	78
4.7	Extraction of a "chunk" by multiple alignment and unification.	79
4.8	An inference that someone other than Socrates, Xanthippe and Sappho is mortal.	80
4.9	A geometric analogy problem.	82
4.10	Multiple alignment in a geometric analogy problem.	83
5.1	Initial part of the tree for language L_3.	96
5.2	Mean denotational learning curves for English robotic corpus.	97
5.3	Mean denotational learning curves for 60 physics word problems in English.	98
5.4	Rate of denotational learning with different values.	99
6.1	A diagram representing a causal theory on five variables.	112
7.1	Causal relations between the result A of a test, the doctor's decision D on whether to prescribe a drug, the outcome B of the treatment, and the unobservable state U of the patient.	126

7.2	Causal relations between smoking D, tar deposits A, lung cancer B, and the unobservable genotype U which may be carcinogenic and may involve inborn craving for nicotine.	127
7.3	Inference tree for $\hat{D}A \to B$ and $\hat{D} \to B$ in the example of Figure 7.1	134
7.4	Inference tree for $\hat{D} \to B$ in the example of Figure 7.2	135
8.1	(a) The Alarm structure. (b) Results for 10000 and 2000 cases.	156
8.2	(a) The big121 problem. (b) Result for 10000 cases.	157
9.1	M-graph given formally (**a**) and expertly (**b**).	163
10.1	Diagrammatic representation that attribute A causes or influences attribute B.	172
10.2	Seven node network showing relationship among three offender characteristics and four victim and crime characteristics.	174
10.3	Ten node network showing relationship among four offender characteristics and six victim and crime characteristics.	179
11.1	The leading graph for lacal grades of disease.	190
11.2	The fragment of the graph of relations for the third day of diphtheria.	195
12.1	Starting from a dependency graph with no loops...	202
12.2	A one-layer BCPNN with three binary input attributes and four output classes.	202
12.3	A multi-layer BCPNN with a hidden layer consisting of one first-order column c and one second-order complex column ab.	203
12.4	A multi-layer BCPNN with a hidden layer consisting of units for the input attributes, plus an overlapping column for a dependency between a and b.	204
12.5	The dependencies accounted for by the two versions of multi-layer BCPNNs.	207
13.1	Pseudo code of a genetic algorithm.	212
13.2	With order assumption: Crossing over two BN structures.	220
13.3	With order assumption: Mutating a BN structure.	220
13.4	Without order assumption: The crossover operator is not a closed operator.	221
13.5	Without order assumption: The mutation operator is not a closed operator.	222
13.6	The ALARM network structure.	222
13.7	BNs coming from two different authors.	226
13.8	BN obtained after fusion.	226

List of Figures

14.1 A simple sufficiency theorem reduction. 243
14.2 Common-knowledge influence diagrams in a simple game. . . . 246
14.3 Using common-knowledge to deduce implied conditional independence. 247
14.4 Reduction using equilibral rationality. 249
14.5 Another example of reduction. 250
14.6 An observer's deductions assuming rationality. 252
14.7 The market at time t. 253
14.8 Π_1's influence diagram of a game played on a partially segmented market. 254

15.1 A valuation network for the optimization problem 262
15.2 The fusion algorithm for the optimization problem using deletion sequence CDABE 268

List of Tables

1.1	 29
8.1	A comparison between the different estimation algorithms. 153
9.1	 165
10.1	Comparison between some of the conditional probabilities. 176
10.2	Prediction accuracy as a percentage of the BBN for a seven node network. 176
10.3	The three most important attributes for prediction of particular offender attributes when no information about the victim is available. 177
10.4	The first three most important attributes when a female victim of age 8–12 has been found. 178
10.5	Prediction accuracy as a percentage of the BBN for a ten node network. 180
11.1	Frame of the scale of patient state severity, therapy and typical time (in order of severity increase). 187
11.2	Main current specific complications of combined diphtheria and their numerical characteristics (in %). 192
12.1	Classification results. 206
13.1	Optimal decomposition. Results presented in [46, 51]. 218
13.2	Optimal decomposition. Best results (Graph Sparse). 218
13.3	Optimal decomposition. Best results (Graph Dense). 219
13.4	Structure learning. Best results. 223
13.5	Structure learning. Best evaluations obtained with the different combinations of genetic operators. Population size 50. 224
15.1	Factors of the Objective Function, $\phi_1, \phi_2,$ and ϕ_3 261
15.2	Numerical computations in the fusion algorithm for the optimization problem 267

18.1 The range of the absolute value of p-quantile for small p 306

List of Contributors

Aitken C.G.G., Department of Mathematics and Statistics, The University of Edinburgh, Edinburgh, EH9 3JZ UK.

Aïvazian S., Institute of Russian Academy of Sciences, Central Economics & Mathematics Institute, ul. Krasikova 32, Moscow 117418 Russia.

Albizuri X., University of the Basque Country, Dept. of Computer Science and AI, PO Box 649 E-20080 San Sebastian, Spain.

Allard C., 22 Mycanae Road, Blackheath, London SE3 7SG UK.

Bailey D., Derbyshire Constabulary Headquarters, Butterley Hall, Ripley, Derbyshire DE5 3RS UK.

Blagoveschensky Y.N., Moscow State University, Soil Science Faculty, Moscow 119899 Russia.

Connolly T., Biomathematics and Statistics Scotland, Scottish Crop Research Institute, Invergowrie, Dundee DD2 5DA UK.

D'Anjou A., University of the Basque Country, Dept. of Computer Science and AI, PO Box 649 E-20080 San Sebastian, Spain.

Gammerman A., Royal Holloway, University of London, Egham, Surrey TW20 0EX UK.

Gordon R., Derbyshire Constabulary Headquarters, Butterley Hall, Ripley, Derbyshire DE5 3RS UK.

Graña M., University of the Basque Country, Dept. of Computer Science and AI, PO Box 649 E-20080 San Sebastian, Spain.

Holst A., Royal Institute of Technology, Studies of Artifical Neural Systems, Department of Numerical Analysis and Computing Science, S-100 44 Stockholm, Sweden.

Kovalenko A, Institute of Russian Academy of Sciences, Central Economics & Mathematics Institute, ul. Krasikova 32, Moscow 117418 Russia.

Kuijpers C.M.H., University of the Basque Country, Dept. of Computer Science and AI, PO Box 649 E-20080 San Sebastian, Spain.

Liang L., Stanford University, Stanford CA 94305, USA.

Lansner A., Royal Institute of Technology, Studies of Artifical Neural Systems, Department of Numerical Analysis and Computing Science, S-100 44 Stockholm, Sweden.

Larrañaga P., University of the Basque Country, Dept. of Computer Science and AI, PO Box 649 E-20080 San Sebastián, Spain.

Suppes P., Stanford University, Stanford CA 94305, USA.

Lozano J.A., University of the Basque Country, Dept. of Computer Science and AI, PO Box 649 E-20080 San Sebastian, Spain.

Meshalkin L.D., Institute of the Russian Academy of Sciences, Central Economics and Mathematics Institute, ul Krasikova, 32 Moscow, 117418 Russia.

Murga R.H., University of the Basque Country, Dept. of Computer Science and AI, PO Box 649 E-20080 San Sebastian, Spain.

Nagaev A., Kopernikus University, Department of Mathematics, 87-100 Torun, ul. Chopina 12/18, Poland.

Oldfield R., Police Research Group, Home Office, Queen Anne's Gate, London SW1H 9AT UK.

Pearl J., University of California at Los Angeles, Cognitive Systems Laboratory, Dept. of Computer Science CA 90024, USA.

Rissanen J., IBM Almaden Research Centre, K52/802, 650 Harry Road, San Jose CA 95120, USA.

Shenoy P.P., The University of Kansas, The School of Business, 330 Summerfield Hall, Lawrence, Kansas 66045-2003, USA.

Smith J.Q., University of Warwick, Department of Statistics, Coventry CV4 7AL UK.

Suppes P., Stanford University, Stanford CA 94305, USA.

Torrealdea F.J., University of the Basque Country, Dept. of Computer Science and AI, PO Box 649 E-20080 San Sebastian, Spain.

Tsybulkin E.K., Pediatric Medical Institute, St. Petersburg, Russia.

Vapnik V., AT&T Bell Labs, Crawfords Road, Holmdel NJ 07733 USA.

List of Contributors

Vovk V., Centre for Advanced Study in Behavioural Sciences, 202 Junipero Serra Blvd., Stanford CA 94305 USA.

Wallace C.S., Monash University, Department of Computer Science, Clayton 3168, Victoria, Australia.

Wedelin D., Department of Computer Science, Chalmers Institute of Technology S-412 96 Gotheburg, Sweden.

Wolff J.G., University of Wales, School of Electronic Engineering & Computer Systems, Dean Street, Bangor, Gwynedd LL57 1UT UK.

Yurramendi Y., University of the Basque Country, Dept. of Computer Science and AI, PO Box 649 E-20080 San Sebastian, Spain.

Zhang G., Department of Computer and Electrical Engineering, Heriot-Watt University, Edinburgh EH14 4AS UK.

Part I
Generalisation Principles and Learning

1

Structure of Statistical Learning Theory

V. Vapnik

The structure of a learning theory as well as the main concepts of the theory are described.

1.1 Function Estimation Model

The general model of learning from examples can be described through three components (Figure 1.1):

1. A generator of random vectors x, drawn independently from a fixed but unknown distribution $P(x)$.

2. A supervisor that returns an output vector y for every input vector x, according to a conditional distribution function[1] $P(y|x)$, also fixed but unknown.

3. A learning machine capable of implementing a set of functions $f(x, \alpha)$, $\alpha \in \Lambda$.

The problem of learning is that of choosing from the given set of functions $f(x, \alpha), \alpha \in \Lambda$, the one which best approximates the supervisor's response. The selection is based on a training set of ℓ independent identically distributed (i.i.d.) observations drawn according to $P(x, y) = P(x)P(y|x)$:

$$(x_1, y_1), \ldots, (x_\ell, y_\ell). \tag{1.1}$$

---

[1]This is the general case which includes a case where the supervisor uses a function $y = f(x)$.

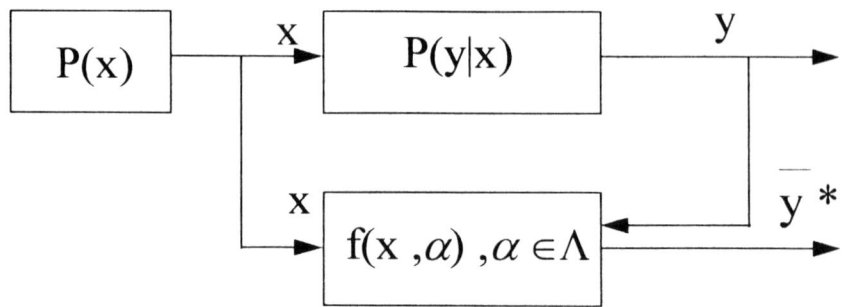

Figure 1.1: Three components of the model of learning processes.

1.2 Problem of Risk Minimization

To choose the best available approximation to the supervisor's response, one measures the *loss* or discrepancy $L(y, f(x, \alpha))$ between the response y of the supervisor to a given input x and the response $f(x, \alpha)$ provided by the learning machine. Consider the expected value of the loss, given by the *risk functional*

$$R(\alpha) = \int L(y, f(x, \alpha)) dP(x, y). \qquad (1.2)$$

The goal is to find the function $f(x, \alpha_0)$ which minimizes the risk functional $R(\alpha)$ (over the class of functions $f(x, \alpha)$, $\alpha \in \Lambda$) in the situation where the joint probability distribution $P(x, y)$ is unknown and the only available information is contained in the training set (1.1).

1.3 Three Main Learning Problems

This formulation of the learning problem is rather broad. It encompasses many specific problems. Consider the main ones: the problems of pattern recognition, regression estimation, and density estimation [10].

1.3.1 The Problem of Pattern Recognition

Let the supervisor's output y take only two values $y = \{0, 1\}$ and let $f(x, \alpha)$, $\alpha \in \Lambda$, be a set of *indicator* functions (functions which takes only two values zero and one). Consider the following loss-function:

$$L(y, f(x, \alpha)) = \begin{cases} 0 & \text{if } y = f(x, \alpha) \\ 1 & \text{if } y \neq f(x, \alpha). \end{cases} \qquad (1.3)$$

For this loss function, the functional (1.2) determines the probability of the different answers: the answers y given by supervisor and the answers given by indicator function $f(x, \alpha)$. (We call the case of different answers a *classification error*). The problem, therefore, is to find the function which minimizes the probability of classification errors when probability measure $P(x, y)$ is unknown, but the data (1.1) are given.

1.3.2 The Problem of Regression Estimation

Let the supervisor's answer y be a real value, and let $f(x, \alpha), \alpha \in \Lambda$, be a set of real functions which contains the *regression function*

$$f(x, \alpha_0) = \int y dP(y|x).$$

It is known that the regression function is the one which minimizes the functional (1.2) with the following loss-function

$$L(y, f(x, \alpha)) = (y - f(x, \alpha))^2. \tag{1.4}$$

Thus the problem of regression estimation is the problem of minimizing the risk functional (1.2) with the loss function (1.4) in the situation where the probability measure $P(x, y)$ is unknown but the data (1.1) are given.

1.3.3 The Problem of Density Estimation

Finally, consider the problem of density estimation from the set of densities $p(x, \alpha), \alpha \in \Lambda$. For this problem we consider the following loss-function

$$L(p(x, \alpha)) = -\log p(x, \alpha). \tag{1.5}$$

It is known that desired density minimizes the risk functional (1.2) with the loss-function (1.5). Thus, again, to estimate the density from the data one has to minimize the risk-functional under the condition where the corresponding probability measure $P(x)$ is unknown but i.i.d. data

$$x_1, \ldots, x_n$$

are given.

1.3.4 The General Setting of the Learning Problem

The general setting of the learning problem can be described as follows. Let the probability measure $P(z)$ be defined on the space Z. Consider the set of functions $Q(z, \alpha), \alpha \in \Lambda$. The goal is to minimize the risk functional

$$R(\alpha) = \int Q(z, \alpha) dP(z), \quad \alpha \in \Lambda \tag{1.6}$$

if probability measure $P(z)$ is unknown but an i.i.d. sample

$$z_1, \ldots, z_\ell \tag{1.7}$$

is given.

The learning problems considered above are particular cases of this general problem of *minimizing the risk functional* (1.6) *on the basis of empirical data* (1.7), where z describes a pair (x, y) and $Q(z, \alpha)$ is the specific loss function (for example, one of (1.3, 1.4, 1.5)). Below we will describe the result obtained for the general statement of the problem. To apply it for specific problems one has to substitute the corresponding loss-functions in the formulas obtained.

1.4 Empirical Risk Minimization Induction Principle

To minimize the risk functional (1.6), for an unknown probability measure $P(z)$ the following induction principle is usually used:

The risk functional $R(\alpha)$ is replaced by the *empirical risk* functional

$$R_{emp}(\alpha) = \frac{1}{\ell} \sum_{i=1}^{\ell} Q(z, \alpha) \qquad (1.8)$$

constructed on the basis of the training set (1.7).

The principle is to approximate the function $Q(z, \alpha_0)$ which minimizes risk (1.6) by the function $Q(z, \alpha_\ell)$ minimizing empirical risk (1.8). This principle is called the Empirical Risk Minimization induction principle (ERM principle).

The ERM principle is quite general. The classical methods for solution of a specific learning problem such as the least squares method in the problem of regression estimation or the maximum likelihood method in the problem of density estimation are realizations of the ERM principle for the specific loss functions considered above.

1.5 Four Parts of Learning Theory

Learning theory has to address the following four questions [11]:

1. *What are (necessary and sufficient) conditions for consistency of the learning process based on the ERM principle?*

   This means to specify the necessary and sufficient conditions for convergence in probability[2] both the values of risks $R(\alpha_\ell)$ for the functions $Q(z, \alpha_\ell)$ minimizing the empirical risk $R_{emp}(\alpha))$ to the minimal possible value of the risk $R(\alpha_0)$ and value of obtained empirical risks $R_{emp}(\alpha_\ell)$

---

[2] Convergence in probability of values $R(\alpha_\ell)$ means that for any $\varepsilon > 0$ and for any $\eta > 0$ there exists a number $\ell_0 = \ell_0(\varepsilon, \eta)$ such, that for any $\ell > \ell_0$ with probability at least $1 - \eta$ the inequality $R(\alpha_\ell) - R(\alpha_0) < \varepsilon$ holds true.

STRUCTURE OF STATISTICAL LEARNING THEORY

to the minimal possible value of the risk [3] $R(\alpha_0)$ as the number of observations is increase (Figure 1.2):

$$R(\alpha_\ell) \xrightarrow[\ell \to \infty]{P} R(\alpha_0),$$

$$R_{emp}(\alpha_\ell) \xrightarrow[\ell \to \infty]{P} R(\alpha_0). \quad (1.9)$$

2. *How fast is the rate of convergence of the learning process?*

3. *How can one control the generalization ability of the learning process?*

4. *How can one construct algorithms that can control the generalization ability?*

The answers to these questions form the four parts of learning theory:

1. Theory of consistency of the learning processes.

2. Non-asymptotic theory of the rate of convergence of the learning processes.

3. Theory of controlling the generalization ability of the learning processes.

4. Theory of constructing the learning algorithms.

1.6 Theory of Consistency of the Learning Processes

1.6.1 The Key Assertion of the Learning Theory

The key assertion on the theory of the learning processes is the following [17]:
Let $Q(z, \alpha)$, $\alpha \in \Lambda$ be a set of totally bounded functions

$$A \leq Q(z, \alpha) \leq B.$$

Then for the ERM principle to be consistent it is necessary and sufficient that the empirical risk $R_{emp}(\alpha)$ converges *uniformly* to the actual risk $R(\alpha)$ over the set $Q(z, \alpha), \alpha \in \Lambda$.

Uniform (one-sided) convergence is defined as

$$\lim_{\ell \to \infty} \text{Prob}\{\sup_{\alpha \in \Lambda} (R(\alpha) - R_{emp}(\alpha)) > \varepsilon\} = 0, \quad \forall \varepsilon. \quad (1.10)$$

---

[3] A particular case of the classical definition of consistency (1.9) where $R(\alpha_0) = 0$ (in this case $R_{emp}(\alpha_\ell) = 0$) is used as the definition of the Probability Approximately Correct (PAC) model of the learning [9]. The restriction $R(\alpha_0) = 0$, however, is too strong. This restriction, makes it impossible to analyze the most important cases of the learning problem, namely the case where the set of functions of the learning machine does not contain the supervisor's rule and the case where data contain noise.

The results obtained in the framework of PAC model are corollary from the results of Statistical Learning Theory where $R(\alpha_0) = R_{emp}(\alpha_\ell) = 0$.

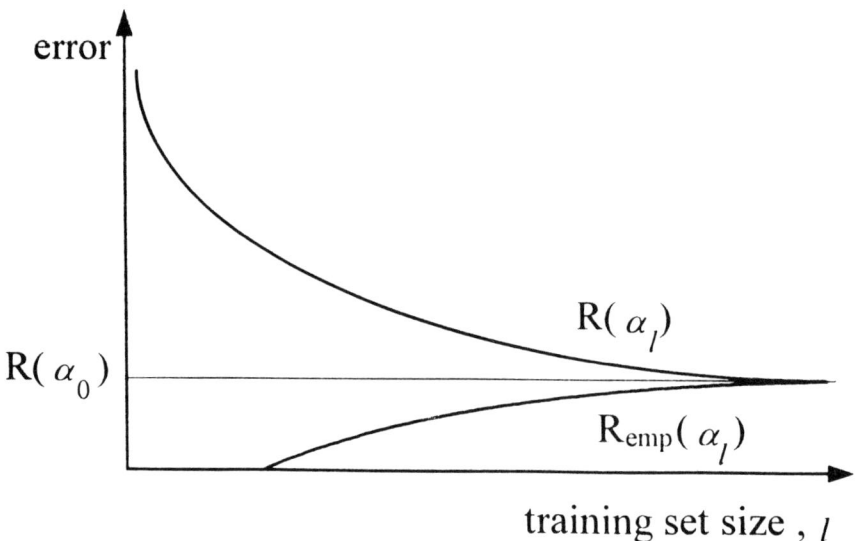

Figure 1.2: Both the values of empirical risks $R_{emp}(\alpha_\ell)$ and the values of risk for functions minimizing the empirical risk $R(\alpha_\ell)$ converge to minimal possible risk $R(\alpha_0)$.

In other words, according to the key assertion the conditions for consistency of ERM principle is equivalent to the conditions for existence of the uniform one-sided convergence (1.10).

1.6.2 The Necessary and Sufficient Conditions for Uniform Convergence

To describe the necessary and sufficient condition for existence of the uniform convergence (1.10), it is necessary to introduce a new concept which is called *the entropy of the set of functions* $Q(z, \alpha)$, $\alpha \in \Lambda$, *on the sample of size* ℓ.

Let $A \leq Q(z, \alpha) \leq B$, $\alpha \in \Lambda$, be a set of bounded loss functions. Using this set of functions and the training set (1.7) one can construct the following set of ℓ-dimensional vectors

$$q(\alpha) = (Q(z_1, \alpha), \ldots, Q(z_\ell, \alpha)), \quad \alpha \in \Lambda. \tag{1.11}$$

This set of vectors belongs to the ℓ-dimensional cube and has a finite ε-net[4] in the metric C. Let $N = N^\Lambda(\varepsilon; z_1, \ldots, z_\ell)$ be the number of elements of the minimal ε-net of this set of vectors $q(\alpha)$, $\alpha \in \Lambda$.

---
[4]The set of vectors $q(\alpha)$, $\alpha \in \Lambda$ has minimal ε-net $q(\alpha_1), \ldots, q(\alpha_N)$ if:

1. There exist $N = N^\Lambda(\varepsilon; z_1, \ldots, z_\ell)$ vectors $q(\alpha_1), \ldots, q(\alpha_N)$, such that for any vector $q(\alpha^*)$, $\alpha^* \in \Lambda$ one can find among these N vectors one $q(\alpha_r)$ which is ε-close to this

The logarithm of the (random) value $N^\Lambda(\varepsilon; z_1, \ldots, z_\ell)$

$$H^\Lambda(\varepsilon; z_1, \ldots, z_\ell) = \ln N^\Lambda(\varepsilon; z_1, \ldots, z_\ell)$$

is called the *random VC-entropy*[5] of the set of functions $A \leq Q(z, \alpha) \leq B$ on the sample $z_1, \ldots, z_\ell$. The expectation of the random VC-entropy

$$H^\Lambda(\varepsilon; \ell) = EH^\Lambda(\varepsilon; z_1, \ldots, z_\ell)$$

is called the *VC-entropy* of the set of functions $A \leq Q(z, \alpha) \leq B$, $\alpha \in \Lambda$ on the sample of the size ℓ. Here expectation is taken with respect to product-measure $P(z_1, \ldots, z_\ell) = P(z_1) \cdot \ldots \cdot P(z_\ell)$.

The main results of the theory of uniform convergence are connected with the equality

$$\lim_{\ell \to \infty} \frac{H^\Lambda(\varepsilon, \ell)}{\ell} = 0, \quad \forall \varepsilon. \tag{1.12}$$

This equality describes the necessary and sufficient conditions for uniform one-sided convergence (1.10) over the set of bounded functions.

According to the key assertion the equality (1.12) implies the necessary and sufficient conditions for consistency of ERM principle.

These necessary and sufficient conditions of consistency of ERM principle were found in [17]. However, previously it was found that equality (1.12) forms the necessary and sufficient condition for uniform two-sided convergence[6]

$$\lim_{\ell \to \infty} \text{Prob}\{\sup_{\alpha \in \Lambda} |R(\alpha) - R_{emp}(\alpha)| > \varepsilon\} = 0, \quad \forall \varepsilon \tag{1.13}$$

over the set of indicator functions $Q(z, \alpha)$, $\alpha \in \Lambda$ [12, 13] and forms the necessary and sufficient conditions for two-sided uniform convergence (1.13) over the set of real bounded functions as well [16].

1.6.3 Three Milestones of Learning Theory

In this section we consider the set of indicator functions $Q(z, \alpha)$, $\alpha \in \Lambda$ (we consider the problem of pattern recognition). Analysis of this set of functions plays crucial part in learning theory. For the set of indicator functions

---

vector (in a given metric). For a C metric that means

$$\rho(q(\alpha^*), q(\alpha_r)) = \max_{1 \leq i \leq \ell} |Q(z_i \alpha^*) - Q(z_i, \alpha_r)| \leq \varepsilon.$$

2. N is minimal number of vectors which possess this property.

[5]Note that VC-entropy is different from classical metrical ε-entropy

$$H_{cl}^\Lambda(\varepsilon) = \ln N^\Lambda(\varepsilon)$$

where $N^\Lambda(\varepsilon)$ is cardinality of the minimal ε-net of the set of functions $Q(z, \alpha)$, $\alpha \in \Lambda$.

[6]Uniform two-sided convergence defines a stronger mode of uniform convergence then uniform one sided convergence.

$Q(z,\alpha), \alpha \in \Lambda$ and for the data $z_1, \ldots, z_\ell$ the minimal ε-net of vectors $q(\alpha)$, $\alpha \in \Lambda$ (see (1.11)) does not depend upon ε if $\varepsilon < 0.5$

$$N^\Lambda(\varepsilon; z_1, \ldots, z_\ell) = N^\Lambda(z_1, \ldots, z_\ell)$$

and is equal to the number of different separations of the data $z_1, \ldots, z_\ell$ by the functions of the set $Q(z, \alpha), \alpha \in \Lambda$.

The VC-entropy for this set of functions also does not depend upon ε

$$H^\Lambda(\ell) = H^\Lambda(\varepsilon, \ell) = E \lg N^\Lambda(z_1, \ldots, z_\ell).$$

Consider two new concepts that are constructed on the basis of the values $N^\Lambda(z_1, \ldots, z_\ell)$: so-called *Annealed VC-entropy*

$$H^\Lambda_{ann}(\ell) = \ln E N^\Lambda(z_1, \ldots, z_\ell)$$

and *Growth function*

$$G^\Lambda(\ell) = \ln \sup_{z_1, \ldots, z_\ell} N^\Lambda(z_1, \ldots, z_\ell).$$

The functions are determined in such a way that for any ℓ the inequalities

$$H^\Lambda(\ell) \leq H^\Lambda_{ann}(\ell) \leq G^\Lambda(\ell)$$

are valid.

On the basis of these functions the main milestones of the learning theory are constructed.

In the previous section we introduced the equation

$$\lim_{\ell \to \infty} \frac{H^\Lambda(\ell)}{\ell} = 0$$

describing the *necessary and sufficient condition* for consistency of ERM principle. This equation is the first milestone in learning theory: any machine minimizing empirical risk should satisfy it.

However, this equation says nothing about the rate of convergence of obtained risks $R(\alpha_\ell)$ to the minimal one $R(\alpha_0)$. It is possible to construct examples where ERM principle is consistent but has arbitrary slow asymptotic rate of convergence.

The question is:

Under which conditions is the asymptotic rate of convergence fast?

We say that the asymptotic rate of convergence is fast if for any $\ell > \ell_0$ the exponential bound

$$P\{R(\alpha_\ell) - R(\alpha_0) > \varepsilon\} < e^{-c\varepsilon^2 \ell}$$

holds true, where $c > 0$ is some constant.

The equation

$$\lim_{\ell \to \infty} \frac{H_{ann}^{\Lambda}(\ell)}{\ell} = 0$$

describes a *sufficient* condition for fast convergence[7]. It is the second milestone of the learning theory: it guarantees a fast asymptotic rate of convergence.

Note that both the equation describing the necessary and sufficient condition for consistency and the one describing the sufficient condition for fast convergence of the ERM method are valid for *given* probability measure $P(z)$ (both VC-entropy $H^{\Lambda}(\ell)$ and VC-annealed entropy $H_{ann}^{\Lambda}(\ell)$ are constructed using this measure). However, our goal is to construct a learning machine for solving many different problems (*i.e.* for many different probability measures).

The question is:

Under which conditions is the ERM principle consistent and rapidly convergant for any probability measure?

The following equation describes the *necessary and sufficient conditions* for consistency of ERM for any probability measure

$$\lim_{\ell \to \infty} \frac{G^{\Lambda}(\ell)}{\ell} = 0.$$

This condition is also sufficient for fast convergence.

This equation is the third milestone in the learning theory. It describes the conditions under which the learning machine that implements the ERM principle has an asymptotic high rate of convergence independently of the problem to be solved.

These milestones form a foundation for constructing both distribution independent bounds for the rate of convergence of learning machines and rigorous distribution dependent bounds.

1.7 Bounds for the Rate or Convergence of the Learning Processes

The non-asymptotic theory of the bounds for the rate of convergence of the learning processes is based on the concept of *VC-dimension* (abbreviation for Vapnik-Chervonenkis dimension) of the set of functions $Q(z, \alpha)$, $\alpha \in \Lambda$. It was developed in the 1960's and 1970's [12, 13, 10].

The concept of the VC-dimension is connected with a remarkable property of the Growth-function $G^{\Lambda}(\ell)$. [12, 13].

---

[7] The necessity of this condition for fast convergence is open question.

1.7.1 The Structure of the Growth-function

Any growth function either satisfies the equality

$$G^\Lambda(\ell) = \ell \ln 2$$

or is bounded by inequality

$$G^\Lambda(\ell) < h(\ln \frac{\ell}{h} + 1),$$

where h is an integer for which

$$G^\Lambda(h) = h \ln 2,$$

$$G^\Lambda(h+1) \neq (h+1) \ln 2.$$

In other words, Growth function can be either a linear function or should be bounded by a logarithmic function. (It cannot be, for example, of the form $G^\Lambda(\ell) = c\sqrt{\ell}$.)

We will say that the VC-dimension of the set of indicator functions $Q(z, \alpha)$, $\alpha \in \Lambda$ is infinite if the Growth function for this set of functions is linear.

We will say that the VC-dimension of the set of indicator functions $Q(z, \alpha)$, $\alpha \in \Lambda$ is finite and equals h if the Growth function is bounded by a logarithmic function with coefficient h.

Therefore finiteness of the VC-dimension of the set of indicator functions implemented by the learning machine forms the necessary and sufficient conditions for consistency of the ERM method independent of the probability measure. Finiteness of the VC-dimension also implies fast convergence.

1.7.2 VC-dimension of the Set of Functions

In this section we give an equivalent definition of the VC-dimension of a set of indicator functions and then we generalize this definition for sets of real functions.

The VC-dimension of a set of indicator functions. The *VC-dimension of a set of indicator functions* $Q(z, \alpha)$, $\alpha \in \Lambda$, is the maximum number h of vectors $z_1, \ldots, z_h$ which can be separated in all 2^h possible ways using functions of this set[8] (*shattered* by this set of function). If for any n there exists a set of n vectors which can be shattered by the set $Q(z, \alpha)$, $\alpha \in \Lambda$, then the VC-dimension is equal to infinity.

VC-dimension of a set of real functions [10]. Let $a \leq Q(z, \alpha) \leq A$, $\alpha \in \Lambda$, be a set of real functions bounded by constants a and A (a can be equal to $-\infty$ and A can be equal to ∞).

---

[8] Any indicator function separates set of vectors into two subsets: the subset of vectors for which this function takes value 0 and the subset of vectors for which it takes value 1.

STRUCTURE OF STATISTICAL LEARNING THEORY

Let us consider along with the set of real functions $Q(z, \alpha)$, $\alpha \in \Lambda$, the set of indicator functions

$$I(z, \alpha, \beta) = \theta \{Q(z, \alpha) - \beta\}, \quad \alpha \in \Lambda \tag{1.14}$$

where $a < \beta < A$ is some constant, $\theta(x)$ is the step-function

$$\theta(x) = \begin{cases} 0 & \text{if } x < 0 \\ 1 & \text{if } x \geq 0. \end{cases}$$

The *VC-dimension of the set of real functions* $Q(z, \alpha)$, $\alpha \in \Lambda$, is defined to be the VC-dimension of the set of indicator functions (1.14).

Example.

1. The VC-dimension of the set of *linear indicator functions*

$$Q(z, \alpha) = \theta \left\{ \sum_{p=1}^{n} \alpha_p z_p + \alpha_0 \right\}$$

in n-dimensional coordinate space $Z = (z_1, \ldots, z_n)$ is equal to $h = n+1$, since using functions of this set one can shatter at most $n+1$ vectors. Here $\theta\{\cdot\}$ is the step-function, which takes value 1 if the expression in the brackets is positive and 0 otherwise.

2. The VC-dimension of the set of *linear functions*

$$Q(z, \alpha) = \sum_{p=1}^{n} \alpha_p z_p + \alpha_0, \quad \alpha_0, \ldots, \alpha_n \in (-\infty, \infty)$$

in n-dimensional coordinate space $Z = (z_1, \ldots, z_n)$ is equal to $h = n+1$, because the VC-dimension of corresponding linear indicator functions is equal to $n+1$ (using $\alpha_0 - \beta$ instead of α_0 does not changes the set of indicator functions).

1.7.3 Distribution Independent Bounds for the Rate of Convergence of the Learning Processes

Consider sets of functions which possess a finite VC-dimension h. We distinguish between two cases:

1. The case where the set of loss functions $Q(z, \alpha)$, $\alpha \in \Lambda$ is a set of *totally bounded functions*, and

2. The case where the set of loss functions $Q(z, \alpha)$, $\alpha \in \Lambda$ is *not necessarily a set of bounded functions*.

Case 1. The set of totally bounded functions. Without restriction in generality, we assume that

$$0 \leq Q(z,\alpha) \leq B, \ \alpha \in \Lambda. \tag{1.15}$$

The main results in the theory of learning machines with the set of totally bounded functions are the following two assertions [10, 11]:

- With probability at least $1 - \eta$, the inequality

$$R(\alpha) \leq R_{emp}(\alpha) + \frac{B\varepsilon}{2}\left(1 + \sqrt{1 + \frac{4R_{emp}(\alpha)}{B\varepsilon}}\right), \tag{1.16}$$

holds true simultaneously for all functions of the set (1.15), where

$$\varepsilon = 4\frac{h(\ln\frac{2\ell}{h} + 1) - \ln\eta}{\ell}. \tag{1.17}$$

- Let $Q(z,\alpha_\ell)$ be the function that minimizes the empirical risk functional (1.8) and let $Q(z,\alpha_0)$ be the function which minimizes the (expected) risk functional (1.6). Then with probability at least $1-2\eta$ the inequality

$$R(\alpha_\ell) - R(\alpha_0) \leq \frac{B\varepsilon}{2}\left(1 + \sqrt{1 + \frac{4R_{emp}(\alpha)}{B\varepsilon}}\right) + B\sqrt{\frac{-\ln\eta}{2\ell}} \tag{1.18}$$

holds true, where ε is determined by (1.17).

For the set of indicator functions, $B = 1$.

The first inequality bounds the risks for all functions of the set (1.15) (including the function $Q(z,\alpha_\ell)$ which minimizes empirical risk (1.8)). The second inequality bounds the non-asymptotic rate of convergence of the sequence of risks for the functions $Q(z,\alpha_\ell)$, $\ell = 1,...$ minimizing empirical risk to the minimal possible risk.

Case 2. The set of unbounded functions. Consider the set of (non-negative) unbounded functions $0 \leq Q(z,\alpha)$, $\alpha \in \Lambda$

It is easy to show (by construction) that without additional information about the set of unbounded functions and/or probability measures it is impossible to obtain an inequality of type (1.16) and (1.18). Below we use the following information: so-called

$$\sup_{\alpha \in \Lambda} \frac{\left(\int Q^p(z,\alpha)dP(z)\right)^{\frac{1}{p}}}{\int Q(z,\alpha)dP(z)} \leq \tau < \infty \tag{1.19}$$

where $p > 1$ is some fixed constant[9].

---

[9] This inequality describes some general properties of the distribution functions of the random variables $\xi_\alpha = Q(z,\alpha)$, generated by the $P(z)$. It describes the "tails of distributions" (the probability of big values for the random variables ξ_α). If the inequality (1.19) with $p > 2$ holds, then the distributions have so-called "light tails" (big values occur not very often). In this case rapid convergence is possible. If, however, the inequality (1.19) holds only for $p < 2$ (big values of the random variables ξ_α occur rather often) then the rate of convergence will be small (it will be arbitrarily small if p is sufficiently close to one).

STRUCTURE OF STATISTICAL LEARNING THEORY

The main results of the theory of learning machines with an unbounded set of functions are the following two assertions which for simplicity we will describe for the case $p > 2$ (the results for case $p > 1$ can be found in [10]):

- With probability at least $1 - \eta$ the inequality

$$R(\alpha) \leq \frac{R_{emp}(\alpha)}{(1 - a(p)\tau\sqrt{\varepsilon})_+}, \quad a(p) = \sqrt[p]{\frac{1}{2}\left(\frac{p-1}{p-2}\right)^{p-1}} \quad (1.20)$$

holds true simultaneously for all functions of the set (1.15), where ε is determined by (1.17), $(a)_+ = max(a, 0)$.

- Let $Q(z, \alpha_\ell)$ be a function which minimizes the empirical risk functional (1.8). Then with probability at least $1 - 2\eta$, the inequality

$$\frac{R(\alpha_\ell) - R(\alpha_0)}{R(\alpha_0)} \leq \frac{a(p)\tau\sqrt{\varepsilon}}{(1 - a(p)\tau\sqrt{\varepsilon})_+} + 0(\frac{1}{\ell}), \quad R(\alpha_0) \neq 0. \quad (1.21)$$

holds true.

As before, the first inequality bounds the risks for all functions of the set (1.15) (including the function $Q(z, \alpha_\ell)$ which minimizes the empirical risk (1.8)). The second inequality bounds the relative rate of convergence of the sequence of risks to the minimal possible risk.

1.7.4 Problem of Constructing Rigorous (distribution dependent) Bounds

To construct rigorous bounds for the rate of convergence one has take into account information about the probability measure. Let $\mathcal{P}_0$ be a set of all probability measures and let $\mathcal{P} \subset \mathcal{P}_0$ be a subset of the set $\mathcal{P}_0$. We say that one has prior information about unknown probability measure $P(z)$ if one knows a set of measures $\mathcal{P}$ that contains $P(z)$.

Consider the following generalization of the Growth function:

$$\mathcal{G}_\mathcal{P}^\Lambda(\varepsilon, \ell) = \lg \sup_{P \in \mathcal{P}} E_P N^\Lambda(\varepsilon; z_1, \ldots, z_\ell)$$

For indicator functions $Q(z, \alpha), \alpha \in \Lambda$ and for the extreme case where $\mathcal{P} = \mathcal{P}_0$ the Generalized Growth function $\mathcal{G}_\mathcal{P}^\Lambda(\varepsilon, \ell)$ coincides with Growth function $G^\Lambda(\ell)$. For another extreme case where $\mathcal{P}$ contains only one function $P(z)$ the Generalized Growth function coincides with annealed VC-entropy.

Rigorous bounds for the rate of convergence can be derived in the terms of the Generalized Growth function. They have the same functional form as the distribution independent bounds (1.16), (1.18) and (1.20), (1.21) but a different expression for ε. The new expression for ε is

$$\varepsilon = \frac{\mathcal{G}_\mathcal{P}^\Lambda(\varepsilon, 2\ell) - \ln \eta}{\ell}.$$

However these bounds are non-constructive because there are no general methods to evaluate bounds for the Generalized Growth function (as there were in the case of the Growth function, where constructive bounds were obtained on the basis of the VC-dimension of the set of function).

To find rigorous constructive bounds one has to find a way for evaluating the Generalized Growth function for different sets $\mathcal{P}$ of probability measures[10].

1.8 Theory for Controlling Generalization Ability of Learning Processes

The theory for controlling the generalization ability of a learning machine is devoted to constructing the induction principle for minimizing the risk functional which takes into account *the size of the training set* (the induction principle for a *"small" sample size*[11]).

1.8.1 Structural Risk Minimization Induction Principle

The ERM principle is intended for dealing with large sample size. Indeed, the ERM principle can be justified by considering the inequalities (1.16) or (1.20). When ℓ/h is large, the second summand in the right hand side of inequality (1.16) becomes small. The actual risk is then close to the value of the empirical risk. In this case, a small value of the empirical risk provides a small value to the (expected) risk.

However, if ℓ/h is small, even small $R_{emp}(\alpha_\ell)$ does not guarantee a small value of risk. In this case the minimization for $R(\alpha)$ requires a new principle, based on the simultaneous minimization of two terms in inequality (1.16) (inequality (1.20)), one of which depends on the value of the empirical risk while the second depends on the VC-dimension of the set of functions. To minimize risk in this case it is necessary to find a method which along with minimizing the value of empirical risk controls the VC-dimension of the learning machine.

The following principle, which is called the principle of Structural Risk Minimization (SRM), is intended to minimize the risk functional with respect to both empirical risk and VC-dimension of the set of functions [14, 10].

Let the set S of functions $Q(z, \alpha)$, $\alpha \in \Lambda$, be provided with *structure*: the nested subsets of functions $S_k = \{Q(z, \alpha),\ \alpha \in \Lambda_k\}$, such that

$$S_1 \subset S_2 \subset ... \subset S_n ..., \tag{1.22}$$

and $S^* = \bigcup_k S_k$.

An *admissible structure* is one satisfying the following three properties:

---

[10]The main problem here is to find a subset $\mathcal{P}$ differing from the set $\mathcal{P}_0$ for which a Generalized Growth function can be evaluated on the basis of some constructive concepts.

[11]The sample size ℓ is considered to be small if ℓ/h is small, say $\ell/h < 20$.

1. The set S^* is everywhere dense in S.

2. The VC-dimension h_k of each set S_k of functions is finite.

3. Any element S_k of the structure contains either:

    A set of totally bounded functions $0 \le Q(z,\alpha) \le B_k$, $\alpha \in \Lambda_k$, or

    A set of functions satisfying the inequality

$$\sup_{\alpha \in \Lambda_k} \frac{(\int Q^p(z,\alpha) dP(z))^{\frac{1}{p}}}{\int Q(z,\alpha) dP(z)} \le \tau_k, \quad p > 2 \qquad (1.23)$$

The SRM principle suggests a two-step procedure:

1. For a given sample size ℓ choose an element S_n of structure (1.22), where

$$n = n(\ell).$$

2. In the set S_n, choose the function $Q(z, \alpha_\ell^n)$ which minimize the empirical risk.

For a given set of observations $z_1, ... z_\ell$ one can choose the element of structure S_n for which the guaranteed risk (1.16) (or (1.20)) is minimal.

The SRM principle actually suggests a *trade-off between the quality of the approximation and the complexity of the approximating function*. (As n increases, the minima of empirical risk are decreased; however, the term responsible for the confidence interval (summand in (1.16) or multiplier in (1.20)) is increased. The SRM principle takes both factors into account.)

For admissible structures the method of structural risk minimization provides approximations $Q(z, \alpha_\ell^{n(\ell)})$ for which the sequence of risks $R(\alpha_\ell^{n(\ell)})$ converge to the best one $R(\alpha_0)$ with *asymptotic rate of convergence*[12]

$$V(\ell) = r_{n(\ell)} + T_{n(\ell)} \sqrt{\frac{h_{n(\ell)} \ln \ell}{\ell}} \qquad (1.24)$$

if the law $n = n(\ell)$ is such that

$$\lim_{\ell \to \infty} \frac{T_{n(\ell)}^2 h_{n(\ell)} \ln \ell}{\ell} = 0. \qquad (1.25)$$

In equation (1.25) $T_n = B_n$ if one considers a structure with totally bounded functions in subsets S_n and $T_n = \tau_n$ if one considers a structure with elements satisfying the equality (1.25), $r_n(\ell)$ is the rate of approximation

$$r_n = \inf_{\alpha \in \Lambda_n} \int Q(z,\alpha) dP(z) - \inf_{\alpha \in \Lambda} \int Q(z,\alpha) dP(z).$$

---

[12] We say that the random variables ξ_ℓ, $\ell = 1, 2, ...$ converge to the value ξ_0 with asymptotic rate $V(\ell)$ if there exists constant C such that

$$V^{-1}(\ell) |\xi_\ell - \xi_0| \xrightarrow{P}_{\ell \to \infty} C.$$

Note that for any given structure (1.22) the parameters of structure h_k, T_k are fixed. Therefore the law satisfying the condition (1.25) can be chosen *a priori*. To find a law $n = n(\ell)$ which provides the best rate of convergence one has to know the *rate of approximation r_n for the chosen structure*. In this case by minimizing the right-hand side of equality (1.24) one can *a priori* find the law $n = n(\ell)$ which provides the best asymptotic rate of convergence.

Example

Let $Q(z, \alpha), \alpha \in \Lambda$ be a set of functions satisfying the inequality (1.19) for $p > 2$. Consider a structure for which $n = h_n$. Let the asymptotic rate of approximation be described by the law

$$r_n = \left(\frac{1}{n}\right)^c$$

(This law describes the typical classical results in approximation theory.) Then the asymptotic rate of convergence achieves its maxima if

$$n(\ell) = \left[\frac{\ell}{\ln \ell}\right]^{\frac{1}{2c+1}}$$

where $[a]$ is integer part of a. The asymptotic rate of convergence equal

$$V(\ell) = \left(\frac{\ln \ell}{\ell}\right)^{\frac{c}{2c+1}}.$$

1.8.2 Examples of the Structure for Neural Nets

The general principle of SRM can be implemented in many different ways. Here we consider three different examples of structures built for the set of functions implemented by a neural network.

1. Structure given by the architecture of the neural network. Consider an ensemble of fully connected neural networks in which the number of units in one of the hidden layers is monotonically increased. The set of implementable functions defines the structure as the number of hidden units is increased.

2. Structure given by the learning procedure. Consider the set of functions $S = \{f(x, w), w \in W\}$ implementable by a neural net of fixed architecture. The parameters $\{w\}$ are the weights of the neural network. A structure is introduced through $S_p = \{f(x, w), ||w|| \leq C_p\}$ and $C_1 < C_2 < ... < C_n$. Under very general conditions on the set of loss function, the minimization of the empirical risk within the element S_p of the structure is achieved through the minimization of

$$E(w, \gamma_p) = \frac{1}{\ell} \sum_{i=1}^{\ell} L(y_i, f(x_i, w)) + \gamma_p ||w||^2$$

STRUCTURE OF STATISTICAL LEARNING THEORY

with appropriately chosen Lagrange multipliers $\gamma_1 > \gamma_2 > ... > \gamma_n$. The well-known "weight decay" procedure refers to the minimization of this functional.

3. Structure given by pre-processing. Consider a neural net with fixed architecture. The input representation is modified by a transformation $z = K(x, \beta)$, where the parameter β controls the degree of degeneracy introduced by this transformation (for instance β could be the width of a smoothing kernel).

A structure is introduced in the set of functions $S = \{f(K(x, \beta), w), w \in W\}$ through $\beta \geq C_p$, and $C_1 > C_2 > ... > C_n$.

To implement the SRM principle using these structures, one has to know (estimate) the VC-dimension of any element S_k of structure and has to be able for any S_k to find the function which minimizes empirical risk. For the pattern recognition problem this is quite a difficult task.

1.8.3 The SRM Principle in the Problem of Polynomial Regression

Consider the problem of approximating an unknown regression function by the polynomial

$$p_k(x, \alpha) = \sum_{i=0}^{k} a_i x^i$$

(note that the regression is not necessary polynomial). This means that for given data one has to determine both the order of the approximating polynomial and the coefficients of the polynomial which provide the best approximation (in $L_2(P)$ metric) to the unknown regression function.

To solve this problem one can use the SRM principle, where element S_r contains the polynomials of order r. The VC-dimension of polynomials of order r equals $r+1$. Setting $\tau = 1$ in (1.20) one has to minimize the functional

$$\Phi(\alpha, r) = \frac{R_{emp}(\alpha, r)}{\left(1 - \sqrt{\frac{(r+1)(\ln \frac{\ell}{(r+1)}+1) - \ln \eta}{\ell}}\right)_+}$$

over both order of polynomials r and coefficients of polynomials α.

1.8.4 Problem of Local Function Estimation

On the basis of the SRM principle one can consider the problem of local function estimation. Let $f(x, \alpha), \alpha \in \Lambda$ be a set of functions in which one would like to approximate a desired function. Let the concept of locality of the vector x_0 be described by the kernel $K(x - x_0, b) \geq 0$ which selects a region of input space of width b, centred at x_0. For example, consider the rectangular kernel,

$$K_r(x - x_0, b) = \begin{cases} 1 & \text{if } |x - x_0| \leq b \\ 0 & \text{otherwise} \end{cases}$$

or a continuous kernel, such as the Gaussian

$$K_g(x - x_0, b) = \exp -\{\frac{(x - x_0)^2}{b^2}\}.$$

The goal is to minimize the local risk functional

$$R(\alpha, b, x_0) = \int L(y, f(x, \alpha))\frac{K(x - x_0, b)}{K(x_0, b)} dP(x, y) \quad (1.26)$$

where probability measure $P(x, y)$ is unknown but we are given data (y_1, x_1), ..., (y_ℓ, x_ℓ). The normalization in the risk-functional $R(\alpha, b, x_0)$, is defined by

$$K(x_0, b) = \int K(x - x_0, b) dP(x).$$

The local risk functional evaluates how the functions of the set $f(x, \alpha)$, $\alpha \in \Lambda$ approximate the desired one in the different neighbourhood of the vector x_0.

The task of local approximation is to find both the best neighbourhood (the best b) and the best approximating function $f(x, \alpha^*)$ for this neighbourhood.

Note that the solution of this problem is based on some trade-off between the size of region of the point of interest x_0 (the bigger the size, the more data can be taken into account) and the quality of the local approximation of the unknown function in the neighbourhood of the point of interest (the lesser the size, the better local approximation can be achieved).

Thus the local risk functional (1.26) is to be minimized over the class of functions $f(x, \alpha)$, $\alpha \in \Lambda$ and over all possible neighbourhoods $b \in (0, \infty)$ centered at point of interest x_0. To solve this problem one obtains the bounds for local risks (which are valid simultaneously for all functions and all regions of the point of interest x_0) and then one uses the SRM principle to minimize this bound [18].

Usually the function that is found is used for estimating of the value of the unknown function near the point x_0.

1.9 Theory of Constructing Learning Algorithms for the Pattern Recognition Problem

To implement both ERM and SRM induction principles in learning algorithms for pattern recognition problem one has to develop *methods for minimizing empirical risk on the different sets of indicator functions*. Minimizing empirical risk on the sets of indicator functions is a rather hard computational task. Developing methods for solving this task is the goal of the theory of constructing learning algorithms for pattern recognition.

STRUCTURE OF STATISTICAL LEARNING THEORY

Generalization of the two ideas for minimizing the empirical risk on the set of linear indicator functions developed in the 1960s constitutes two main branches in the theory of constructing learning algorithms.

1.9.1 Sigmoid Approximation of Indicator Functions and Neural Nets

Consider first the problem of minimizing empirical risk on the set of *linear indicator functions*

$$f(x, w) = \theta \left\{ \sum_{i=0}^{n} w_i x^i \right\}, \quad w \in W. \tag{1.27}$$

Let

$$(x_1, y_1), \ldots, (x_\ell, y_\ell)$$

be a training set, where $x_j = (x_j^1, \ldots, x_j^n)$ is a vector, $y_j \in \{0, 1\}$, $j = 1, \ldots, \ell$.

The goal is to find the parameters $w = (w^1, \ldots, w^n)$ (weights) which minimize the empirical risk functional

$$R_{emp}(w) = \frac{1}{\ell} \sum_{j=1}^{\ell} (y_i - f(x_j, w))^2. \tag{1.28}$$

For minimizing this functional it is impossible to use regular *gradient-based* methods of optimization. (The gradient of the indicator function $R_{emp}(w)$ is either equal to zero or undefined.) The idea is to approximate the set indicator function (1.27) by so-called *sigmoid functions*

$$\bar{f}(x, w) = S \left\{ \sum_{i=0}^{n} w_i x^i \right\} \tag{1.29}$$

where $S(u)$ is a smooth monotonic functions such that $S(-\infty) = 0$, $S(+\infty) = 1$, for example,

$$S_1(u) = \frac{1}{1 + \exp^{-u}}, \quad S_2(u) = \frac{2 \operatorname{arctg} u + \pi}{2\pi}.$$

For the set of sigmoid functions, the empirical risk functional

$$R_{emp}(w) = \frac{1}{\ell} \sum_{j=1}^{\ell} (y_i - \bar{f}(x_j, w))^2 \tag{1.30}$$

is smooth in w. It has a gradient $\operatorname{grad} R_{emp}(w)$ and therefore can be minimized using standard gradient-based methods, for example, the *gradient descent method*:

$$w_{new} = w_{old} - \gamma(\cdot) \operatorname{grad} R_{emp}(w_{old})$$

where $\gamma(\cdot) = \gamma(n) \geq 0$ is a value which depends on the number n of iteration. For convergence of the gradient descent method to local minima it is enough that $\gamma(n)$ satisfy the conditions:

$$\sum_{n=1}^{\infty} \gamma(n) = \infty, \quad \sum_{n=1}^{\infty} \gamma^2(n) < \infty$$

Thus, the idea is to use the sigmoid approximation at the stage of estimating the coefficients, and use the indicator functions with these coefficients at the stage of recognition.

The generalization of this idea leads to Learning Neural Nets. In this generalization, instead of the set of linear indicator functions (single neuron), one considers a set of functions which are the super-position of the several linear indicator functions (neural nets) [7]. All indicator functions in this super-position should be replaced by sigmoid functions.

A method for calculating the gradient of the empirical risk for the sigmoid approximation of neural nets, the so-called *back-propagation method*, was found [8, 5]. Using this gradient, one can calculate the corresponding coefficients (weights) of the neural net on the basis of standard gradient-based procedures [4].

The main problems with the back-propagation approach are:

1. The empirical risk functional has many local minima. Standard optimization procedures guarantee convergence to some local minimum. The function which will be found from the gradient-based procedure can provide a local minimum which is far from the best one. The quality of the approximation obtained depends on many factors, in particular on the initialization parameters of the algorithm.

2. Convergence to the local minima can be rather slow (due to high dimensionality of the weight-space).

3. The sigmoid function has a scaling factor which affects the quality of approximation. To choose the scaling factor one has to make a trade-off between quality of approximation and the rate of convergence.

1.9.2 Method of Optimal Separating Hyperplane and Support Vectors Networks

The second idea for constructing learning algorithms is a generalization of the idea of constructing the *optimal separating hyperplane*[13] [14, 2, 3, 11]. Below we describe this idea for the particular case where two subsets of the training set (belonging to two different classes) can be separated by a hyperplane

---

[13] The optimal separating hyperplane is the one which, from among all separating hyperplanes, has a maximal distance (maximal margin) between the hyperplane and the vector closest to it.

without errors. However, the technique described is generalized for the non-separating case as well [3, 11].

The support-vector network implements the following idea: it maps the input vectors into a very high dimensional feature space Z through some non-linear mapping chosen *a priori*. In this space a linear decision surface is constructed with special properties that ensure high generalization ability of the network.

Example

To construct a decision surface corresponding to a polynomial of degree two, one can create a feature space, Z, which has $N = \frac{n(n+3)}{2}$ coordinates of the form:

$$z_1 = x_1, \ldots, z_n = x_n , \qquad n \text{ coordinates,}$$

$$z_{n+1} = x_1^2, \ldots, z_{2n} = x_n^2 , \qquad n \text{ coordinates,}$$

$$z_{2n+1} = x_1 x_2, \ldots, z_N = x_n x_{n-1} , \qquad \tfrac{n(n-1)}{2} \text{ coordinates,}$$

where $x = (x_1, \ldots, x_n)$. The separating hyperplane constructed in this space is the separating second degree polynomial in the input space.

Two problems arise in the above approach: one conceptual and one technical. The conceptual problem is how to find a separating hyperplane that will generalize well: the dimensionality of the feature space will be huge, and not all hyperplanes that separate the training data will necessarily generalize well.

The technical problem is how computationally to treat such high-dimensional spaces: to construct a polynomial of degree 4 or 5 in a 200 dimensional space it is necessary to construct hyperplanes in a billion dimensional feature space.

The conceptual part of this problem was solved in 1965 [14] by constructing *optimal hyperplanes* for separable classes. An optimal hyperplane is defined as the linear decision function with maximal margin between the vectors of the two classes, see Figure 1.3.

It was observed that to construct the optimal hyperplanes one only has to take into account a small amount of the training data, the so-called *support vectors*, which determine this margin. It was shown that if the training vectors are separated without errors by an optimal hyperplane the expectation value of the probability of committing an error on a test example is bounded by the ratio between the expectation value of the number of support vectors and the number of training vectors:

$$E[\Pr(\text{error})] \leq \frac{E[\text{number of support vectors}]}{\text{number of training vectors} - 1} . \qquad (1.31)$$

This bound does not explicitly contain the dimensionality of the space of separation. It follows from this bound, that if the optimal hyperplane can be

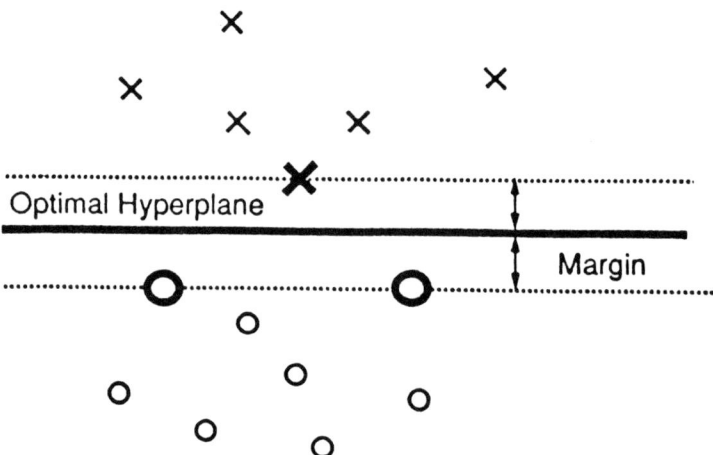

Figure 1.3: The optimal hyperplane is the one which has a maximal margin. For constructing this hyperplane one can takes into account only support vectors (denoted in bold face).

constructed from a small number of support vectors relative to the training set size, the generalization ability will be high — even in an infinite dimensional space.

Let
$$(w_0 \cdot z) + b_0 = 0$$
be the optimal hyperplane in feature space. The vector of weights w_0 for the optimal hyperplane can be written as a linear combination of support vectors

$$w_0 = \sum_{\text{support vectors}} \alpha_i z_i \ . \tag{1.32}$$

The linear decision function $I(z)$ in the feature space will accordingly be of the form:

$$I(z) = \text{sign}\left(\sum_{\text{support vectors}} \alpha_i (z_i \cdot z) + b_0 \right), \tag{1.33}$$

where $(z_i \cdot z)$ is the dot-product between support vectors z_i and vector z in feature space. To find both the support vectors z_i and coefficients α_i one has to solve the simplest quadratic programming problem (to maximize some quadratic form in the non-negative quadrant).

However, even if the optimal hyperplane generalizes well and theoretically it can be found, the technical problem of how to treat the high dimensional feature space remains.

In 1992 it was shown [2] that for constructing the optimal separating hyperplane in the feature space one can use the general expression for the dot product in the Hilbert space

$$(z_i \cdot z) = K(x, x_i),$$

where z is image in transformed space of the vector x in input space. $K(x, x_i)$ can be any symmetric function satisfying some very general properties [1].

This leads to the opportunity to construct nonlinear decision functions in input space

$$I(\mathbf{x}) = \text{sign}\left(\sum_{\substack{support \\ vectors}} \alpha_i K(\mathbf{x}_i \cdot \mathbf{x}) + b_0\right), \qquad (1.34)$$

that are equivalent to the linear decision functions (1.33) in the feature space.

The same method for constructing the optimal hyperplane provides both the pre-images x_i (in the input space) of support vectors z_i (in the feature space) and weights α_i.

The learning machines which construct decision functions of the type (1.34) are called *Support Vectors Networks*.[14]

Using different dot-products $K(x, x_i)$ one can construct different learning machines with arbitrary types of (nonlinear in input space) decision surfaces. For example, to specify polynomials of different order d one can use the following functions of the dot-product in corresponding feature space

$$K(x, x_i) = ((x \cdot x_i) + 1)^d.$$

Radial Basis Function machines with decision functions of the form

$$f(x) = \text{sign}\left(\sum_{i=1}^{n} \alpha_i \exp\left\{-\frac{|x - x_i|^2}{\sigma^2}\right\}\right)$$

can be implemented by using function of the type

$$K(x, x_i) = \exp\left\{-\frac{|x - x_i|^2}{\sigma^2}\right\}.$$

In this case the SVN machine will find both the centres x_i and the corresponding weights α_i.

SVN possess some useful properties:

- The optimization problem for constructing an SVN has a unique solution.

---

[14] This name stresses that, for constructing this type of machine, the idea of expanding the solution on support vectors is crucial. In SVN the complexity of construction depends on the number of support vectors rather than on the dimensionality of the feature space.

- The learning process for constructing an SVN is rather fast.

- Simultaneously with constructing the decision rule, one obtains the set of support vectors.

- Implementation of a new set of decision functions can be done by changing only one function which defines the dot product in Z-space.

1.9.3 Why Can Neural Networks and Support Vectors Networks generalize?

The generalization ability of the both neural networks and support vectors networks is based on the factors described in the theory for controlling the generalization ability of the learning processes. According to this theory, to guarantee a high level of generalization ability of the learning process one has to construct a structure

$$S_1 \subset S_2 \subset, \ldots, \subset S$$

on the set of decision functions $S = \{Q(z, \alpha), \alpha \in \Lambda\}$ and then choose both an appropriate element S_k of the structure and a function $Q(z, \alpha_\ell^k) \in S_k$ in this element that minimizes bound (1.16). The bound (1.16) can be rewritten in the simple form

$$R(\alpha_\ell^k) \leq R_{emp}(\alpha_\ell^k) + \Omega\left(\frac{\ell}{h_k}\right) \tag{1.35}$$

where the first term is an estimate of the risk and the second is the confidence interval for this estimate.

Designing a neural network (with the rules for initialization, the stopping rules and the rules for constructing the desire functions), one determines a set of admissible functions with some VC-dimension h^*. For a given amount ℓ of training data the value h^* actually determines the confidence interval $\Omega(\frac{\ell}{h^*})$ for the network. Choosing the appropriate element of structure is therefore a problem of designing the network for a specific amount of data.

During the learning process this network minimizes the first term in the bound (1.35) (number of errors on the training set[15]).

If it happens that at the stage of designing the network one constructs too complex a network (for the given amount of training data), the confidence interval $\Omega(\frac{\ell}{h^*})$ will be big. In this case, even if one could minimize the empirical risk down to zero the number of errors on the test set could be big. This case is called *over fitting*.

To avoid over fitting (to get a small confidence interval) one has to construct networks with a small VC-dimension. On the other hand, in the set of

---

[15] The same is true even one uses a weight-decay learning method. The coefficient for the term with the weight-decay determines the construction of the learning machine. During the learning process one minimizes the empirical risk for this construction of the learning machine.

STRUCTURE OF STATISTICAL LEARNING THEORY

functions with small VC-dimension it is difficult to approximate the training data (to get a small value for the first term in (1.35)). To obtain a small approximation error one has to choose an architecture of network which reflects a prior knowledge about problem at hand.

Thus, to solve a problem by neural networks one must first find the appropriate neural network architecture (which is result of a trade-off between over fitting and poor approximation), and second, find in this network the function that minimizes the number of error on the training data. This technology of minimizing the right hand side of inequality (1.35) can be described as follows:

Keep a fixed confidence interval (by choosing appropriate network) and minimize the empirical risk.

The Support Vectors Networks use another technology of minimizing the right hand side of inequality (1.35). It can be described as follows:

Keep fixed the value of empirical risk (say equal to zero) and minimize the confidence interval.

Indeed, according to the example given in the Section 1.7, the VC-dimension of the set of linear discriminant functions

$$f(z) = \theta((w \cdot z) + b) \tag{1.36}$$

is equal to N, the number of free parameters. Now let $||z|| \leq 1$ and consider the subset

$$f(z) = \theta((w \cdot z) + b), \quad ||w|| \leq C \tag{1.37}$$

of the set of functions (1.36), where coefficients w of hyperplane is normalized with respect to vectors $Z = z_1, \ldots, z_r$ (on which this hyperplane is defined):

$$\min_{z_i \in Z} |((w \cdot z) + b)| = 1.$$

The VC-dimension of this subset of function defined on Z is bounded as follows:

$$h \leq \min(C^2, N) + 1.$$

Let us consider the structure on the set of linear functions where element S_k contains the functions with $||w|| \leq C_k$ (the smaller C_k the less VC-dimension of the subset S_k). Let us minimize the right hand side of (1.35) by making the empirical risk equal to zero and minimizing the confidence interval. To do this we have to separate, without errors, the training set using the hyperplane with the smallest norm of weights $||w||$, which is the optimal hyperplane[16].

---

[16]To construct a hyperplane that separates the vectors z_i belonging to the first class, of the training data from the vectors z_j^* belonging to the second class one has to find the

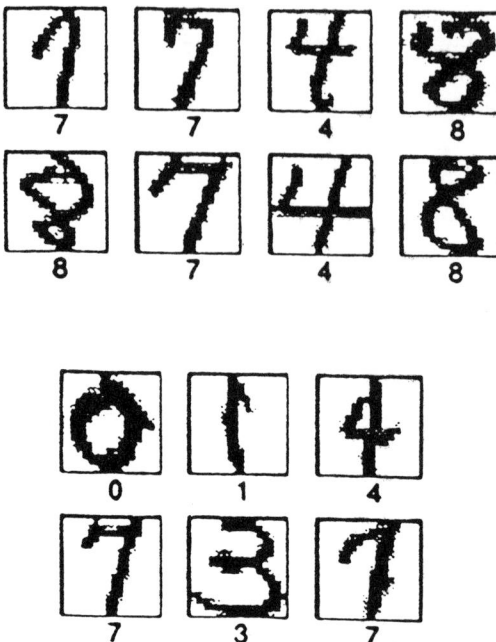

Figure 1.4: Examples of a pattern with labels from a US Postal Service database.

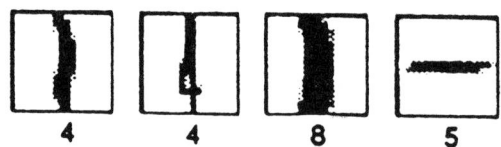

Figure 1.5: Labelled examples of training errors for the 2nd degree polynomial.

We conclude this overview of the learning and generalization theory with

vector w and value b such that inequalities

$$(z_i \cdot w) + b \geq 1,$$
$$(z_i^* \cdot w) + b \leq -1.$$

hold. To construct the *optimal* separating hyperplane one has to find the vector of coefficients w_0 and constant b_0 that satisfy these inequalities and has the minimal norm $||w||$.

STRUCTURE OF STATISTICAL LEARNING THEORY

Table 1.1:

Degree of polynomial	Raw error, %	Support vectors	Dimensionality of feature space
1	12.0	200	256
2	4.7	127	~ 33000
3	4.4	148	$\sim 1 \times 10^6$
4	4.3	165	$\sim 1 \times 10^9$
5	4.3	175	$\sim 1 \times 10^{12}$
6	4.2	185	$\sim 1 \times 10^{14}$
7	4.3	190	$\sim 1 \times 10^{16}$

an example describing experiments on learning the digit recognition. In these experiments a U.S. Postal Service database was used [6]. It contains 7300 training patterns and 2000 test patterns. The resolution of the database is 16 × 16 pixels, therefore the dimensionality of the input space is 256. Some typical examples are shown in Figure 1.4.

Table 1.1 describes the results of the experiments on digit recognition using 10 decision polynomials (one per class) constructing by SVN.

The training data are not linearly separable. The number of mis-classifications on the training set averages 34 per classifier for the linear case (total number of mis-classifications on the training set is equal to 340). For 2nd degree polynomial classifiers the total number of mis-classifications on the training set is down to four. These four examples are shown in Figure 1.5. Starting with polynomials of degree 3 the training data are separable.

The number of "support vectors" shown in the table is a mean value per classifier.

Notice that the number of support vectors increases very slowly with the order of polynomials. The 7 degree polynomial has only 30% more support vectors than the 3rd degree polynomial[17].

The dimensionality of the feature space for a 7 degree polynomial is, however, 10^{10} times larger than the dimensionality of the feature space for a 3d degree polynomial classifier. Note that performance almost does not change

---

The hyperplane with w_0 and b_0 provides the maximal margin

$$\rho = \frac{2}{||w_0||}$$

between separating vectors (see Figure 1.3).

[17] The relatively high number of support vectors for the linear separator is due to non-separability: the number 200 includes both support vectors and mis-classified data.

with increasing dimensionality of the space — indicating no over-fitting problems.

The performance of the neural net (LeNet 1) specially constructed for this task is 5.1% raw error [6].

References

[1] Aizerman E.A., Braverman E.M. and Rozonoer L.I. (1970) *Method of Potential Functions in the Theory of Pattern Recognition*. Nauka, Moscow.

[2] Boser B., Guyon I. and Vapnik V.N. (1992) An training algorithm for optimal margin classifiers. In *Fifth Annual Workshop on Computational Learning Theory*, Pittsburgh. ACM.

[3] Cortes C. and Vapnik V. (1995) Support Vector Networks. *Machine Learning* 20(3): 1–25.

[4] Hertz J., Krogh A. and Palmer R. (1991) *Introduction to the Theory of Neural Computation*. Addison-Wesley.

[5] Le Cun Y. (1986) Learning processes in an asymmetric threshold network. In Beinenstock E., Fogelman-Soulie F. and Weisbuch G. (eds) *Disordered systems and biological organizations*. Les Houches, France, Springer-Verlag.

[6] LeCun Y., Boser B., Denker J.S., Henderson D., Howard E.R., Hubbard W. and Jackel L.J. (1990) Handwritten digit recognition with back-propagation network. In *Advances in Neural Information Processing Systems*, (2). Morgan Kaufman, San Mateo.

[7] Minsky M.L. and Papert S.A. (1969) *Perceptrons*, MIT Press, Cambridge.

[8] Rumelhart D.E., Hinton G.E. and Williams R.J. (1986) Learning internal representations by error propagation. In *Parallel distributed processing: Explorations in macrostructure of cognition* (I) 318–362. Bradford Books, Cambridge, MA.

[9] Valiant L.G. (1984) A Theory of Learnability. *Comm. ACM* (27)11: 1134–1142.

[10] Vapnik V.N. (1979) *Estimation of Dependencies Based on Empirical Data*. Nauka, Moscow. (English translation: (1982) Springer-Verlag, New York.)

[11] Vapnik V.N. (1995) *The Nature of Statistical Learning Theory*. Springer-Verlag, New-York.

[12] Vapnik V.N. and Chervonenkis A.Ja. (1968) On the uniform convergence of relative frequencies of events to their probabilities. *Reports of Academy of Science USSR* (181)4:.

[13] Vapnik V.N. and Chervonenkis A.Ja. (1971) On the uniform convergence of relative frequencies of events to their probabilities. *Theory Probab. Apl* 16: 264–280.

[14] Vapnik V.N. and Chervonenkis A.Ja. (1974) *Theory of Pattern Recognition*. Nauka, Moscow.

[15] Vapnik V.N. and Chervonenkis A.Ja. (1979) *Theory of Pattern Recognition*. German translation: *Theorie der Zeichenerkennung* Akademia-Verlag, Berlin.

[16] Vapnik V.N. and Chervonenkis A.Ja. (1981) Necessary and sufficient conditions for the uniform convergence of the means to their expectations. *Theory Probab. Apl.* 26: 532–553.

[17] Vapnik V.N. and Chervonenkis A.Ja. (1989) The necessary and sufficient conditions for consistency of the method of empirical risk minimization. *Yearbook of the Academy of Sciences of the USSR* on *Recognition, Classification, and Forecasting* (2) 217–249. Nauka, Moscow. (English translation: (1991) *Pattern Recogn. and Image Analysis*, (1)3: 284–305.)

[18] Vapnik V.N. and Bottou L. (1993) Local Algorithms for Pattern Recognition and Dependencies Estimation. *Neural Computation* (5)6: 893–908.

2

Stochastic Complexity — an Introduction

J. Rissanen

2.1 General

Model building (or modelling for short) refers to the activity aimed at finding a description of the machinery that has generated a given set of data, "the go of it" in the words of Maxwell. In traditional statistics this has often been misunderstood as the problem of estimating the "true" underlying cause for the data, which moreover is taken to be a mathematical one, namely, a probability measure. In very simple cases, such as coin flipping data, there indeed seems to be only one meaningful probabilistic model, and little harm is done if it is estimated as the "true" data generating machinery. But most often no such preferred explanation exists. Consider, for example, the problem in which we wish to have an algorithm for detecting "scenes" in a grey scale image, where intuitively by a "scene" we mean a closed region with a roughly uniform grey level. Here, it makes little sense to talk about the "true" number of scenes, because we have trouble even to specify precisely what we mean by a "scene". We can, of course, define an "epsilon-scene", where the grey levels vary within an epsilon from a constant, but then the question arises how to pick the epsilon. Now, one may dismis this problem as not belonging to statistics at all, but the best algorithms are found with probability models, and besides the same considerations apply to other similar and clearly statistical problems such as the so-called unsupervised clustering problem. In general, then, an arbitrary choice of the "truth" leads into a biased result, because the objects to be estimated and the very way the estimation is done depends upon what the "truth" was assumed to be. And since there is no general way in traditional statistics to compare different assumptions about the "truth", the recipes offered degenerate to a bag of tricks and *ad hoc* choices.

There is, however, another quite different approach to modelling problems and, in fact, to all other problems of statistical inference as well. The basic premise is not to estimate any "truth" at all, but to fit models, which are just ways to express the constraints in the data. Why this, in itself hardly a surprising nor novel view, works is that there is a way to measure the strength of the constraints represented by the models, which depends only upon the data and hence needs no assumption of underlying "true" generators of the data. Perhaps the easiest way to explain the ideas involved is to recall the algorithmic notion of information or, synonymously, complexity, which is defined to be the length of the shortest program for a universal computer with which the data can be reproduced. Because such a program must take advantage of all the constraints that the data have, it may be taken as the best model of the data and the generating machinery as well — provided of course that the data set is large enough to reflect its properties. It is quite clear to this author that this is all one can ask for unless, of course, there is prior knowledge; *i.e.* additional knowledge not in the data, which in this view is to be taken advantage of in selecting the types of models to be fitted. Unfortunately, the shortest program or even its length, the celebrated Kolmogorov complexity, cannot be found by algorithmic means, which has the devastating implication that, even though we can estimate it from the above in a computable manner with the error going to zero, we cannot assess the goodness of the estimate. And this puts an end to the dreams of basing inductive inference on the Kolmogorov complexity.

While the Kolmogorov complexity is not the appropriate concept as a foundation of statistics, it nevertheless incorporates the important idea of measuring the strength of constraints by code length. Indeed, a program for generating a data string is clearly a code, of which the string can be decoded with help of the computer. But this being the case we do not need the algorithmic framework, and we can consider any set of codes with which to encode the given data string. By a strange coincidence the idea of a code, the so-called *prefix code*, where the codewords are self-containing in the sense that no commas are needed to separate them when they are concatenated, is tantamount to a discrete probability measure. Indeed, with a small idealization we may equate the length of the codewords for strings $x^n = x_1, \ldots, x_n$, from which a prefix code can be constructed, with $-\log P(x^n)$, where the logarithm is to the base two. Hence, we have the convenient unification that a computer, a prefix code, and a probability measure $P(.)$, all may be regarded as a *model* of the string, or, perhaps better, of the machinery that generates it. In case the computer is a universal one, it will, of course, act as a universal model in the sense that any computable model may be represented in it.

Pursuing this line of thought, we may now consider a class of probability density functions as models, $\mathcal{M}_k = \{f(x^n|\theta^k)\}$, or even their union over

k, where the parameters $\theta^k = \theta_1, \ldots, \theta_k$, also written just as θ, range over some subset of the k-dimensional euclidean space. Often, each member is required to satisfy the marginality requirement for a random process so that the conditional densities $f(x_{n+1}|x^n, \theta)$ can be defined and the future data linked with the current ones. Hence, we now equate a *model* with a parametric density function $f(x^n|\theta)$. A bit more generally, by taking a model class as $\mathcal{M} = \bigcup_k \{f(y^n|x^n, \theta^k)\}$, where y^n denotes another data string, we capture essentially all models that actually can be fitted to data. To define density functions as models represents an abstraction, because only discrete data can be encoded with a finite code length. The use of such abstractions is, however, meaningful in view of the fact that they define probabilities for data written with any finite precision, which, in turn, define finite code lengths for them.

If we try to define the shortest code length for a string relative to a parametric model class, which is smaller than the set of all computable models, to avoid the noncomputability problem, we face a difficulty: we can no longer look for the code length which is literally the shortest for every string. Rather, we must define it as the shortest in a weaker probabilistic sense, which nevertheless is strong enough to guarantee that it cannot be beaten for practical purposes. Once this is done the problem becomes to compute, or at least estimate, this shortest code length, which we call the string's *stochastic complexity*, relative to the model class considered. This can be done in a number of ways. For "smooth" model classes we can give a quite accurate formula for it. For many other cases an excellent approximation can be obtained predictively by an algorithm. Finally, complex model classes can be handled by breaking them up into simpler ones, for which the code length can be computed more easily, and adding the code length needed to link them together. All this may still not be straightforward, and the estimation of code lengths can be a challenge. However, such model classes are generally so complex that traditional statistical techniques have little to offer, and the code length approach at least provides a clearly defined goal what to look for.

Once the stochastic complexity for the data, relative to a number of suggested model classes, is calculated, it serves as a criterion for selecting the optimal one, namely, the one with the smallest stochastic complexity. This is the *MDL (Minimum Description Length)* principle for model selection. The code length criterion has the distinguished property that it admits three *data* dependent interpretations: first, as a code length, then, as the probability or density assigned to the data sequence, and finally, often also as a prediction error — not as the usual estimate of the mean prediction error, but the actual errors committed as the given data string is predicted. Clearly, then, there is no need to make the untenable assumption that any of the selected models has generated the data. Indeed, we may paraphrase Laplace' famous phrase thus "the assumption of 'truth' is not needed in the application of the principle of MDL to statistical problems".

Such a framework broadens the scope of traditional estimation and modelling. Not only does the stochastic complexity act as a criterion for model selection but it also defines a *universal model*, which represents the entire class in question and in effect reduces the class into a single probability measure. When it receives a data string generated by any model in the class, its behaviour imitates the data generating process. Clearly, for this to be possible the string must be long enough to represent accurately the statistical properties of the generating process. Moreover, ultimately, the universal process ought to be optimal in the sense that no other process whatsoever can imitate the data generating process better. For some model classes we actually have algorithms for implementing such optimal universal models, and for them there is little that remains to be added, see [11]. In general, though, the computation of the universal process is a challenge with much work remaining.

2.2 Stochastic Complexity

In this section we give without derivation the formula for the stochastic complexity for model classes which are "smooth", essentially in the sense that the maximum likelihood estimates of the parameters satisfy the Central Limit Theorem. For the derivation we refer to [8]. First we define

$$\hat{L}(x^n|\mathcal{M}_k) = -\ln \hat{f}(x^n) = -\ln \frac{f(x^n|\hat{\theta}(x^n))}{\int_{\hat{\theta}(y^n)\in\Gamma} f(y^n|\hat{\theta}(y^n))dy^n} \qquad (2.1)$$

to be the *stochastic complexity* of x^n, relative to $\mathcal{M}_k$, where $\hat{\theta} = \hat{\theta}(y^n)$ is the maximum likelihood estimate, and Γ is so chosen that the integral is finite. This is true if Γ is an open bounded subset of the k-dimensional euclidean space such that its closure contains no singular points of the following version of the Fisher information matrix

$$I(\theta) = \lim_{n\to\infty} -n^{-1}\left\{E_\theta \frac{\partial^2 \log f(x^n|\theta)}{\partial \theta_i \partial \theta_j}\right\}, \qquad (2.2)$$

the limit satisfying $0 < C_0 \leq I(\theta) \leq C_1 < \infty$ for all $\theta \in \Gamma$, where C_0 and C_1 are positive definite matrices. For such model classes we can give an accurate estimate of the denominator in equation (2.1) with the resulting code length

$$\hat{L}(x^n|\mathcal{M}_k) = -\ln f(x^n|\hat{\theta}) + \frac{k}{2}\ln\frac{n}{2\pi} + \ln\int_\Gamma \sqrt{|I(\theta)|}d\theta + o(1). \qquad (2.3)$$

The term $o(1)$ goes to zero as $n \to \infty$, which for instance for the Bernoulli class takes place fast and monotonically starting with the value 0.674 for $n = 0$, after which it declines to 0.119 at $n = 40$.

The way we have defined it, the stochastic complexity acts as the Kolmogorov complexity and does not define a random process without further normalization. It nevertheless gives a lower bound for general processes, and we have the first justification for the stochastic complexity as the shortest code length in the mean [3, 4],

Theorem 2.1 *Let $\mathcal{M} = \bigcup_k \mathcal{M}_k$ be a class of random process models such that for each k the maximum likelihood estimates $\hat{\theta}(x^n)$ of $\theta = \theta_1, \ldots, \theta_k$ satisfy the central limit theorem for $\theta \in \Gamma$, a bounded open subset of R^k. Further, let $g(x^n)$ be any density function defining a random process; i.e. for each n, $g(x^n) = g_n(x^n)$ defines a probability measure such that the marginality conditions are satisfied. Then for all k, all ϵ, and all $\theta \in \Gamma$, except in a subset $A_g(n)$ whose volume goes to zero as $n \to \infty$,*

$$E_\theta \ln \frac{f(X^n|\theta)}{g(X^n)} \geq \frac{k-\epsilon}{2} \ln n. \qquad (2.4)$$

The expectation E_θ is taken with respect to $f(x^n|\theta)$.

We have additionally under weak smoothness conditions

$$\ln \frac{f(x^n|\theta)}{g(x^n)} \geq \frac{k-\epsilon}{2} \ln n \qquad (2.5)$$

for all but finitely many n in $f(.|\theta)$ - probability 1, and for all θ except in a subset of Γ of Lebesgue measure 0.

In light of the second result, in particular, there is little chance to find a way to encode sequences, generated by some parametric model in the class, with a shorter code length than the stochastic complexity, unless, of course, the sequences are so short that they do not reflect the data generating distribution. To avoid a frequent misunderstanding we emphasize that these are analytic results, and when we apply them to compare model classes we make no assumption that the observed data have been generated by any distribution in any of our model classes. All we claim is that the winning model class permits a shorter encoding of our particular data set than the losing one, or, what amounts to the same thing, the winning model class captures better the "laws" that restrict the data than the losing one. This leaves open the possibility that an even better model class could some day be found. For now, however, this is the best we have — and nobody can teach us more about the properties of the data than what we have learned so far without suggesting another better model class. In fact, this is the most one can ever ask for.

There is another way to interpret the formula in equation (2.3). It can be viewed to within terms of order $o(1)$ as a two-part code length as follows:

$$\hat{L}(x^n|\mathcal{M}_k) = -\ln \frac{f(x^n|\hat{\theta})}{f(\hat{\theta})} - \ln \pi(\hat{\theta}) + o(1), \qquad (2.6)$$

where

$$f(\theta) = \int_{\hat{\theta}(y^n)=\theta} f(y^n|\hat{\theta}(y^n)) dy^n,$$

and π is Jeffreys' prior

$$\pi(\theta) = \frac{|I(\theta)|^{1/2}}{\int_{\eta \in \Gamma} |I(\eta)|^{1/2} d\eta}.$$

What distinguishes this from the earlier two-part codes [5, 10] is the normalization provided by the denominator, which is needed to make $\hat{f}(x^n)$, defined by $\hat{L}(x^n|\mathcal{M}_k)$ to integrate to unity. This, of course, is the very first requirement for a code to be efficient. Also, there is no longer any optimal precision for the parameters, which allowed us to put it as infinite. Another consequence of the same normalization is that the minimization of $\hat{L}(x^n|\mathcal{M}_k)$ over the number of parameters is no longer equivalent with the maximization of the posterior probability, proportional to the joint density $f(x^n|\theta)\pi(\theta)$. This means that the MDL principle cannot be regarded as an instance of Bayesianism, which of course should be evident anyway.

For many model classes, even when they satisfy the required smoothness conditions, the calculation of the integral of the square root of the Fisher information can be difficult. However, if we just estimate the Fisher information by replacing θ in equation (2.2) by $\hat{\theta}$ and dropping the expectation operation, we can get an idea of whether the integral is larger than unity. If it is, as normally is the case, then by leaving out the integral term in equation (2.3) we get a lower bound for the stochastic complexity. This can be important, because any coding method will give an upper bound, and hence we can bracket it. For instance, by truncating the maximum likelihood estimate $\hat{\theta}$ to an optimal precision $\hat{\theta}^d$ the two-part code length $-\ln f(x^n|\hat{\theta}^d) + L(\hat{\theta}^d)$, where the second term is any prefix code length, obtained by a prior, say, will be an upper bound for the stochastic complexity.

Example 1. We conclude this section by evaluating equation (2.3) for the important class of independent processes in m symbols $\{0,\ldots,m-1\}$; the same formula extends immediately to discrete Markov chains. Let the probability of the occurrence of symbol i be θ_i. The Fisher information is given by $|I(\theta)| = \prod_i \theta_i^{-1}$. Its square root is integrable over the open set of the parameters, defined by the simplex $0 < \theta_i < 1, \sum_i \theta_i = 1$, and the result is given by Dirichlet's integral

$$\int \prod_i \theta_i^{-1/2} d\theta_i = \frac{\pi^{m/2}}{\Gamma(m/2)}.$$

When this is substituted into equation (2.3) for $k = m-1$ we get the stochastic complexity,

$$\hat{L}(x^n) = n \ln n - \sum_i n_i \ln n_i + \frac{m-1}{2} \ln \frac{n}{2\pi} + \ln \frac{\pi^{m/2}}{\Gamma(m/2)} + o(1), \qquad (2.7)$$

where n_i denotes the number of times symbol i occurs in x^n.

Importantly, it follows from the work in [2] that formula (2.7) can be evaluated predictively as follows:

$$\hat{L}(x^n) = -\sum_{t=0}^{n-1} \ln \frac{n(x_{t+1}|x^t) + 1/2}{t + m/2} + o(1), \qquad (2.8)$$

where $n(x_{t+1}|x^t)$ denotes the number of times symbol $x_{t+1} = i$ occurs in the past string x^t. Although such a near-perfect coincidence is rare, the general predictive code length, which we introduced and analyzed around 1982 (see Section IV in [3]), and which was discovered independently in [1] as the "prequential principle", coincides with equation (2.3) up to terms of order $o(\ln n)$. It therefore provides an important way to estimate the stochastic complexity even when the formula in equation (2.3) is not valid. We reproduce in the next section a version of the predictive code length, in which the start-up difficulty, otherwise plaguing this method, is avoided.

2.3 Predictive Coding

Consider the following sequential coding: first, order the data set in any manner, unless already done, say as $x_1, x_2, \ldots, x_t, \ldots, x_n$. Next, subdivide the data into $M + 1$ segments of length d, except the last, which may be shorter. The segment length d will be optimized. Encode the first d numbers $x_1, \ldots, x_d$ any way agreed with the decoder, say by adjoining to the model class a special distribution $f(x^n|\lambda)$, where λ represents the empty parameter. Then, recursively, let $\hat{\theta}(x^{md})$ denote the maximum likelihood estimate, determined from the "past" sequence x^{md} for $m \geq 1$, and encode the numbers in the next segment with the help of the conditional distributions $f(x_{md+i}|x^{md+i-1}, \hat{\theta}(x^{md}))$, $i = 1, \ldots, d$, which can be calculated from the members of the model class. The resulting code length for the data is then given by

$$PMDL(x^n) = -\sum_{m=0}^{M} \sum_{i=1}^{d} \ln f(x_{md+i}|x^{md+i-1}, \hat{\theta}(x^{md})), \qquad (2.9)$$

where $\hat{\theta}(x^0) = \lambda$ and the terms for $t > n$ are zero. We may minimize this criterion with respect to the block length d, although to be precise we should then add the code length required to encode the numbers d, or about $\log d + 2 \log \log d$. Results of such a procedure when applied to neural nets appear in [7].

This type of recursive calculation for $d = 1$ has often been done to simplify the computation of the maximum likelihood estimate $\hat{\theta}(x^n)$ or in data compression to have a convenient encoder. However, such a purpose misses a much bigger point, namely, that the so-obtained code length is actually a good approximation of the stochastic complexity, and hence it can be used as a criterion for finding even the number of parameters and for model selection in general. Despite the fact that there is no explicit code length added for the parameters, because they are determined by an algorithm from the past data, the predictively calculated code length nevertheless includes it. To see it, rewrite the formula as

$$PMDL(x^n) = -\ln f(x^n|\hat{\theta}(x^n)) + \ln \frac{f(x^n|\hat{\theta}(x^n))}{f(x^n|\hat{\theta}(x^{Md}))} + \sum_{m=0}^{M} \ln \frac{f(x^{md}|\hat{\theta}(x^{md}))}{f(x^{md}|\hat{\theta}(x^{(m-1)d}))},$$

and we see that the code length for the parameters, taken as the model complexity, consists of the second term and the last sum, which are certainly nonnegative. Moreover, one can prove that under suitable conditions the model complexity behaves asymptotically as $\frac{k}{2}\ln n$, just as the corresponding term in the stochastic complexity.

Whenever there are enough data so that the initialization problem can be overcome the predictive code length provides a powerful way to approximate the stochastic complexity. This is particularly important in neural net models, where it is very difficult to estimate the stochastic complexity well by other means. We illustrate the computations with an example, which is a variant of one analyzed in [9].

Example 2. The data consist of pairs (y_t, x_t) for $t = 1, 2, \ldots, n$. The range of y_t is $\{0, 1\}$ and that of x_t is a bounded interval of the real line, say the unit interval, which is achieved by normalization of the weights. Such data could result from polling n men, in which case x_t is the weight of the tth man, and y_t is 1 if he thinks himself to be fat; otherwise, $y_t = 0$. The objective is to learn a conditional binary probability distribution $P(y = 0|x)$ as a function of the weight x, representing the probabilistic concept of a man's perception of his obesity. We model the data for different values for t as independent.

Consider the growing ordered sequence of weights $x^t = x_1, \ldots, x_t$ for $t = 0, 1, \ldots n$. For m equal-width bins partitioning the unit interval let $t(i|x^t, m)$ denote the number of points in the sequence x^t that fall within the ith bin, and let $t_j(i|y^t, x^t, m)$ be the number of times the y-component of these points is j. Using the predictive formula, equation (2.8), we define the conditional probabilities

$$P(Y_{t+1} = j|y^t, x^{t+1}, m) = \frac{t_j(i(x_{t+1})|y^t, x^t, m) + 1/2}{t(i(x_{t+1})|x^t, m) + 1}, \quad (2.10)$$

where $i(x)$ denotes the index of the bin in which x falls. Next, form the convex mixture

$$P_M(Y_{t+1} = j|y^t, x^{t+1}) = \frac{1}{M} \sum_{m=1}^{M} P(Y_{t+1} = j|y^t, x^{t+1}, m). \quad (2.11)$$

Here we used a uniform "prior" $\mu_M(m) = 1/M$, as it were, even though we are guided by the provably good properties of mixtures rather than by any consideration of prior knowledge. Finally, to get rid of the dependence on M, let $\hat{M}_t = \hat{M}(y^t, x^t)$ be the smallest value which maximizes

$$\prod_{s=0}^{t-1} P_M(y_{s+1}|y^s, x^{s+1}).$$

The final conditional probability function, after given the data (y^n, x^n), is $P(y|x) = P_{\hat{M}_n}(y|x, y^n, x^n)$. We are indebted to the referee for the remark that one could obtain a better result by using a more sophisticated prior which gives more weight to those values of m that give a shorter code length for the data.

References

[1] Dawid A.P. (1984) Present Position and Potential Developments: Some Personal Views, Statistical Theory, The Prequential Approach. *J. Royal Stat. Soc. A* (147) 2: 278–292.

[2] Krichevsky R.E. and Trofimov V.K. (1983) The Performance of Universal Coding. *IEEE Trans. on Information Theory* (IT-27) 2: 199–207.

[3] Rissanen J. (1983) A Universal Data Compression System. *IEEE Trans. Inform. Theory* (IT-29) 5: 656–664.

[4] Rissanen, J. (1986) Stochastic Complexity and Modeling. *Annals of Statistics* (14) 1080–1100.

[5] Rissanen J. (1987) Stochastic Complexity. *J. Royal Statistical Society, Series B* (49) 3: (with discussions) 223–265.

[6] Rissanen J. (1989) *Stochastic Complexity in Statistical Inquiry*. World Scientific Publ. Co., Suite 1B, 1060 Main Street, River Edge, New Jersey.

[7] Rissanen J. (1992) Information Theory and Neural nets. In *Mathematical Perspectives of Neural Networks*, Smolensky P., Mozer M. and Rumelhart D. (eds), Lawrence Erlbaum, (to appear).

[8] Rissanen J. (1993) Fisher Information and Stochastic Complexity. *IEEE Trans. Information Theory*. (Accepted.)

[9] Rissanen J. and Yu B. (1992) Learning by MDL (to appear).

[10] Wallace C.S. and Freeman P.R. (1987) Estimation and Inference by Compact Coding. *Journal of the Royal Statistical Society, Series B* (9) 3: 240–251, and 252–265 (with discussions).

[11] Weinberger M.J., Rissanen J. and Feder M. (1995) A Universal Finite Memory Source. *IEEE Trans. on Information Theory* (IT-41) 3: 643–652, May.

3

MML Inference of Predictive Trees, Graphs and Nets

C.S. Wallace

3.1 A Critique of Computational Learning Theory

The term "learning" here is used in a restricted sense. In human terms, we use the term to cover a number of related but distinct activities, including at least:

Rote learning: getting fixed into the memory a set of essentially arbitrary facts, *e.g.* the names of the days of the week, the names of the students in a class, the genders of French nouns, *etc.*

Inductive learning: detecting and perhaps formally expressing an apparent regularity or pattern in a body of data about many particular instances or events. Included in this meaning of learning are human activities such as learning a language simply by osmosis from one's parents, the discovery of "scientific" laws, the classification of things and phenomena into classes and the attachment of names to these classes, learning patterns of behaviour through punishment and reward, *etc.*

Learning "how to": learning how to walk, to waterski, to integrate functions, *etc.* Obviously, there is an element of inductive learning in learning how to do things, but the induction is usually not formally expressed, and in some cases may be almost incapable of verbal expression.

Learning from instruction: usually an activity designed to bypass the need for inductive learning. A pattern, rule, scientific law or classification previously learnt by another person or persons is expressed, learned initially by rote, but then "internalized" as the learner appreciates how the pattern or rule does indeed fit the data available to the learner.

Learning deductions: *e.g.* learning theorems such as Pythagoras's. Here, the learning may be guided by instruction, or the unaided discovery of the theorem. In the latter case, some element of inductive reasoning may well have entered into the process, but in both cases the process is essentially deductive.

We will here restrict attention to inductive learning, and consider how this activity may be initiated by computers.

3.1.1 Computational Learning Theory

The theoretical study of inductive learning has been called computational learning theory. Two well-known theoretical analyses were initiated by Gold and Valiant, respectively and will be briefly outlined. Unfortunately, neither stream of thought has born much fruit except theorems and papers.

3.1.2 Gold's Model

Gold suggests that all inductive learning can be regarded as special cases of a single general induction problem: to learn the grammar of a language given sentences from that language. This model indeed seems general enough to cover all the learning activities of interest here, and has been widely accepted.

A distinction can be drawn between two learning situations. In one, the learner is given only a set of sentences from the lanaguage. In the other, the learner is given two sets of sentences, one set from the language and the other containing sentences which are not in the language (negative instances). While there may be pragmatic differences between the two situations, there is little distinction in principle. The task of learning a language L_1 given both positive and negative instances is equivalent to the task of learning a slightly different language L_2 from positive instances only, where the sentences of L_2 have the form

$$< \text{sentence from } L_1: \text{``Yes''} >$$
$$\text{or}$$
$$< \text{sentence not in } L_1: \text{``No''} >.$$

For simplicity, we will discuss learning from positive instances only.

3.1.3 Gold's Theory

Without loss of generality, sentences in any alphabet (or sensory modality) may be recoded as binary strings, and the set of finite binary strings mapped onto the natural numbers. The theory developed by Gold and his followers regards a language as a (possibly infinite) set of finite sentences, and hence as a subset of the natural numbers. The theory addresses only languages for which the subset is effectively computable. Theoretical work has concentrated on characterizing those languages which are "learnable in the limit", *i.e.* languages such that the learner can, given sufficient instances, be guaranteed ultimately to discover just this effective computation.

Clearly, the unrestricted problem here posed is insoluble, since, whatever finite set of sentences (and negative instances) has been seen by the learner, there are infinitely many languages compatible with this data.

The theory therefore considers that the language is drawn from some defined subset or class of languages, and asks which classes are such that any language within the class can be "learnt" given sufficient data and the knowledge that the language belongs to the class. This assumption is sufficient to make any conclusions of the theory almost useless, since in the real world we rarely know a *priori* that what we wish to learn lies in some well-defined subset of possible grammars.

Even with this Procrustean simplification, theory based on Gold's model remains unable to say much of interest. Only a rather small number of restricted classes of language can be shown to be learnable in Gold's terms, and in many of these classes negative as well as positive instances are needed.

Gold's model is an acceptable model of inductive inference, and, we believe, can be made more productive by one small change. Rather than viewing a grammar as a subset of the natural numbers, and learning as a process for discovering this subset, it is both more realistic and more useful to regard a grammar as a probability distribution over the natural numbers, and learning as a process for approximating this distribution.

3.1.4 Valiant's Model

Valiant and his followers, perhaps dismayed by the virtual impossibility of learning in Gold's model, consider a much less ambitious framework. Rather than taking the learning of a language as the archetype problem, they consider the much simpler problem of "learning" a Boolean function of a known, finite number of variables, given a data set comprising many instances of variable values and the resulting function value. Of course, if the data set is exhaustive, containing all possible combinations of variable values, the problem is trivial. Instead, it is assumed that the number of variables K is fairly large, and the number of data instances $N \ll 2^K$. Clearly, one can no longer expect any learning process to guarantee discovery of the "correct" Boolean function, so instead the theory asks whether the learning process will probably learn a function which is approximately in agreement with the true function, *i.e.* in agreement for the great majority of possible combinations of variable values. If the number of known function values N is less than half the number 2^K of possible variable value combinations, the problem is still in general insolvable even by the relaxed Probably Approximately Correct (PAC) criterion.

Valiant's scheme, like Gold's, attempts to make the unsolvable learning problem solvable by imposing an *a priori* restriction on the set of possible functions, and allowing knowledge of this restriction to be used in the learning process. But in Valiant's model, the finite nature of the thing to be learnt has the result that, if the restriction on the set of possible functions is strong enough to make PAC learning possible, then the learning is in principle trivial. Almost all functions within the permitted set and which are compatible with the known data *must* agree with the target function for the great majority of cases. Thus, "learning" in principle requires nothing more than testing the permitted functions until one is found compatible with the data.

The theory based on Valiant's model is, however, not mathematically trivial. Rather than considering which restricted sets of functions are PAC-learnable in principle, attention has focused on what restricted sets are PAC-learnable in polynomial time. As with Gold's model, the theoretical results are not encouraging. Relatively few function sets appear to be tractable.

3.1.5 The Deductive Fallacy

The kinds of computational learning theory outlined above have given little insight into human or machine learning, and have difficulty accounting for apparently successful learning by humans and algorithms such as C4.5. Their failure suggests that their underlying models of the learning process are flawed.

Some obvious limitations of their models are:

Learning, as they define it, is possible only if that which is to be learnt is known *a priori* to be in a restricted set;

There is no inherent preference order among theories all compatible with the data;

There is only limited consideration of erroneous data;

The theories almost totally neglect the only well-proven, if limited, technique for systematic induction, *viz.*, the theory of statistical inference and estimation.

Most importantly, learning is treated by Valiant as a form of deductive reasoning. In his model, what is learnt is considered learnt only if it can be deduced from the data, or (in PAC learning) a probability statement about it can be deduced from the data.

As has been sufficiently demonstrated by Popper, deduction can play only a negative role in learning. Deductive reasoning at most allows the data to rule out some possible languages, functions, *etc.* but never allows the data to support what we learn. By concentrating on provable theories about what can be proved from the data, computation learning theory has totally missed the point.

3.2 Learning from Data

3.2.1 Learning is NOT Deduction

Inductive reasoning is reasoning from data about particular instances to a general conclusion intended to apply to all similar instances, including ones not in the data. Such reasoning is an essential component of learning as here defined. Unlike deduction, induction leads to a conclusion which:

  a. is not provably correct;

  b. is (usually) not exactly true;

  c. is expected to be revised, refined or possibly overthrown in the light of new data;

  d. is probabalistic rather than definite.

Since, in practical cases, the given data is rarely wholly reliable and exact, the conclusion of an induction will, if only tacitly, be understood to make only a probabalistic assertion about data instances. It is thus possible to arrive at a conclusion which, when stated absolutely rather than probabalistically, appears to be contradicted by some of the given data. It is entirely reasonable to draw the inductive conclusion

$$\text{Force} = \text{Mass} \times \text{Acceleration}$$

from a body of instances where no instance exactly agrees with the conclusion. The conclusion should really be read as asserting something like

> "Measured force is randomly distributed with a Normal distribution of mean (Measured mass x measured acceleration) and a standard deviation of 1% of the mean."

3.2.2 Learning as a Competition

If we accept the above rather modest ambitions for what is learnt, there will usually be many conclusions which could be drawn from the data, and the above criteria must be augmented by some way of choosing among competing conclusions. (Having emphasized the necessarily tentative character of inductive conclusions, we will hereafter call them "theories".)

The components from which a criteria of choice among competing theories may be built are usually considered to be:

a. Degree of fit to the known data. A theory under which some data instances seem unlikely is not favoured, one which predicts the data instances to be likely is favoured;

b. Complexity. Other things being equal, simple theories are to be preferred to complex ones (Occam's Razor);

c. *A priori* plausibility. A theory is not favoured if it seems at odds with prior knowledge. In a way, this may be interpreted as meaning that the theory should fit not only the given data, but also any previous data which seems relevant;

d. Specificity. A theory which fits the data well only if some of its details (*e.g.* unknown parameters) are given very precise values is less favoured than one which fits the data well over a wide range of detailed assumptions;

e. Generality. Other things being equal, a theory is preferred if it is applicable to a wide range of (as yet unseen) possible instances. A theory which makes an assertion only about just those instances in the known data is not much use.

3.2.3 Explanations

Before advancing our proposed criterion of choice among competing theories, we introduce the notion of an "explanation".

Recall that the available data may be considered to be encoded as a binary string D.

An *explanation* of the data is a recoded binary string, also encoding the data, but having a particular form. An explanation is a two-part string. The first part (A) encodes a theory or inductive conclusion θ drawn from the data. The second part (B), encodes the data using a code which would be optimal were θ true. Here, optimality means minimizing the expected length of B.

The theory of optimal codes is well developed and based on Shannon's theory of Information. For our purposes, we need to know that:

a. A string optimally encoding a datum of probability p has length $-log_2 p$ (we neglect fractional parts);

b. A string optimally encoding a set of data drawn from the probability distribution assumed in constructing the optimal code has the statistical properties of a random binary stream;

c. Well-known techniques exist for the construction of an optimal code for any given probability distribution. These techniques produce code strings at most one digit longer than the theoretical optimum $-log_2 p$. Common techniques are Huffman codes and Arithmetic codes (Langdon [23]). We will not be concerned with the details of these code constructions.

3.3 The Message Length Criterion

We prefer that theory giving the shortest explanation of the data.

From the above, it is clear that the length of the second part B of an explanation is

$$-\log[\text{Prob (data) assuming } \theta \text{ is true}] = -\log P(D \mid \theta)$$

where θ is the "theory" asserted in the first part (A).

Note that Length (B) is small if θ fits the data well, large if θ fits the data poorly.

The encoding of part A of the explanation is slightly trickier. Formally, if we define

$$\text{Prior } (\theta)$$

to be the probability we would assign to the theory θ before the data is seen, then the length of part A, using an optimal code, is just

$$\text{length } (A) = -\log \text{ Prior } (\theta)$$

whence the total length of the explanation E is

$$\text{length } (E) = -\log \text{ Prior } (\theta) - \log P(D \mid \theta) = -\log \text{ Prob } (\theta \text{ and } D).$$

Usually, it is not immediately obvious how we might assign a prior probability to each competing theory. Indeed, if the range of theories being contemplated includes a continuum, *i.e.* a family of theories parameterized by real numbers, it is not obvious that a particular theory with particular parameter values can be assigned any prior probability at all!

However, we can usually attack the question of prior probability from the other end. Instead of asking how to assign prior probabilities to theories, we ask how can we describe a theory? That is, what kind of language would be reasonable for describing a theory?

For instance, if we are considering trying to learn a Boolean function, as in the Valiant model, we might reasonably decide that the theory, *i.e.* the hypothesized function, could be represented in a code more or less as we would represent it on paper: as a string of variable labels and Boolean operators:

$$A.B \text{ or (not } C \text{ or } D). \text{ not } A.$$

It is quite easy to see how, knowing the number of variables on which the function might depend, a fairly efficient binary encoding of such formulas can be designed. We may take such a code as defining a prior probability distribution over functions. In effect, the code assigns a low probability to functions represented by long, complex formulae, and high probability to functions having simple formulae.

3.3.1 Parameter Coding

When the hypothesized theory has one or more real-valued parameters, prior expectations will not usually assign finite prior probability to any specific value. Rather, prior knowledge will define a prior probability *density* over the continuum of possible values. (Often, it is rather the case that the lack of prior knowledge, together with symmetry arguments, will lead us to adopt some form of "uninformative" prior density, *e.g.* uniform.)

Having defined a prior density, the coding of a parameter value in part A of an explanation involves a compromise. Typically, the length of part B will be minimized by some single value of the parameter, so we would like part A to state this value. If we state it to high precision (many binary places), the length of B is minimized, but the coded value in A will be long. If we state it imprecisely, rounding off to a few binary places, A becomes short but in general B will be longer than its minimum possible length.

The best compromise states a parameter value to a precision δ or maximum roundoff error $\delta/2$, where

$$\delta = \frac{1}{\sqrt{\kappa_k^k F}}$$

where F is the Fisher information, or the expected value of the second derivative of $-\log \text{Prob}(D \mid \theta)$ with respect to the parameter, and κ_k is a constant depending on the dimensionality k of the parameter (*i.e.* on the number of free scalar parameters).

This choice of precision is essentially the precision to which we expect to be able to estimate the parameter.

3.3.2 Bayesian Parallel

In general terms, given data D, a family of possible theories parameterized by the k-dimensional parameter θ, and a prior density $h(\theta)d\theta$, the length of an explanation asserting theory $\hat{\theta}$ is

$$L = -\log \frac{h(\hat{\theta})}{\sqrt{\kappa_k^k F(\hat{\theta})}} - \log \text{Prob}(D \mid \hat{\theta}) + k/2$$

where κ_k is a geometric constant, the last term arises from round-off error in stating $\hat{\theta}$, and logs are to base e. (So the length is in units of $\log_2(e)$ binary digits.)

Choosing the theory $\hat{\theta}$ to minimize L is equivalent to maximizing

$$\frac{h(\hat{\theta})}{\sqrt{F(\hat{\theta})}} \cdot \text{Prob}(D \mid \hat{\theta})$$

$$\text{Now } h(\hat{\theta})\delta = \frac{h(\hat{\theta})}{\sqrt{\kappa_k^k F(\hat{\theta})}}$$

is approximately the total prior probability mass assigned to all theories within $\pm \delta/2$ of the stated theory $\hat{\theta}$. Since the data does not allow us to distinguish among these theories, we may regard $h(\hat{\theta})\delta$ as being the finite prior probability of $\hat{\theta}$, and indeed the length of part A, stating $\hat{\theta}$, is just
$-\log(h(\hat{\theta})\delta)$.

Thus, minimizing L is equivalent to choosing $\hat{\theta}$ to maximize

$$\text{Prob } (\hat{\theta}) \cdot \text{Prob } (D \mid \hat{\theta}) = \text{Prob } (D) \cdot \text{Prob } (\hat{\theta} \mid D),$$

i.e. equivent to choosing $\hat{\theta}$ to maximize the Bayesian posterior probability

$$\text{Prob } (\hat{\theta} \mid D).$$

Thus our proposed criterion is a close parallel to Bayesian statistical inference. However, conventional Bayesian analysis has no general, principled method of choosing among a continuum of theories: it yields only a posterior probability density over the continuum. Further, conventional Bayesian analysis does not use coding theory to assign prior probabilities to structural "theories" such as Boolean functions.

3.4 Theoretical Results

Based on the modified Bayesian formulation of Minimum Message Length learning, the following results have been shown.

3.4.1 Sufficiency

The statement of the "theory" in part A contains (within 1 bit) all the information relevant to the theory that was available in the data. (Wallace & Freeman [16], Rissanen [24]). Were this not so, part B could not be optimally coded.

3.4.2 Avoidance of Overfitting

Part A contains (within 1 bit) no information from the data which was not relevant to the theory (Wallace & Freeman [16]).

Note that the above results do not show that part A is a good representation of the available information, only that it contains the information, perhaps in a misleading form (*e.g.* parameter estimates might be biased).

3.4.3 Consistency

As the amount of data increases, the "theory" stated in part A will converge in probability to the correct theory, iff the code used in part A can represent the correct theory (Schwartz [22], Barron and Cover [3]).

3.4.4 Oracular Imitation

Suppose two general learning processes X and Y are tested on a sequence D of data sets, not necessarily all relating to problems of the same kind. They produce respectively two sequences of inferences or theories, T_x and T_y. Now suppose that we are asked to choose the better of the two processes X and Y, using knowledge of D, T_x, T_y and the circumstances (sample size, domain, prior expectations *etc.*) of each data set. The choice must be made without knowledge of the "correct" theories, or of the internal workings of X or Y.

Whatever criterion of choice is used, it should be fair, *i.e.* have no in-built preference for either X or Y.

It can be shown (Wallace [11]) that:

- a. There exist inference processes, called false oracles, such that if X is a false oracle and Y an oracle, no fair criterion will be expected to distinguish between them;
- b. False oracles are stochastic: they may yield different inferences from repetitions in D of the same data set;
- c. The MML inference process is deterministic, and hence not a false oracle. However, it imitates a false oracle in the same sense that a good pseudo-random number generator imitates a random process.

This curious result is a consequence of the following:

If several data sets (*e.g.* the sequence D) are encoded as a single explanation, the part A will state a sequence of inferences, one per data set, then part B will encode all the data. When this is done, it turns out that the shortest explanation is obtained when the inference made on one data set is allowed to depend (slightly) on other data sets. The simplest illustration of this phenomenon is when the sequence D contains two unrelated samples from two different Normal distributions of unknown means μ_1 and μ_2. Suppose that, for both data sets, the sample size is N and the standard deviation is known to be 1.0. Then, part A should state estimates $\hat{\mu}_1, \hat{\mu}_2$ each with precision about $\sqrt{12/N}$, *i.e.* differing from the respective sample means m_1, m_2 by at most about $\sqrt{3/N}$. The expected increase in the length of part B is proportional to the squares of the roundoff-errors $(\hat{\mu}_1 - m_1), (\hat{\mu}_2 - m_2)$. This is minimized when the set of $(\hat{\mu}_1, \hat{\mu}_2)$ pairs which can be stated in part A is chosen to be an hexagonal lattice rather than a rectangular lattice. This is a simple and well-known result in the theory of quantizing lattices.

Given the data, the explanation will choose the $(\hat{\mu}_1, \hat{\mu}_2)$ pair closest to (m_1, m_2). Were the lattice rectangular, this would give the $\hat{\mu}_1$ closest to m_1, and the $\hat{\mu}_2$ closest to m_2. But in an hexagonal lattice, the choice of $\hat{\mu}_1$ will depend in part on m_2.

For a long sequence of data sets, it can be shown that this kind of optimal encoding has the result that the "theory" chosen for one data set amounts to a pseudo-random choice of a theory from the posterior distribution over theories given that data set, where the data from other data sets play the role of a source of pseudo-random digits. This enables MML to imitate a false oracle.

3.4.5 Invariance

Unlike most Bayesian methods of theory selection and estimation, MML is invariant under arbitrary monotonic transformations of parameter spaces.

3.4.6 Kullback-Liebler Distance

The "false oracle" result applies to situations where many inferences are made at once. For the more common situation, where a single theory with a few free parameters is to be learnt from a single, finite data set, general theoretical results are few. Sufficiency, lack of overfitting and invariance are nice, but say nothing about bias or estimation error. In the Bayesian context, perhaps the most general criterion for assessing the success of an inference, when the "true" model is known, is the Kullback-Liebler distance between the data distribution $P(D \mid \hat{\theta})$ implied by the inferred model and the data distribution $P(D \mid \theta)$ of the true data source.

$$KL(\hat{\theta}, \theta) = \int_{\text{all } D} P(D \mid \theta) \log \frac{P(D \mid \theta)}{P(D \mid \hat{\theta})} dD.$$

An ideal Bayesian inference from data D might then be that model $\hat{\theta}$ hat which minimizes the expected KL distance $E[KL(\hat{\theta}, \theta)]$, the expectation being taken over the posterior distribution of θ

$$EKL(\hat{\theta} \mid D) = C \int_\theta h(\theta) P(D \mid \theta) \int_{\text{all } D'} P(D' \mid \theta) \log \frac{P(D' \mid \theta)}{P(D' \mid \hat{\theta})} dD' d\theta$$

where C is a constant independent of $\hat{\theta}$. Unfortunately, minimization of EKL is extraordinarily difficult except in very simple cases.

However, it can be shown that MML inference approximately minimizes EKL, and in simulation tests involving model selection and estimation in a number of domains, MML has been found on average to do better at minimizing $KL(\hat{\theta}, \theta)$ than some inference methods (*e.g.*, AIC, Akaike Information Criterion) specifically intended to minimize EKL.

3.5 The Kolmogarov-Chaitin Formulation

Hitherto, we have discussed MML in a Bayesian context, assuming a prior probability distribution or density over possible inferences, even if this distribution was defined by a choice of language rather than *vice versa*. There is another very general way of considering the encoding of data via explanations.

Recall that the data is available as a binary string D. Given some universal Turing machine T_1, with one-way binary input and output tapes, we consider input tapes (binary strings) I such that T_1, given input I, will write D and then try to read more input. Any such I, of length shorter than D, is considered an explanation of D, and the shortest such I is preferred.

Actually, we impose a technical requirement on the form of I, *viz.*:

$$I = I_A : I_B$$

where

  a. Output from I_A is null;

  b. Output from $I_A : I_B = D$;

  c. For any string X.

Output from $I_A : I_B : X = D :$ (Output from $I_A : X$). (For any I having such a form, the division into the parts I_A, I_B is unique.) This requirement allows us to identify I_A as a statement of an inference, and I_B as an encoding of D assuming the truth of the inference. Note that the choice of UTM T_1 in effect expresses a "prior" on inferences.

3.5.1 Theoretical Results from Kolmogorov-Chaitin Form

3.5.1.1 Significance

If D is random, no explanation can be found (Chaitin [25]).

3.5.1.2 Convergence

If D is drawn from a computable probability function, then as more data is added, with probability 1 the input I_A will be a program for computing this function, and will not change with further data (Barron & Cover [3]).

If D is drawn from a non-computable function which is computable given a finite number of real values, then the function computed by I_A will converge to this function in the limit. The rate of convergence is (usually) $1/\sqrt{\text{datasize}}$ (Barron & Cover [3]).

3.6 Prediction vs. Induction

So far, we have discussed ways of inductively inferring a model of the real world from a set of data. That is, we discussed how to arrive at a general proposition about the source of the data.

If a good inductive inference is drawn, it may be used to make predictions about future data from the same source. In the Bayesian formulation, the model found to give the best explanation of the data is typically a probability distribution over the population of all similar data, and so can be used directly to attach probabilities to possible future observations and outcomes. As mentioned in Section 3.4.6, the MML inferences approximately minimize the expected KL distance between the inferred model and the real word. Hence, they approximately minimize the degree to which future data will surprise us.

Sometimes a more precise objective may be stated for predictions. In these cases, we may know the cost of predicting A when in fact B is the case, cost $= C(A, B)$. This cost function may be strongly asymmetric, and not closley related to KL distance. In such cases, conventional Bayesian minimum-cost estimation (or prediction) may be useable, and should be preferred to inductive inference if feasible. Here, we choose to make that prediction or estimate, or choose that action, which has minimum expected cost. The expected cost of prediction A is the expected value of $C(A, B)$, taken over the posterior distribution of B given the data

$$EC(A \mid D) = \int d\theta h(\theta) P(D \mid \theta) C(A, \theta) P(D).$$

This way of making predictions or decisions is theoretically unchallengeable if its premises are true. Note that it involves no commitment to any one model of the real world, no inductive inference and no learning other than the accumulation of data. It is a purely deductive process using the axioms of probability theory.

Unfortunately, true minimum-cost inference is often infeasible, and cost functions are often unknown at the time a prediction must be made. We resort to induction, the (*pro tem*) commitment to one model of reality, because we are often unable to make minimum-cost inferences.

Predictions made on the basis of inductively-learned models will therefore rarely be optimal with respect to a given cost function. It is sometimes possible to improve such predictions by using not just the one best inductive inference, but by averaging over a small number of reasonably good models.

3.6.1 Prediction in the Kolmogarov-Chaitin Formulation

Using the notation of Section 3.5, an "inference" I_A is made from data D using Turing machine T_1. Then a probability $P(X)$ cam be attached to any possible future observation, given by

$$P(X) = \sum_Y 2^{-\text{length}(Y)}$$

where Y is any binary string such that the output of T_1 given input $I_A : Y$ is X.

Often, the sum is dominated by a single term due to the shortest such input Y.

This formulation of predictive induction was first stated by Ray Solomonoff.

3.7 Applications

This is a small sample of applications of MML inference. Under each head is an outline of the message structure used in explanations, giving both the part 1 structure describing the model, and the part 2 encoding the data.

3.7.1 Learning the Class Structure of Populations ("unsupervised learning"). Boulton & Wallace [4, 5]

Part 1: Number of classes. For each class:

> Fraction of population;
>
> Parameters of statistical distributin of each variable, or "insignificant" flag.

Part 2: For each member of sample:

> Class to which it belongs; Variable values encoded assuming distribution parameters of variable in class.

3.7.2 Learning the Shape Patterns of Megalithic Stone Circles. Patrick and Wallace [8], Patrick [21]

(A comparison of Thom's Theory of megalithic geometry *vs.* a "null thoery" based on Fourier series of near-circular shapes.)

Part 1: (Thom)

> Dense encoding of Thom's family of ellipes circles and "flattened circles";
>
> Centre position, orientation, size encoded using Thom's "megalithic yard".

Part 1: (Null)

> Centre position, orientation, size;

Up to 3 co-efficients of Fourier series for radius as function of angle.

Part 2: (Both theories)

Radial position of stone encoded as Gaussian deviation from theory position (Angular positions assumed known).

3.7.3 Learning Simple Grammars. Georgeff & Wallace [5]

Part 1: A description of a probabalistic finite state automaton, as follows:

Number of states
For each state:
For each alphabet symbol:
Probability of occurrence;
Destination state.

Part 2: Initial state

For each sentence symbol:
Huffman code specifying symbol using symbol probability in current state.

3.7.4 Learning Rules in the Form of Decision Trees. Quinlan & Rivest [10], Patrick & Wallace [19]

Part 1: (Case attribute values assumed known. Explanation encodes classes.)

A description of tree, being a prefix-ordered recursive description of nodes. Node description is:
leaf or non-leaf flag;
If leaf, probability distribution over classes;
If non-leaf:
Decision attribute followed by description of child node for each attribute value.

Part 2: For each case:

Class encoded using probability distribution in leaf implied by case attribute values.

3.7.5 Learning Decision Graphs. Oliver [6, 7]

Explanation as for 6.4, but part 1 elaborated by inclusion of "join" nodes at which two branches of tree join. A superior model for decision rules including disjunction.

3.7.6 Inferring Relationships among DNA Strings. Allison & Yee [1, 2]

For any pair of strings:

Part 1: Parameters of a finite state "mutation engine". Parameters include probabilities of single-base mutation, insertion or deletion, and multi-base insertion or deletion.

Part 2: A list of "instructions" to the mutation engine for changing one string into the other. Instructions coded using part 1 parameters.

For a set of related strings:

Part 1: An inferred "evolutionary tree" and an assumed "ancestral string".

Then for each arc of the tree, a mutation engine part 1 as above.

Part 2: For each arc of tree:

Mutation instructions as above.

3.7.7 Learning Existence and Effects of Hidden Variables in Linear Models. Wallace & Freeman [17]

Part 1: Number of latent variables.

Parameters of classical factor model.

Estimated factor scores for each case.

Part 2: Case variable values encoded as Gaussian deviations from expected values implied by factor loads and scores.

3.7.8 Inference of Prolog Programs. Muggleton [26]

Part 1: A logic program.

Part 2: A sequence of deductive steps.

3.8 Coding Techniques

In using MML as a principle of induction, it is not necessary actually to construct the binary string which is the explanation. Rather, one need only be able to calculate the length which this explanation would have, were it to be constructed. If actual explanation strings were in fact constructed, it would be found that the length of an explanation, necessarily an integer number of binary digits, would be a discontinuous and very complicated function of the parameters of the hypotheses used in the explanation. For practical use of MML, it is not only permissible but highly desirable that expressions be found which give smooth, ideally differentiable, approximations to the actual explanation length. All MML applications discussed here make use of such approximations, and we discuss below some of them.

It is important to note that while use of approximations to message lengths are essential on pragmatic grounds, the approximations should not be too sloppy. Recall that the difference in the lengths of two competing explanations of the same data is more or less the logarithm of the posterior odds ratio in favour of the shorter explanation. Thus a length difference of 10 binary digits expresses *a posterior* odds ratio of over 1000:1, which would normally be regarded as a very strong preference for the shorter explanation. Since such rather small length differences have highly significant implications, one needs to be sure that errors in calculating message lengths are kept very small, ideally only one or two bits. Use of sloppy approximations can easily lead to length differences which appear highly significant, but which are merely due to approximation errors.

Note that some kinds of approximation error can lead to errors in message length which are proportional to the message length. Thus, even a 0.1% approximation error can lead to an error of 10 bits in estimating the length of an explanation of around 10,000 bits (not unusual in a large data set.) But it is the *absolute* difference in explanation lengths, not relative difference, which expresses the significance of a preference between competing explanations. One must therefore either

(a) use very good approximations, or

(b) be satisfied that the approximation errors will affect competing hypotheses equally, or

(c) take message length differences with several grains of salt, and accept that one hypothesis is definitely to be preferred to another only if it gives an explanation shorter by a substantial number of bits. In practice, all three of these precautions should be followed where possible.

3.8.1 Encoding Real-Valued Data

Suppose some component of the data is a set of values thought to be independently-drawn random values from a probability density, where the parameters of the density distribution are unknown, and form part of the hypothesis to be inferred. Let the data values be $\{x_1, x_2, \ldots, x_n, \ldots, x_N\}$ and the hypothesized density be $f(x \mid \hat{\theta})$ with estimated parameter $\hat{\theta}$ stated in part 1 of the explanation. Then the length of the message fragment encoding, say, datum

x_n in part 2 of the message will depend on $f(x_n \mid \hat{\theta})$, and will be minus the log of the probability of drawing this value from the density $f(x \mid \hat{\theta})$.

If the datum $x_n = 1.31126$ is taken at face value as a real number, then the probability is zero: a probability *density* assigns probabilities only to intervals of variable values, not to point values. In fact, the datum must be interpreted as the inequality $1.311255 \leq x_n < 1.311265$ which, given $\hat{\theta}$, has the probability $f(1.31126 \mid \hat{\theta}) \times 0.00001$ approximately, provided $f(x \mid \hat{\theta})$ changes little over this interval. It can be seen that the *precision to which the data is stated* comes into the calculation of message length. Datum x_n, if stated to precision $\pm \delta/2$ will require a message fragment of length

$$-\log(\delta f(x_n \mid \hat{\theta})) = -\log f(x_n \mid \hat{\theta}) - \log \delta.$$

Since the precision δ of the data is a given of the data, and not a function of any inferred hypothesis, the term $-\log \delta$ will be a constant component of the explanation length, and hence will not (directly) appear in the length difference between competing hypotheses. In this case, it seems as if consideration of δ can be ommitted from the inductive process, and indeed it often, legitimately, is.

Why, then, is the precision of data worth mentioning?

(a) It is useful to remind oneself that data is always finite: it is always expressed in a form of finite length, and never contains arbitrary real values, only finite interval values and rationals;

(b) The real length of an explanation does depend on δ. If two different forms of explanation are attempted, one of which encoded a scaled or transformed form of the data, their explanation lengths will not be comparable unless account is taken of how δ is also scaled or transformed.

(c) Sometimes, the model density $f(x \mid \hat{\theta})$ can change singificantly over an interval of size δ. For instance, the age distribution in some human group may be modelled by a Normal density $N(\mu, \sigma^2)$, but the data may state ages only to the year. If some group arising in the hypothesis has a narrow spread of ages, say $\hat{\sigma} \approx 5$ years, then the probability of drawing the value 28 from $f(x) = N(35, 5^2)$ is not well approximated by $\delta f(28)$ with $\delta = 1. f(x)$ changes quite a lot over the interval 27.5 to 28.5. Accurate calculation of the probability requires a better approximation, e.g. Sheppards approximation Prob $(27.5 < x < 28.5) \approx \delta f^*(x)$ where $f^*()$ is the density $N(35, 5^2 + \delta^2/12)$. Use of this approximation is advised whenever the expected scale parameter of a density is not orders of magnitude greater than the data precision. Failure to use it (or analogous approximations for densitites of forms other than Normal) can result in silly estimates of scale parameters, "probabilities" greater than one, and negative message lengths.

3.8.2 Coding Real Parameters

In the first part of an explanation it is often necessary to state estimates of one or more real-valued parameters of the inferred model. As noted in Section 3.3.1, such estimates should be stated to a certain precision to minimize the explanation length, and this precision depends on the sensitivity of the log likelihood of the model to small perturbations in its parameter values. Suppose we write the data as D, the vector of unknown parameters as θ, and the probability law of the model as Prob $(D \mid \theta) = f(D \mid \theta)$. Then the estimated vector $\hat{\theta}$ should be rounded-off to a precision represented by a volume *delta* of the parameter space, where

$$\delta = \frac{1}{\sqrt{\kappa_k^k F(\hat{\theta})}}$$

where k is the dimension of the parameter vector, and $F(\theta)$, the Fisher Information, is the determinant of the matrix of expected second partial derivatives of $-\log f(x \mid \theta)$ with respect to the components of θ, evaluated at $\hat{\theta}$. That is, if we write θ as $(\theta_1, \theta_2, \ldots, \theta_k)$, then the (i,j)th element of the matrix is

$$-E \frac{\partial^2}{\partial \theta_i \partial \theta_j} \log f(y \mid \theta) = -\int dy f(y \mid \theta) \frac{\partial^2}{\partial \theta_i \delta \theta_j} \log f(y \mid \theta).$$

Note that $F(\hat{\theta})$ is a function of the estimate $\hat{\theta}$ but not of the data. Thus, a part-1 message stating $\hat{\theta}$ can be encoded to be intelligible to a receiver who does not (as yet) know the data D.

There is no problem in allowing the precision for stating $\hat{\theta}$ to depend on $\hat{\theta}$. All this means is that the discrete set of estimate values which may be stated in part 1 are not necessarily evenly-spaced in the parameter space.

If we think of the discrete set of possible parameter values as a set of points in the parameter space, we have a choice, in choosing this set and designing a code for part 1, of how best to arrange the points in parameter space. It turns out that, in a region of space where the ratio $h(\theta)/\sqrt{F(\theta)}$ is slowly-varying, the estimate points should be arranged so as to minimize the average squared distance of any point in the region to the nearest estimate point. Such an arrangement is called an "optimum quantizing lattice". The constant κ_k relates to the resulting mean squared distance for a unit density of estimate points. In one dimension, $\kappa_1 = 1/12$, and the estimate points are just laid out along the line. In two dimensions, the optimum arrangement is an hexagonal lattice. In three dimensions, it is believed to be the same as the lattice of centres of close-packed spheres. For higher dimensions, the best arrangement is unknown, but quite tight bounds are known for κ_k. See TR 89/128. Curiously, for large numbers of parameters (large k) a random arrangement of estimate points is asymptotically optimal.

The uncertainty about the exact value of κ_k for $k > 3$ fortunately leads to an uncertainty in calculating explanation lengths of less than one bit.

Note that the best arrangement is never a rectangular lattice. There is an odd but important consequence. Suppose θ has two components, θ_1 and θ_2, and that some of the data, say D_2 is irrelevant to estimation of θ_1. That is, suppose that

$$Prob(D_2 \mid \theta) = Prob(D_2 \mid \theta_1, \theta_2) = Prob(D_2 \mid \theta_2).$$

Then, nonetheless, the estimated value $\hat{\theta}_1$ of θ_1 will be found to depend upon D_2. As D_2 is varied, $\hat{\theta}_1$ may take a range of values, all perfectly reasonable but differing slightly. In fact, for many parameters, it can be shown that as data irrelevant of θ_1 is varied, $\hat{\theta}_1$ behaves like a value randomly selected from the Bayesian posterior distribution of θ_1.

3.8.3 Coding Discrete Parameters

Where the hypothesis family used in the explanation has a small number of discrete, or structural, parameters, the best one can do is try to design a code for these parameters which is non-redundant, and which encodes most briefly those models or hypotheses which are intuitively simplest. For instance, if the hypothesis involves a binary tree of unknown size and shape, one would expect to use a code which encodes trees with few nodes in a short string, and uses long codes for trees with many nodes. See Sections 3.7.3, 3.7.4.

From experience with real-valued parameters, we know that one should try to encode one's hypothesis no more precisely than is warranted by the data. With small numbers of structural or discrete parameters, it is not clear how one could encode their values "imprecisely", and one must simply accept that MML may lead to choosing an hypothesis whose structure is stated without qualification, even though the data is reasonably compatible with other structures.

However, there are cases in which "imprecise" coding of discrete parameters is possible, and should be done. An example is fully described in the reference of 3.7.1, and the theory elaborated in TR 89/128. Suppose that the hypothesis contains many discrete unknown parameters, and that the given data suggest that many of them could well be given two or more different values without much affecting the message length. Essentially, the idea is this: If it does not much matter whether discrete parameter θ_i is stated to be A or B, (where A and B are two possible discrete values), but the specification of an hypothesis requires some value to be stated, then base the choice between A or B on some binary digit which will appear later in the message, and which occurs after all parts of the message relevant to θ_i. Then that binary digit may be omitted from the message, since, by the time the receiver requires it, she has all information relevant to θ_i, can hence deduce that the transmitter could have chosen either of values A or B indifferently, and knowing the coding conventions used, can then recover the missing digit from the transmitter's choice of value for $\hat{\theta}_i$.

MML INFERENCE OF PREDICTIVE TREES

Although this may seem a most contrived device, using it not only shortens the explanation but also can remove bias in the estimation of other parameters.

The idea generalizes to choice among any number of possible discrete values, and its use in MML does not require actual construction of code strings using the trick. It can be shown that when the trick is used, the resulting explanation, although stating a definite value for θ_i, has a length given by

$$-\log(\exp(-L_A) - + \exp(-L_B))$$

where L_A and L_B are the lengths the explanations would have were this coding device not used, and θ_i stated to have value A or B, respectively.

3.8.4 Prior and Progressive Codes

Suppose some part of the data is a sequence of binary values, and it is thought that these are uncorrelated, each value having the same, unknown, probability p of being one. The normal MML approach would be to include in part 1 an estimate $\hat{p}$ of p, and then in part 2 encode each binary value in sequence, using a code word of length $-\log \hat{p}$ for a one, and a word of length $-\log(1-\hat{p})$ for a zero. Suppose further that *a priori*, p is considered equally likely to have any value between 0 and 1, *i.e.* $h(p) = 1 (0 < p < 1)$.

Consider data containing N binary values, of which k are ones. Then the standard MML techniques give:

$$\hat{p} = (k+1/2)/(N+1).$$

Part 1 length $= \dfrac{1}{2}\log(N/12) - 1/2\log(\hat{p}) - 1/2\log(1-\hat{p})$.

Part 2 length $= -k\log\hat{p} - (N-k)\log(1-\hat{p}) + \dfrac{1}{2}$.

Total length $= \dfrac{1}{2}\log(N/12) - (k + 1/2\log(\hat{p})) - (N - k + \dfrac{1}{2}\log(1-\hat{p}) + \dfrac{1}{2}$.

It may well be that the value of p is not a question of great interest in the inference, and one might instead consider coding the sequence thus:

Encode the first value with one binary digit.

After having encoded $n < N$ values, of which say m were ones, encode the next value with a word of length

$-\log(m+1)/(n+2)$ if it is one, or

$-\log(n-m+1)/(n+2)$ if it is zero.

It is easily shown that using this "progressive code", the total length needed to encode call N values is

$$\log(N+1) + \log\binom{N}{k}$$

which may be interpreted also as the length of a different message which first encodes the number of ones (all numbers in the range 0 to N being equally likely *a priori*) and then states which of the $\binom{N}{k}$ possible arrangements of the ones in the sequence in fact occurred. Both codes are strictly optimal: the message length is exactly minus the log of the *a priori* probability of getting the sequence observed, and so the latter code may be termed a "prior" code. Its length is always shorter than that obtained by MML, just because the "prior code" message includes no estimate of the unknown probability p. However, the difference is small, less than one bit. (The difference may be interpreted as the log of the posterior probability of the MML estimate, given the data.)

In some MML applications, it may well be more convenient to calculate the length of the explanation of such a sequence by using the formulae for progressive or prior codes. While these lengths underestimate the MML length, the difference is small and may be neglected, especially if there is little interest in the unknown parameter p.

More generally, for data D, model probability function $f(D \mid \theta)$, and unknown (vector) parameter θ with prior density $h(\theta)$:

The length of a prior or progressive encoding of D is $-\log r(D)$ where $r(D) = \int h(\theta) f(D \mid \theta) d\theta$ (that is, $r(D)$ is the marginal prior probability of D), and the length of an MML explanation of D will exceed this length by approximately $\frac{1}{2} \log(\pi k)$ where k is the dimensionality of θ. This approximation is valid whenever the likelihood $f(D \mid \theta)$ has a single mode, and approximately quadratic behaviour about the mode. It may sometimes be more convenient to approximate the MML length by $-\log r(D) + \frac{1}{2} \log(\pi k)$ rather than by direct use of MML techniques.

References

[1] Allison L., Wallace C.S. and Yee C.N. (1992) Minimum Message Length Encoding, Evolutionary Trees and Multiple-Alignment. *Proc. of 25th Hawaii Int. Conf. on System Sciences*, Kauai, Hawaii.

[2] Allison L., Wallace C.S. and Yee C.N. (1992) Finite-State Models in the Alignment of Macro Molecules. *Journal of Mol. Evol.*, (35) 77–89.

[3] Barron A.R. and Cover T.M. (1991) Minimum Complexity Density Estimation. *IEEE Transactions on Information Theory* 37: 1034–1054.

[4] Boulton D.M, and Wallace C.S. (1970) A Program for Numerical Classification. *The Computer Journal* (13)1: 63–69, February.

[5] Georgeff M.P, and Wallace C.S. (1993) A General Selection Criterion for Inductive Inference. *Proceedings 6th European Conference on Artificial Intelligence*, September 1984, 473-482.

[6] Oliver J.J. (1993) Decision Graphs - An Extension of Decision Trees. *Proceedings of the Fourth International Conference on A.I. and Statistics*, Ford Lauderdale, Miami, U.S.A., Jan.

[7] Oliver J.J. and Wallace C.S. (1991) Inferring Decision Graphs. *IJCAI '91 Workshop 8*, Sydney.

[8] Patrick J. and Wallace C.S. (1977) Stone Circles: A Comparative Analysis of Megalithic Geometry. *48th ANZAAS Conf. Melbourne*.

[9] Quinlan J.R. (1992) *C4.5: Programs for Machine Learning*. Morgan Kaufmann, San Mateo, CA.

[10] Quinlan J.R. and Rivest J.L. (1989) Inferring Decision Trees Using the Minimum Description Length Principle. *Information and Computation* 80: 227–248.

[11] Wallace C.S. (1989) False Oracles and SMML Estimators. Technical Report 89/128, Department of Computer Science, Monash University, June.

[12] Wallace C.S. and Boulton D.M. (1968) An Information Measure for Classification. *Computer Journal* (11)2: 185–194.

[13] Wallace C.S. and Boulton D.M. (1975) An Invariant Bayes Method for Point Estimation. *Classification Society Bull.* (3)3: 11–34

[14] Wallace C.S. and Dowe D.L. (1993) MML estimation of the von Mises concentration parameter. Technical Report 93/193, Department of Computer Science, Monash University, Clayton 3168, Australia.

[15] Wallace C.S. and Dowe D.L. (1994) Intrinsic classification by MML – the Snob program. *Proc. 7th Australian Joint Conference on Artificial Intelligence*, UNE, Armidale, NSW, Australia, November 1994, 37–44.

[16] Wallace C.S. and Freeman P.R. (1987) Estimation and Inference by Compact Coding. *The Journal of the Royal Statistical Society, Series B, Methodology* (49)3: 252–265.

[17] Wallace C.S. and Freeman P.R. (1992) Single Factor Analysis by MML Estimation. *J.R. Statist. Soc. B* (54)1: 195–209.

[18] Wallace C.S. and Georgeff M.P. (1984) A General Selection Criterion for Inductive Inference. *ECAI '84; Advances in Artificial Intelligence*, Elsevier Amsterdam 1–18.

[19] Wallace C.S. and Patrick J.D. (1993) Coding Decision Trees. *Machine Learning* 11: 7–22.

[20] Gold E.M. (1967) Language identification in the limit. *Information and Control* 10: 447–474.

[21] Patrick J.D. (1979) An Information Measure Comparative Analysis of Megalithic Geometries, Ph.D.

[22] Schwarz G. (1978) Estimating the Dimension of a Model. *Ann. Stat.* (6)2: 461–464.

[23] Langdon G.G. An Introduction to Arithmetic Coding. (1994) *IBM J. Res. Dev.* (28)2: 135–149.

[24] Rissanen J. (1987) Stochastic Complexity. *Journal of the Royal Statistical Society* Series B (49) 3: 233–239.

[25] Chaitin G.J. (1975) A Theory of Program Size Formally Identical to Information Theory. *Journal of the Association for Computing Machinery* 22: 329–340.

[26] Muggleton S. (1994) Bayesian inductive logic programming. *COLT-94: 3-11, The Association for Computing Machinary*, New York.

[27] Solomonoff R.J. (1964) A formal theory of inductive inference. *Information and Control* 7: 376–388.

4

Learning and Reasoning as Information Compression by Multiple Alignment, Unification and Search

J.G. Wolff

4.1 Introduction

This chapter presents the tentative idea that "multiple alignment" in a sense which is close to the use of that term in bio-informatics, together with the full or partial merging or "unification"[1] of patterns, and a process of "search", is a framework within which learning and reasoning may be integrated.

This thinking is part of a programme of research aiming to develop the "SP" conjecture ("computing as compression") that *all kinds of computing and formal reasoning may usefully be understood as information compression by pattern matching, unification and search* (PMUS), and to develop a "new generation" computing system based on the theory [19, 20, 21, 24].

Learning and reasoning are both large subjects. In the space of one short chapter it is not possible to do more than present a few examples to suggest how these things may be seen in terms of multiple alignment, unification and search. Relevant issues will be discussed more fully elsewhere.

Research on "inductive logic programming" (ILP) is also concerned with learning and reasoning but the focus is different from the SP programme. In

---

[1] The term *unification* is used in this chapter to mean a simple merging of multiple instances of any pattern to make one. This idea is related to, but simpler than, the concept of "unification" as it is used in logic.

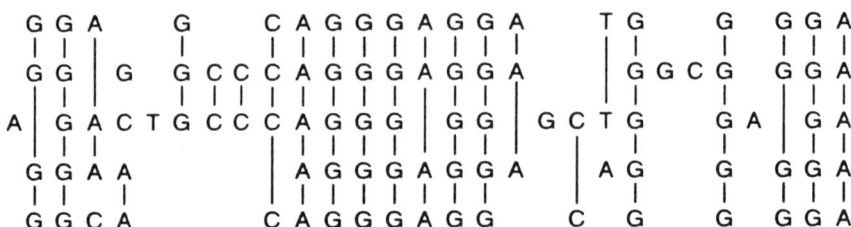

Figure 4.1: An alignment amongst five DNA sequences (adapted from Figure 6 in [13], with permission from Oxford University Press).

ILP (see, for example, [12]) the emphasis is on (supervised) learning within a framework of logic, whereas the SP programme seeks to integrate (unsupervised) learning and reasoning (and other aspects of computing) within a more general framework.

In this chapter, the term *pattern* is used to mean any sequence of atomic symbols including subsequences within a larger sequence, where the symbols in the subsequence are not necessarily contiguous within the larger sequence.

4.2 Multiple Alignment Problems

The term "multiple alignment" is normally associated with the computational analysis of DNA sequences or sequences of amino acid residues as part of the process of elucidating the structure, functions or evolution of the corresponding molecules.

The general idea is to examine two or more sequences to find one or more alignments of matching bases or amino acid residues which are, in some sense, optimal. An example is presented in Figure 4.1.

4.2.1 What is an "Optimal" Alignment?

Intuitively, an "optimal" or "good" alignment amongst two or more sequences of symbols is one which is relatively long, involves a relatively large number of sequences, has few gaps (or no gaps at all) and, where there are gaps, these should be relatively short.

LEARNING AND REASONING   69

These kinds of measures are used directly in many studies of multiple alignment. But it seems that our intuitions in this area may also be formalized in terms of probabilities and compression. These two kinds of analysis are complementary and not in conflict — they are two sides of the information theory coin. Recent studies in this tradition include [1, 2, 4, 16, 22].

4.2.2 Searching for Optimal Alignments

In this area of research, it is widely recognized that, when searching for good alignments between two or more sequences, the space of possible alignments is normally too big to be examined exhaustively. For realistically large examples it is necessary to use "heuristic" techniques (*e.g.* "hill climbing", "beam search") which may be described generically as "metrics-guided search". Existing methods are reviewed in [3, 4, 7, 15]. Also relevant is the method for finding alignments between two sequences which is described in [22].

Given the variety of techniques which already exist to find multiple alignments, one might expect to find one that could be incorporated in a model of learning, reasoning and other aspects of computing as discussed in this and other publications in this research programme. However, an examination of the literature has shown that existing techniques do not precisely meet the needs of the SP programme. Given the differences in objectives between biochemistry and the SP programme, this is not very surprising.

Work is now proceeding on a new search method within a new SP model. The new model is not yet sufficiently mature or robust to demonstrate all the points made in this chapter. However, it is good enough to solve the geometric analogy problem described in Section 4.5.7 and other simple examples.

The new model "broadcasts" symbols systematically to make yes-no matches throughout the set of sequences (in this respect, it is similar to SP20 [20]). The new model then builds alignments by pairwise conjunction of "hits" and groups of hits, guided at all stages by how much compression each alignment may achieve. In accordance with the principles of metrics-guided search, low scoring alignments are eliminated at each stage.

4.3 Unifications and Tags

In the context of biochemistry, multiple alignment means pattern matching and search but it does not normally include unification of patterns, the "U" of "PMUS".

In the context of "computing as compression", the merging or unification of two or more matching patterns to make one is the step which realizes compression. A full discussion is outside the scope of this chapter. It is pertinent here only to note a few points relating to later sections of the chapter.

If subsequences of one or more patterns are unified, non-redundant information about the *locations* of the deleted patterns is lost unless steps are taken to preserve it.

The most widely used technique for preserving this non-redundant information about locations is to insert a short "identifier", "code" or "tag" wherever a pattern has been deleted. Of course, the tags must be shorter than the patterns they replace to achieve a net compression. Tags may be newly-minted patterns which replace deleted patterns. It also possible to delete only part of a redundant pattern and leave the residue of the pattern as a tag to mark the position. In either case, it seems necessary to include with the tag some kind of termination or scope marker, as described briefly later in this chapter.

For the purposes of the SP programme, this simple idea of marking the positions of deleted patterns using tags has the attraction that pattern matching and unification can be applied to tags themselves, just like any other information. This means that PMUS mechanisms can be used for de-referencing tags and de-compressing information as well as for the original compression of information (see [19], Section 5.2).

4.4 Learning as Multiple Alignment with Unification and Search

In this discussion, "learning" will be taken to mean the kind of unsupervised assimilation and organization of knowledge which appears to be required for the learning of a first language by children or the "discovery", "inference" or "abstraction" of a grammar from a sample of an artificial or natural language. It seems that lessons learned in these areas may be applied to tasks like rule induction for expert systems, "data mining" and so on. The language learning task may be analysed into (at least) the following sub-tasks:

- Learning segmental (conjunctive) structure in unsegmented samples (*e.g.* words, phrases and sentences);

- Learning contextually-equivalent disjunctive groupings (*e.g.* word classes – nouns, verbs, adjectives *etc.* — and similar groupings of other linguistic structures);

- Generalization of grammatical rules and the correction of overgeneralizations;

- Learning semantic structures and their associations with syntactic forms.

The SNPR model [24, 25] developed in the tradition of "Minimum Length Encoding" (MLE, [5, 14]), is an unsupervised model of language learning which performs the first three of these tasks recursively to build hierarchical, context-free phrase-structure grammars (CF-PSGs) which may have conjunctive and disjunctive groupings at any level. The model is capable of discovering plausible grammars from unsegmented samples of artificial languages, forming generalizations and plausible corrections for overgeneralizations without correction by an external "teacher" or the provision of negative samples or the "grading" of samples in any way (cf. [11]). The MK10 model (see [25]), which is a sub-set of the SNPR model dedicated to the learning of segmental structures, is capable of discovering word structure in unsegmented samples of English.

Notwithstanding the known shortcomings of CF-PSGs for representing the organization of natural languages, the SNPR model exhibits several of the features of language learning which have been observed in empirical studies of children [25]. These are by-products of the organization of the model, not *ad hoc* features which have been bolted on.

Although the SNPR and MK10 models are largely applications of PMUS mechanisms, they were developed with objectives which were different from the SP programme and it is not altogether surprising that neither of them precisely meet current requirements. However, it is now possible to see in outline how these models may be re-organized to meet these broader objectives. More specifically, the first two of the tasks identified above may now be seen in terms of multiple alignment, unification and search.

4.4.1 Learning Segmental Structure

Figure 4.2 shows an alignment amongst a set of sentences generated by a simple grammar. Of all the possible alignments of these sentences, the one shown in the figure appears to be the best. This judgement is based on the compression that may be achieved by the unification of matching sequences of letters within the alignment.

The point of interest here is that, in the alignment, the patterns which we conventionally recognize as "words" stand out as coherent subsequences. The words in the sentences will fall out as discrete patterns when matching subsequences within the alignment are unified. The words are picked out despite the fact that, in the sentences, there are no explicit markers showing where each word ends and the next begins.

It is not hard to see that the same kind of process operating on sequences of symbols for grammatical categories may serve to identify segments larger than words such as phrases and sentences (cf. [26]).

Of course, the example is relatively simple and is artificial in the sense that, in a typical sample of natural language, it is unlikely that one would find a set of inter-related sentences neatly presented as shown. However, the principles illustrated by this example appear to be general and should apply even in the more haphazard environment of a natural language. Words and other segments may be identified from samples of language containing frag-

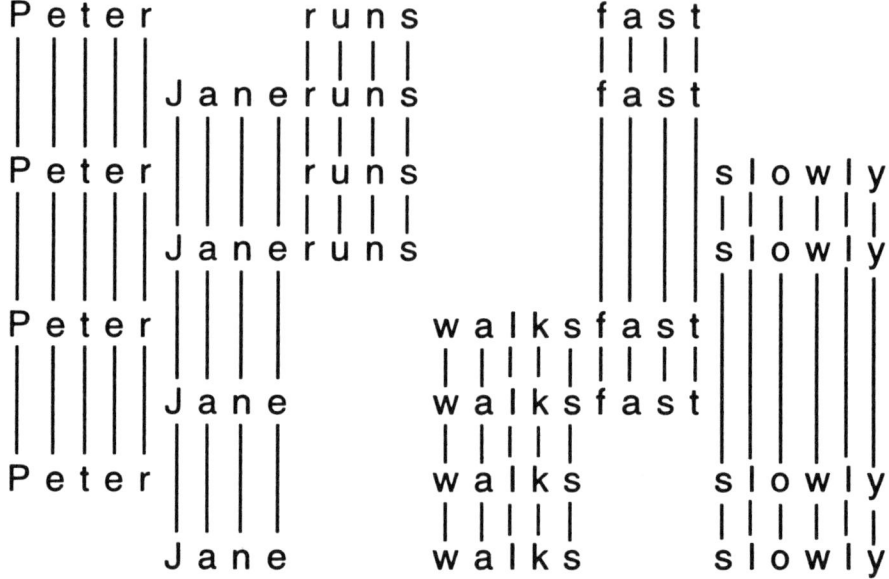

Figure 4.2: An alignment amongst some sentences.

ments smaller than sentences. For example, it only requires three fragments, "r u n s s l o w l y", "r u n s f a s t", "w a l k s f a s t" for one to make a preliminary judgement that "r u n s", "f a s t" and "w a l k s" are discrete segments.

4.4.2 Learning Disjunctive Classes

A key idea in "distributional" or "structural" linguistics (see, for example, [8]) was that distributionally-equivalent classes of structure might be identified on the basis of shared context. For example, the words "accountant" and "clerk" may be placed in the same group (normally known as "nouns") because they may appear in the same position in sentences like:

> The ... remembered the tax.

Distributional linguistics relied mainly on judgements by native speakers to determine these privileges and equivalencies. But if one attempts to discover these groupings by any kind of automatic process, there is a problem: it is

```
N P e t e r ;

N J a n e ;

V r u n s ;

V w a l k s ;

Q f a s t ;

Q s l o w l y ;

N ; V ; Q ; ;
```

Figure 4.3: Unified patterns from the alignment in Figure 4.2, with word class symbols as tags.

rare to find pairs of sentences which are identical except for two words in the one position.

The SNPR model solves this problem in the way it combines the search for disjunctive groups with the building of progressively larger segmental structures. It is not necessary to use whole sentences for a preliminary classification of words or other segments: "accountant" and "clerk" may be assigned to the same class because they both occur in the context "the ..." — and pairs of fragments like that are much more common than the required pairs of sentences. When labels for classes have been introduced, they may themselves provide the contexts from which other classes may be derived. And because labels for classes are more frequent than their constituent structures, it becomes relatively easy to find those shared contexts.

In Figure 4.2, "P e t e r" and "J a n e" share the same set of contexts and may thus be classified together syntactically as "nouns". In a similar way, "r u n s" and "w a l k s" may be classified as "verbs" and "f a s t" and "s l o w l y" may be classified as "adverbs" or "qualifiers".

There is a natural affinity between the kinds of classification just described and "tags" of the kind described briefly in Section 4.3: word classes and categories of that kind are based on shared contexts and tags provide the means of marking the contexts of unified patterns. Thus, the alignments shown in Figure 4.2 may be compressed into the form shown in Figure 4.3 by the unification of matching patterns and the use of word class symbols as tags to mark the contexts in which those patterns belong.

In Figure 4.3, each unified pattern is terminated by a semi-colon (";") and instances of the same symbol appear in the sequence "N ; V ; Q ; ;" (at the bottom of the figure) which records the sequence of word classes in the sentences shown in Figure 4.2. This symbol has been introduced to ensure that the "correct" alignments and unifications are made when the tags are de-referenced. An example showing how tags may be de-referenced by alignment and unification is presented in Section 4.5.6.

In effect, the set of patterns shown in Figure 4.3 is a simple grammar corresponding to the set of sentences in Figure 4.2. The sequence of tags at the bottom of Figure 4.3 represents the structure of the sentence at a relatively abstract level and the other sequences provide lower-level details of word classes and individual words.

4.5 Reasoning as Multiple Alignment with Unification and Search

In this chapter, "reasoning" is taken to mean deductive reasoning as discussed in standard treatments of logic (*e.g.* [6]) and also probabilistic forms of reasoning — abductive, inductive and analogical reasoning. A working hypothesis is that deduction may be cast in a probabilistic frame in which probability values are extreme (close to 0 or 1).

4.5.1 Recognition and Reasoning

A prominent feature of human (and animal) perception is the ability to recognize a pattern from a sub-set of its features: we may recognize something as a car even though one end is obscured by a building and another part is obscured by a tree; we may recognize a piece of music from hearing only a part.

This kind of recognition may be seen as a form of reasoning: from the observed features we reason that the unobserved features are "really" there. It is tempting to believe that all forms of human reasoning are founded on this ability to recognize a whole pattern from a sub-set of its parts.

This kind of recognition/reasoning may also be seen in terms of the alignment and unification of patterns: sensory data is a pattern to be aligned and unified with one or more patterns in memory.

4.5.2 *Modus Ponens*

> *If something is a mammal then that something is warm blooded.*
> *Tibby is a mammal.*
> *Therefore, Tibby is warm blooded.*

Alignment:

```
N  Tibby  ;  mammal                              ;
|         |  |                                   |
N         ;  mammal  warm-blooded  furry  ...    ;
```

Unification:

```
N  Tibby  ;  mammal  warm-blooded  furry  ...  ;
```

Figure 4.4: *Modus ponens* reasoning by alignment and unification.

In logical notation, this kind of *modus ponens* reasoning may be expressed as:

$\forall x$: mammal$(x) \supset$ warm-blooded(x)
mammal(Tibby)
∴ warm-blooded(Tibby).

A strict interpretation of the implication symbol ("⊃") as "material implication" is outside the scope of these proposals because the concept of material implication includes a concept of negation and no serious attempt has yet been made to interpret negation in the proposed framework. A "naturalistic" concept of implication is intended here, close to the everyday meaning of the word.

Figure 4.4 shows how this kind of reasoning may be re-cast in terms of the alignment and unification of patterns. The first pattern expresses the idea that there is something whose name is "Tibby" which is a mammal. The second pattern expresses the association between being a mammal, having warm blood, being furry and other characteristics of mammals. There is an unfilled "slot" for the name of an individual mammal.

The two patterns may be unified as shown. The unified pattern may be read as "Tibby is a mammal and, as such, is warm-blooded, furry and so on". This contains the required conclusion together with some other conclusions and a re-statement of the "premises".

Any unification like this needs to be "implicit" or "transitory", otherwise the pattern which describes mammals in general, and which, presumably, needs to be preserved, would quickly lose its generality.

4.5.2.1 Modelling Named Variables and the Universal Quantifier

In [24], Chapter 5, it was suggested that, via some manipulation of structures similar to the casting of logical structures into clausal form, the universal quantifier might be modelled in an SP system by a variable like an unnamed variable in Prolog (represented in both systems by the underscore symbol "_"). The effect of a named variable might be achieved by matching and unification applied to the combination of an unnamed variable with a symbol acting as a name, *e.g.* (X _).

More recent thinking has led to the tentative view that there is no need for an unnamed variable with special properties and that an equivalent effect may be achieved by the use of a symbol like ";" to mark the scope of patterns.

In each of the two patterns in Figure 4.4, the symbol ";" serves to mark the scope of the "N" field and it is also a termination marker for each pattern. Apart from its usage as a scope marker, the ";" symbol has no special properties or status and enters into matching and unification like any other symbol.

In the example, the sub-sequence "N ;" behaves like a named variable (*e.g.* "x" in logical notation). The effect of matching and unification is to create the sub-sequence "N Tibby ;" which may be seen as an "assignment" of "Tibby" to the "named variable".

"Quantification" in this example may be seen as "universal" because the same inference would be made for any pattern of the form "N ... ; mammal ;".

Further discussion of the way in which the concept of a variable and types for variables may be modelled in the proposed framework may be found in [16, 17, 18].

4.5.3 *Modus Tollens*

The *modus tollens* form of reasoning is similar to *modus ponens* except that it uses negation:

$$\forall x : p(x) \supset q(x)$$
$$\neg q(a)$$
$$\therefore \neg p(a).$$

Unless negation is to be accommodated with a closed-world assumption in the manner of "negation as failure" (with the associated pitfalls), there appears to be a need for a special negation operator within the proposed framework of compression by PMUS. As noted above, this has not yet been attempted.

Alignment:

```
N Chirpy ;            warm-blooded              ;
| |                   |                         |
N       ; mammal warm-blooded furry ... ;
```

Unification:

```
N Chirpy ; mammal warm-blooded furry ... ;
```

Figure 4.5: Abductive reasoning by alignment and unification.

As suggested previously [23, 19], the case for including negation as a primitive within the SP framework is strengthened by the observation that negation can be a powerful aid to compression: it is more economical to encode this week's shopping list as "last week's, *not* soap" than to write it out in full.

Within a framework of alignment and unification of patterns, we may suppose that a pattern which has been "negated" may be matched and unified like any other pattern but that any resulting "inferences" are negated. These are things to be explored in the future.

4.5.4 Abduction

Instead of inferring that something is warm-blooded because it is a mammal (Section 4.5.2), we may (with some risk) infer abductively that something is a mammal because it is warm-blooded. In this case, of course, there is a good chance that the creature might be a bird. Although abduction is risky, it is a prominent feature of everyday reasoning and can be very useful.

Figure 4.5 shows how this kind of reasoning may be seen as alignment and unification. The pattern "N Chirpy ; warm-blooded ;" would unify equally well with a pattern specifying the features of a bird and the corresponding probabilities (that Chirpy is a mammal or a bird) should be approximately even. However, additional information, *e.g.* that Chirpy has feathers or knowledge about typical pet names for birds, would tilt the probabilities in favour of Chirpy being a bird. In general, there is a probability associated

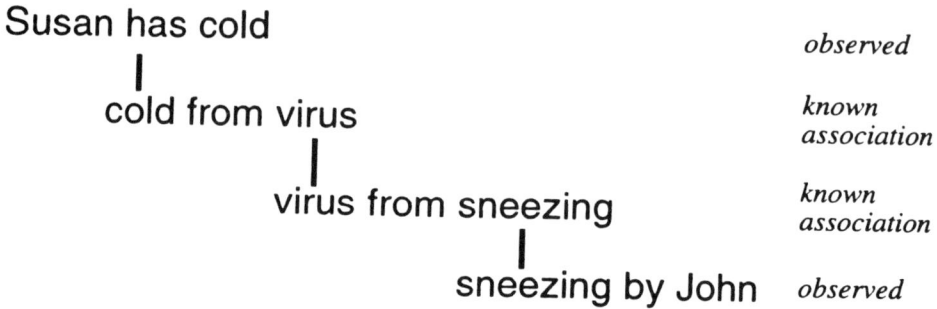

Figure 4.6: Multiple alignment in a chain of inference. The status of each pattern in the alignment is shown on the right.

with each possible alignment of patterns and this probability depends on the range of patterns which are available for alignment.

4.5.5 Chains of Reasoning

A familiar feature of everyday reasoning and of systems for automatic reasoning (*e.g.* Prolog) is the formation of "chains" of two or more inferences, each one except the first depending on those that precede it.

Figure 4.6 shows how this kind of chaining may be seen in terms of multiple alignment. Susan has a cold and John has been sneezing. Knowing that a cold is caused by a virus and that sneezing spreads viruses we may infer that Susan caught her cold from John.

4.5.6 Inductive Reasoning

> *Socrates is human and is mortal.*
> *Xanthippe is human and is mortal.*
> *Sappho is human and is mortal.*
> *Therefore probably all humans are mortal.*

Inductive reasoning, as illustrated in this example from [6], is related to learning as discussed in Section 4.4. The top part of Figure 4.7 shows how the meanings of the first three sentences may be represented as patterns and it shows alignments of symbols within those patterns.

Alignment:

    Socrates human mortal ;
            | | |
    Xanthippe human mortal ;
            | | |
    Sappho human mortal ;

Unification:

    N Socrates ; ;
    N Xanthippe ; ;
    N Sappho ; ;
    N ; human mortal ;

Figure 4.7: Extraction of a "chunk" by multiple alignment and unification.

The sub-sequence "human mortal ;" is a repeating "chunk" of information which may be extracted in much the same manner as words or other segments at the syntactic level (cf. Section 4.4). At the bottom of Figure 4.7, this unified chunk is shown together with modified versions of the three original patterns. In each modified pattern, "N" and ";" have been placed around the "variable" parts of those patterns. The same two symbols are attached to the "chunk" containing the invariant pattern "human mortal".

Although "human mortal" has been deleted from each of the three original patterns, it should be clear that alignment and unification along the lines described in Section 4.5.2 can restore this chunk to each context whence it came.

Compression has been achieved in Figure 4.7 because there are fewer characters in the unified patterns than in the originals. Normally, one would com-

Alignment:

```
N  David  ;  human                ;
|            |      |
N         ;  human  mortal        ;
```

Unification:

N David ; human mortal ;

Figure 4.8: An inference that someone other than Socrates, Xanthippe and Sappho is mortal.

pute optimal compression taking account of the probabilities of symbols. But if compression is to be achieved exclusively by PMUS, it seems more appropriate to use simpler and more direct measures in terms of the numbers of characters or bits used in patterns.

4.5.6.1 Inductive Generalization

How do we reach the inductive generalisation that "probably all humans are mortal"? Within the framework of multiple alignment, unification and search, it seems that this may be achieved in much the same manner as deductive or abductive inferences (Sections 4.5.2 and 4.5.4).

If, for example, we know that someone called "David" is human, we may infer that he is mortal because of the regular association between humanness and mortality. Figure 4.8 shows how this may be achieved by alignment and unification of relevant patterns. Since this kind of inference may be made for any entity which is described as human, and since all these inferences would have a high probability, we may conclude that "probably all humans are mortal".

Why should we not also conclude that "probably all mortals are human"? As was indicated in Section 4.5.4, the probability associated with any alignment depends on the range of patterns in the database. If the database recorded only the association between humanness and mortality then it seems that the conclusion that all mortals are human would be reasonable. But in a database containing a more realistic variety of information, there would be many other associations including, for example, the facts that cows are mortal, birds are mortal, fish are mortal and so on. In this case, the probability associated with the inference that any given mortal being was human would be relatively low.

4.5.7 Analogical Reasoning

It seems that analogical reasoning, as exemplified by geometric analogies of the kind used in puzzles and IQ tests, may be understood in terms of MLE [9, 10]. This chapter makes the more specific proposal that this kind of reasoning, or aspects of it, may be understood in terms of multiple alignment and unification.

To interpret geometric analogies in these terms, it seems necessary to translate the geometrical forms into sequences of symbols ("patterns") like those presented elsewhere in this chapter. In the example shown in Figure 4.9, item A may be translated as "small star left-of large square", B may be translated as "small star above large square" and so on.

This translation is a significant problem in itself, well outside the scope of this chapter except for one aspect which deserves to be noted: as discussed elsewhere [10], the case for an MLE interpretation of geometric analogies requires that textual descriptions are in the "right" form. For example, it makes a difference whether A in Figure 4.9 is described as "small star left-of large square" or as "large square right-of small star". To be more precise, there must be consistency within the scope of one problem in the way geometric forms are described.

The "correct" solution to the problem shown in Figure 4.9 is clearly item 3. This solution is shown in Figure 4.10 as an alignment amongst the textual descriptions of A, B, C and 3.

Unification with an example like this would require the use of tags along the lines described in Section 4.5.7. A full discussion is not appropriate here. However, it should be reasonably clear that the alignment which is shown is better than the best alignment that one might obtain with any of items 1, 2 or 4 in Figure 4.9, and that unification of the alignment in Figure 4.10 should give more compression than any of the alternatives.

4.6 Conclusion

In this chapter I have tried to show the potential of multiple alignment with unification and search as a framework within which learning and reasoning may be integrated. Within the space available it has been possible only to

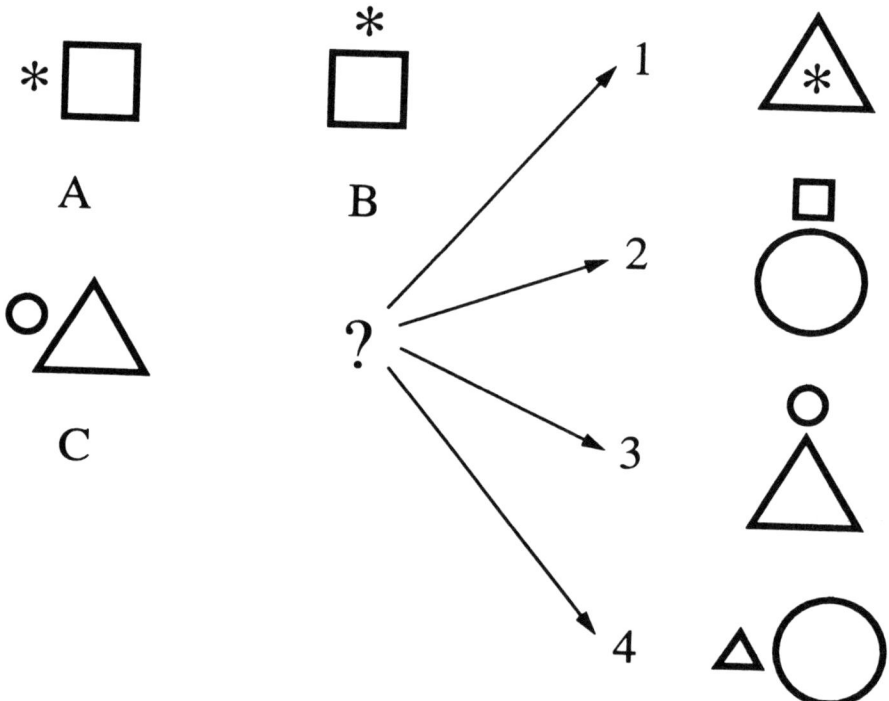

Figure 4.9: A geometric analogy problem.

sketch out the possibilities. These ideas are still quite tentative and there are clearly many issues and avenues to be explored.

In the light of previous and current work, it appears that multiple alignment, unification and search may also provide a framework for understanding several other aspects of computing including information retrieval and pattern recognition, parsing, the execution of functions, planning and problem solving. Relevant discussion may be found in [16, 17, 18].

The possible integration of diverse kinds of computing, including learning and reasoning, within one simple framework provides a powerful motive for developing these ideas. Apart from potential benefits in terms of theory, there is the prospect of developing a "new generation" computing system with practical advantages compared with conventional systems [19].

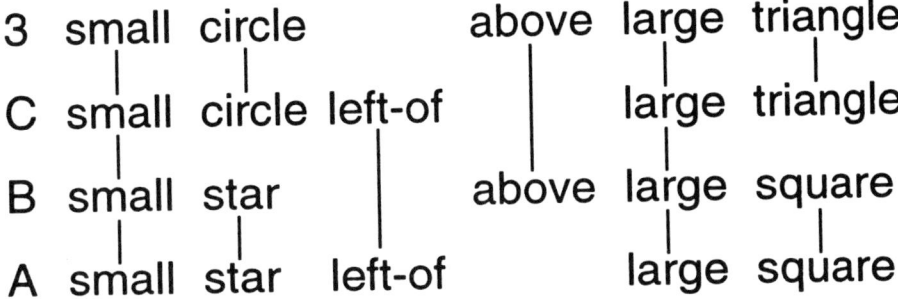

Figure 4.10: Multiple alignment in a geometric analogy problem.

Acknowledgements

I am grateful to Alex Gammerman for useful discussion and for copies of relevant papers; to Lloyd Allison of the Department of Computer Science, Monash University, for sending me articles about multiple alignment and compression; and to an anonymous referee for constructive comments.

References

[1] Allison L. and Yee C.N. (1990) Minimum message length encoding and the comparison of macromolecules. *Bulletin of Mathematical Biology* (52) 3: 431–453.

[2] Allison L., Wallace C.S. and Yee C.N. (1992) Finite-state models in the alignment of macromolecules. *Journal of Molecular Evolution* 35: 77–89.

[3] Barton G.J. (1990) Protein multiple sequence alignment and flexible pattern matching. *Methods in Enzymology* 183: 403–428.

[4] Chan S.C., Wong A.K.C. and Chiu D.K.Y. (1992) A survey of multiple sequence comparison methods. *Bulletin of Mathematical Biology* (54) 4: 563–598.

[5] Cook C.M. and Rosenfeld A. (1976) Some experiments in grammatical inference, in *Computer Oriented Learning Processes*, Simon J.C. (ed) 157–174, Leyden: Noordhoff.

[6] Copi I.M. (1986) *Introduction to Logic*, MacMillan, New York.

[7] Day W.H.E. and McMorris F.R. (1992) Critical comparison of consensus methods for molecular sequences. *Nucleic Acids Research* (20) 5: 1093–1099.

[8] Fries C.C. (1952) *The Structure of English*, New York: Harcourt, Brace & World.

[9] Gammerman A.J. (1991) The representation and manipulation of the algorithmic probability measure for problem solving. *Annals of Mathematics and Artificial Intelligence* 4: 281–300.

[10] Gammerman A.J. (1993) Geometric analogy problems by minimum length encoding. *Proceedings of the 4th Conference of the International Federation of Classification Societies, Paris (IFCS-93)*, August-September.

[11] Gold M. (1967) Language identification in the limit. *Information and Control* 10: 447–474.

[12] Kietz J-U. and Lübbe M. (1994) An efficient subsumption algorithm for Inductive Logic Programming. *Proceedings of the Fourth International Workshop in Inductive Logic Programming (ILP-94)*, 97–105.

[13] Roytberg M.A. (1992) A search for common patterns in many sequences. *Cabios* (8) 1: 57–64.

[14] Solomonoff R.J. (1964) A formal theory of inductive inference, Parts I and II. *Information and Control* 7: 1–22, 224–254.

[15] Taylor W.R. (1988) Pattern matching methods in protein sequence comparison and structure prediction *Protein Engineering* (2) 2: 77–86.

[16] Wolff J.G. (submitted). Computing as compression by multiple alignment, unification and search (1).

[17] Wolff J.G. (submitted). Computing as compression by multiple alignment, unification and search (2).

[18] Wolff J.G. (submitted). Multiple alignment and computing.

[19] Wolff J.G. (1995) Computing as compression: an overview of the SP theory and system. *New Generation Computing* 13: 187–214.

[20] Wolff J.G. (1995) Computing as compression: SP20. *New Generation Computing* 13: 215–241.

[21] Wolff J.G. (1994) Computing and information compression: a reply. *AI Communications* (7) 3/4: 203–219.

[22] Wolff J.G. (1994) A scaleable technique for best-match retrieval of sequential information using metrics-guided search. *Journal of Information Studies* (20) 1: 16–28.

[23] Wolff J.G. (1993) Computing, cognition and information compression. *AI Communications* (6) 2: 107–127.

[24] Wolff J.G. (1991) *Towards a Theory of Cognition and Computing*, Chichester: Ellis Horwood.

[25] Wolff J.G. (1988) Learning syntax and meanings through optimization and distributional analysis. In *Categories and Processes in Language Acquisition*, Levy Y., Schlesinger I.M. and Braine M.D.S. (eds) 179–215, Hillsdale, N.J.: Lawrence Erlbaum (reproduced in Ref. 24, Chapter 2).

[26] Wolff J.G. (1980) Language acquisition and the discovery of phrase structure. *Language and Speech* 23: 255–269.

5

Probabilistic Association and Denotation in Machine Learning of Natural Language

P. Suppes, L. Liang

We have taken what we believe is a new tack in the approach to machine learning by using in a very explicit way principles of association and generalization derived from classical psychological principles. The principles we used were, however, much more specific and technically developed.

The fundamental role of association as a basis for conditioning is thoroughly recognized in modern neuroscience and is essential to the experimental study of the neuronal activity of a variety of animals. For similar reasons its role is just as central to the learning theory of neural networks, now rapidly developing in many different directions. We have not, however, made explicit use of neural networks, but have worked out our theory of language learning at a higher level of abstraction. In our judgment the difficulties we face need to be solved before a still more detailed theory is developed.

The classical psychological principles of learning used here have been thought by linguists to be wholly inadequate as the basis for a theory of language learning. Nothing could be further from the truth. Skinner's [6] naive formulation of the problems of language learning was rightly attacked by Chomsky [5], but no serious alternative learning theory has been offered by linguists even today.

In the first section we briefly describe our approach to machine learning of natural language. In the second section we focus on the problem of denotation that is important in our use of probabilistic association of words and their meaning. In the third section we outline the backgound cognitive and perceptual assumptions of our machine learning work. In the fourth section we formulate explicitly our two general axioms of association and denotation, but do not state the additional axioms describing the full learning process. These may be found in our previous publications, with some changes being made over time [1, 2, 3]. In the fifth section we formulate and analyze two related but distinct denotational learning models. Finally, in the sixth section we present some empirical results.

5.1 Our Approach to Machine Learning

Without going into all the details, we want to convey a rather clear intuitive sense of the process of learning of natural language in terms of the various events that happen when an utterance is given to a robot. (In this and succeeding sections we shall refer to robots, but it should be understood that the basic program of machine learning would apply without serious modification to other applications, for example, machine learning of physics word problems, which we briefly consider later.) Also, following standard learning usage, we shall often speak of trials where, of course, we mean that the trial begins with a command in the form of an utterance to be executed by the robot.

The most important way to describe conceptually the learning process our program embodies is in the description of the state of memory of a robot at the beginning of each trial. There are four aspects of this memory that are changed due to learning. The first is the association relation between words of a given language and internal symbols that have as denotations actions, objects, properties and relations in the robot's world. A central problem is to learn in each language what word is properly associated with a given internal symbol. A second aspect of the memory that changes is the denotational value of a given word, which will affect its probability of being associated. It is this denotational-value aspect of the memory that we describe in some detail, beginning in the next section. The third part that changes is the short-term memory that holds a given verbal command for the period of the trial on which it is effective. This memory content decays and is not available for access after the trial on which a particular command is given. This means that at the beginning of the trial, before a command is given, this short-term buffer is empty. What we have said thus far, could, with some stretching, fit into classical theories of association, but for language learning it is quite evident that the association relation and some simple features of short-term memory are certainly not enough.

The fourth aspect is the important one of learning grammatical forms. Consider the verbal command *Get the nut*. This would be an instance of the grammatical form *A the O*, where *A* is the category of actions and *O* is the category of objects. This form actually represents a mild oversimplification, because we do not have just a single category of actions. There are several subcategories, depending upon the number of arguments required, and certain other natural semantical requirements as well. The example will illustrate how things work, however. The grammatical forms are derived by generalization only from actual instances of verbal commands given to the robot. No prior knowledge of any sort of the grammar of the natural language to be learned is available to the robot. Also important is the fact that associated with each grammatical form as it arises from generalization are the associations of the words which have been the basis for the generalization, along with their internal representations. For example, if *Get the nut* were the occurrence in which the grammatical form just stated was generated, then also stored with that grammatical form would be the associations $get \sim \$g$ and $nut \sim \$n$, where $\$g$ and $\$n$ are internal symbols whose denotations are known to the robot. When incorrect associations are deleted by further learning, the grammatical forms based on such associations are also deleted.

5.2 Problem of Denotation

In the probabilistic theory of machine learning of natural language which we have been developing, we have encountered in a new form a standard problem in the analysis of the semantics of natural language, namely, how to handle words that are nondenoting. We do not mean nondenoting in some absolute sense, but relative to a fixed set of semantic categories. These categories in the robotic case are, roughly speaking, the categories of actions, objects, properties and relations. It may well be that in some elaborate set-theoretical semantics of natural language, nondenoting words like the definite article *the* denote a complicated set-theoretical function, but the relevance of such an elaborate semantics to language learning is doubtful. In the robotic context, we have something simpler and closer to the common man's view of what denotations are. We take as denoting words color and object words, common nouns, familiar concrete action words, *etc.*. We take ordinary prepositions in English and sometimes other devices in other languages to denote relations in most cases. On the other hand, our computational semantics centered on physics words problems, with an internal equational language of physical quantities, is further removed from common sense ideas.

When a child learning a first language or an older person learning a second language first encounters utterances in that new language, there is no uniform way in which nondenoting words are marked. There is some evidence that various prosodic features are used in English and other languages to help the child. For example, in many utterances addressed to very young children, the definite or indefinite article is not stressed but rather the common noun it modifies, as in the expression *Hand me the cup.* But such devices do not seem uniform and in any case are not naturally available to us in our machine-learning research, where we use written input of words without additional prosodic notation.

As has already been made clear, a central feature of our approach to machine learning is the probabilistic association between words of the natural language being learned and denoting symbols of the internal language. It is appropriate that at the beginning all words are treated equally, and so the associations are formed from sampling based on a uniform distribution. On the other hand, after many words have been learned and a good deal of language has been acquired by the robot, it is very unnatural, and also inefficient, if the robot is now given, for example, the esoteric command *Get the astrolabe*, to have the internal symbol ast be associated with equal probability with the definite article *the* and *astrolabe* — we assume here that the association of *get* is already correctly fixed. After much experience, what we want is that there is very little chance of associating the definite article *the* with any denoting symbol.

To incorporate such learning many variant models are easily formulated. We have restricted ourselves to two which bring out the most salient distinction to consider. Should the denotational value of a word change only when it occurs in an utterance that is responded to incorrectly or not at all, or should the denotational value change every time it occurs regardless of whether a correct response is given? Model I is of the first type — learning only from errors. Model II is of the second type — learning occurs on all trials.

These two types of models have a long history of application in the study of human and animal learning. In such studies the empirical question centres on which kind of model approximates more closely the learning that is taking place in the organism. In machine learning this question is answered by fiat, that is, by implementing in the machine-learning program one of the models. But empirical questions remain. Having implemented a model, we study how well it works in providing dynamically changing denotational values for words of a given language. In particular, we analyze in detail, by extensive computation, the error rates of association arising for different parameter values in the different models and for different languages. In this chapter, we present data for English and Chinese.

5.3 Background Cognitive and Perceptual Assumptions

Before explicitly formulating the learning principles of association and denotation we use, we first state informally assumptions we make about the cognitive and perceptual capacities of the class of robots, albeit as yet quite limited, we work with.

1. *Internal language.* The robot has a fully developed internal language, which it does not learn. It is technically important, but not conceptually fundamental, that in our case this language is LISP. When we speak here of the internal language we refer only to the language of the internal representation, which is itself a language at a higher level of abstraction, relative to the

concrete movements and perceptions of the robot. It is the language of the internal representations held in memory that provides the direct interface to the natural-language learning. In fact, most of the machine learning of a given natural language can take place through simulation of the robot's behaviour by using just the language of the internal representation. The first associations learned are between the internal representation in memory of a coerced action and a contiguous verbal utterance in the natural language being learned.

The fundamental importance of this internal representation of a coerced action can be recognized by considering a parallel case of animal learning. When a dog is trained to *Get the paper* or *Get the ball* by being led through the desired action or by some related technique, the residue in memory of what we term the coerced action is surely drastically abstracted from the perceptually rich context of the demonstrated action desired, and it is that abstracted internal representation in memory that must be associated to the verbal stimulus in order for the dog later to perform the desired action upon hearing the verbal command. We are a long way from knowing even the general structure of the dog's internal representation in memory of the action. In this limited sense, life with a robot is much easier, for we ourselves create the form of its internal representation.

2. *Objects, relations and properties.* We furthermore assume the robot begins its natural-language learning with all the basic cognitive and perceptual concepts it will have. In other words, our first-language learning experiments are pure language learning. Any learning of new concepts is delayed to another phase. For example, we have assumed that the spatial relations frequently referred to in all, or at least all the languages we consider in detail, are already known to the robot. This is quite contrary to human language learning. For example, probably in no widely used natural language at least, do children at the age of thirty-six months use or fully understand the relations of left and right. To avoid misunderstanding, we emphasize that we consider it an important future task to have the robot also learn the familiar spatial and temporal relations.

3. *Actions.* What was just said about objects and relations applies also to actions, represented in English by such verbs as *pick up, get, place, screw, etc.*. The English, of course, must be learned, but not the underlying actions.

4. *Associations and grammatical forms.* Before stating any formal principles of learning, we feel it is desirable to describe as informally and intuitively as possible the learning setup we use. Consider the English command *Pick up the screw*, no part of which has as yet been learned by the robot. The learning steps may be roughly schematized as follows:

(i) By coercion, or simulation of coercion, the robot creates in memory an internal representation of the coerced action of picking up the screw;

For statement of learning principles we show this internal representation, not as a LISP expression, but just as a schematic function *I(...)* of the denoting terms in the LISP expression. Here, by *denoting terms* we mean the names in the internal language of the actions, objects, properties and relations mentioned. The internal representation of *Pick up the screw* is then *I($p,$u,$s)*, where $p = the action of picking, $u = the direction up and $s = screw.

(ii) By contiguity the robot associates the verbal utterance and the internal representation

$$Pick\ up\ the\ screw \sim I(\$p,\$u,\$s),$$

where $\sim$ is the symbol we use for association;

(iii) By probabilistic association, the robot associates the internal denotations with the English words, with one possibility the following incorrect result:

$$pick \sim \$s,\ up \sim \$p,\ screw \sim \$u.$$

We need to observe the following:

  a. We assume from the beginning the robot knows word boundaries, as delineated by the typed input. This is an example of an assumption that is natural for robots, but clearly false for very young children;

  b. For our simple example, there are 24 possible ways of associating the three internal symbols to the four denoting words in the English utterance. We initially assign to each of these 24 possibilities equal probability, but as trials contine, modify the probability by dynamic changes in denotational values, as is explained later in detail.

(iv) After the associations are made, by the principle of generalization, which we call the category generalization, each word is assigned the category of its associated internal symbol. In the present case *pick* $\in O$ — the category of objects, $up \in A$ — the category of actions and *screw* $\in R$ — the category of relations. A grammatical form is then also generalized from the verbal command:

$$O\ A\ the\ R$$

which, like the assigned categories is wrong for English, but remember that this is just the starting point of learning. With this grammatical form is associated its internal representation *I(A, R, O)* which characterizes its meaning.

(v) A new command is presented as the next step, say *Pick up the nut*. By coercion the internal representation *I($p, $u, $n)* is created (see (i) above). The robot then first searches its memory to see if any of the words uttered are associated to one of the internal denotations. Here the result is $up \sim \$p$, and also the classification of *the* as a nondenoting word is found. There are then six possibilities of probabilistic association for *pick, the* and *nut*. Note that the earlier incorrect association of *pick* with $\$s$ does not appear here, which means that at this stage of learning it will be changed. So, let us suppose the new associations are

$$pick \sim \$u, \; nut \sim \$n.$$

We also have as a new grammatical form

$$R \; A \; the \; O$$

which though incorrect, now has only the confusion of the associations of *pick* and *up* as its source. To correct these associations we must separate the constant pairing of *pick* and *up*, which is what we do. In any case, we form at once the association to the internal representation:

$$R \; A \; the \; O \sim I(A, R, O).$$

(vi) Learning stops whenever the following steps of interpretation can be successfully completed upon giving the robot a verbal command:

   a. An association to an internal denotation or a nondenoting classification is found in memory for each word;

   b. The category of each word is found in memory;

   c. The grammatical form resulting from (b) is found with an associated internal representation in memory;

   d. The command is correctly executed on the basis of the internal representation.

5.4 The General Axioms of Association and Denotation

We state the axioms in a general form, to be made more specific later, but we assume already that each word a of the target natural language has a denotational value $d_n(a)$ on each trial. This value changes from trial to trial according to the two different models presented in the next section.

I. *Probabilistic association.* On any trial n, let a natural language sentence s be associated to σ, its internal representation, let $\{a_i\}$ be the set of words of s not associated to any internal denoting symbol of σ, let $d_n(a_i)$ be the current denotational value of each such a_i and let $\{\alpha_j\}$ be the set of internal denoting symbols not currently associated with any word of s. Then:

(i) an element α_j is uniformly sampled without replacement from $\{\alpha_j\}$;

(ii) at the same time an element a_i is sampled without replacement from $\{a_i\}$ with the sampling probability

$$p_n(a_i) = \frac{d_n(a_i)}{\sum_{\{a_i\}} d_n(a_i)};$$

(iii) the sampled pairs are associated, *i.e.* $a_i \sim \alpha_j$;

(iv) sampling continues until either the set $\{a_i\}$ or the set $\{\alpha_j\}$ is empty.

II. *Denotational value computation.* If at the end of a trial a word a in the presented sentence is associated with some internal symbol α, then $d(a)$, the denotational value of a, increases and if a is not so associated $d(a)$ decreases. Moreover, if a word a does not occur on a trial, then $d(a)$ stays the same unless the association of a to an internal symbol α is broken on the trial, in which case $d(a)$ decreases.

5.5 Denotational Learning Models

We now turn to the description of the two models we consider. We begin with Model I — learning denotational values only from errors. Only two axioms are needed, but the exact description of the conditions for application of the learning operators are rather complicated. We describe these conditions informally rather than introduce a great deal of formal notation that we subsequently would not make much use of.

Model I

1. *If on trial n a wrong response or no response is given to the verbal command uttered at the beginning of the trial, and if after the coercion and probabilistic association of some (perhaps all) words of s to internal symbols of the internal representation σ of the coerced action, a word a_i is now associated to an internal symbol α_j of σ, i.e. $a_i \sim \alpha_j$, then*

$$d_{n+1}(a_i) = (1-\theta)d_n(a_i) + \theta,$$

and if a_i is not so associated with any denoting internal symbol

$$d_{n+1}(a_i) = (1-\theta)d_n(a_i),$$

where the learning parameter θ satisfies the constraint $0 < \theta \leq 1$, and so does the initial value $d_1(a_i)$, which is the same for all words a_i.

2. *If on trial n a correct response is given to the verbal command s, then no denotational values of words are changed, i.e. for all words a_i*

$$d_{n+1}(a_i) = d_n(a_i). \tag{5.1}$$

Moreover, if a word a_i does not occur on trial n, equation (5.1) also holds unless the association of a_i to an internal symbol α_j is broken on trial n, in which case

$$d_{n+1}(a_i) = (1-\theta)d_n(a_i).$$

There is much to be said for learning only when errors are made. But learning is not a monolithic phenomenon. Many kinds of skills improve with practice even after explicit errors are no longer made. Children, for example, continue to learn to read faster or to solve elementary mathematical problems faster long after their error rates are insignificant. In these and similar cases the evidence is substantial that learning, as reflected in speed of response, is taking place on trials with correct responses as well as on those with errors. The axioms for Model II reflect this kind of learning.

Model II

If, at the end of trial n, a word a_i in the presented verbal stimulus is associated with some denoting internal symbol α_j of the internal representation σ of s at the end of the trial, then

$$d_{n+1}(a_i) = (1-\theta)d_n(a_i) + \theta,$$

and if a_i is not so associated,

$$d_{n+1}(a_i) = (1-\theta)d_n(a_i).$$

Moreover, if a word a_i does not occur on trial n, then

$$d_{n+1}(a_i) = d_n(a_i),$$

unless the association of a_i to an internal symbol α_j is broken on trial n, in which case

$$d_{n+1}(a_i) = (1-\theta)d_n(a_i).$$

It should be apparent that Models I and II have the same two parameters, namely, the initial denotational value d_1 for all words of a language and the learning parameter θ. The general empirical question concerns what values of the parameters work best for each model for a given language, or more explicitly, for a given finite sample of utterances of the language, the one used for learning. Of course, there is no unique best criterion of parameter performance. We have used the following, although many others are easy to formulate.

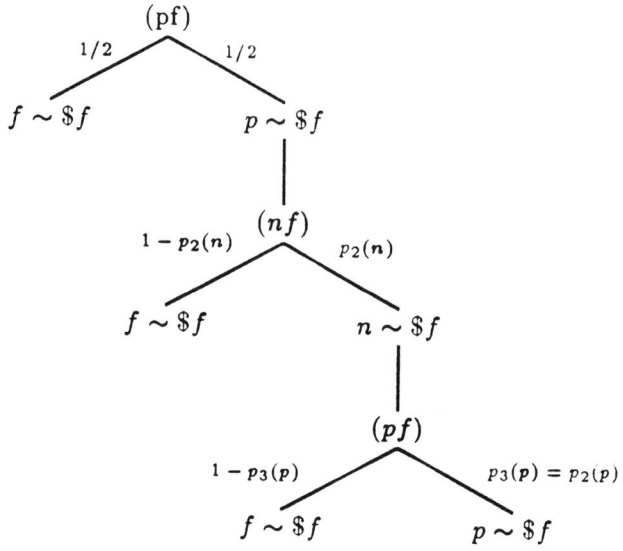

Figure 5.1: Initial part of the tree for language L_3.

First, to introduce some notation, useful now and later, let
$E_\delta(d_n)$ = the expectation or mean of d_n on trial n for all denoting words (indicated by the subscript δ),
$E_\nu(d_n)$ = the expectation of d_n on trial n for all nondenoting words.

The criterion we use is the first trial m^* on which for the finite sample used

$$E_\nu(d_{m^*})/E_\delta(d_{m^*}) < \epsilon. \tag{5.2}$$

In most of the analysis reported below we have chosen $\epsilon = 0.01$.

It is easy to see that if we classify externally words as denoting or nondenoting, as we naturally do on learning trials, then in simple examples m^* may not exist. Consider the language L_3 with three words and two commands, *Please forward!* and *Now forward!* with *please* and *now* being nondenoting and *forward* having the same denotation as *$f*. Let the learning parameter $\theta = 1$. Then suppose on trial 1 with the utterance *please forward* given at the beginning of the trial, as the result of sampling *please* is associated with *$f*, and *forward* has no association. Then with $\theta = 1$ we have at the end of this first trial, ready for trial 2, $d_2(please) = 1$, $d_2(forward) = 0$, $d_2(now) = d_1(now)$. Then on the presentation of (*now forward*) at the beginning of the next trial, the internal symbol *$f* is associated with neither *now* nor *forward*. After coercion, the probability of association with *now* is

$$\frac{d_2(now)}{d_2(now) + d_2(forward)} = \frac{1}{1+0} = 1,$$

and obviously $prob_2$ (*forward* ~ *$f*) = 0. At the end of this trial, trial 2, $d_3(please) = d_2(please) = 1$, $d_3(forward) = d_2(forward) = 0$, $d_3(now) = 1$.

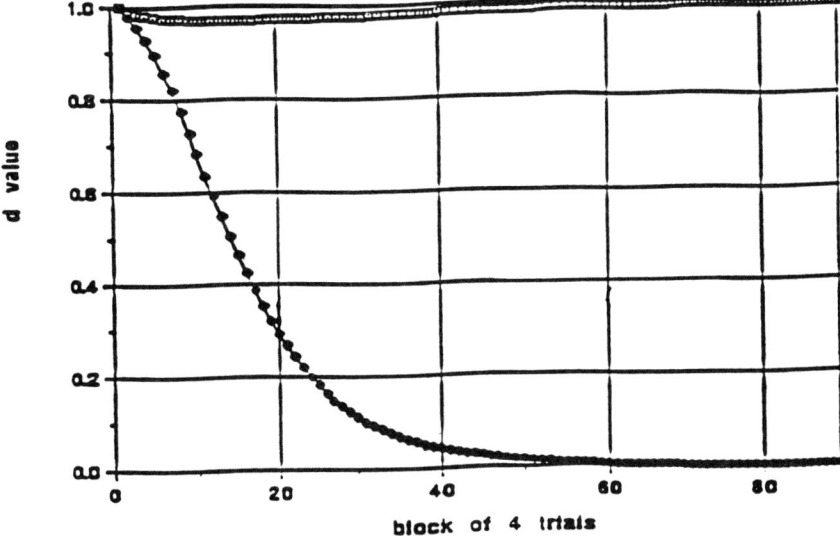

Figure 5.2: Mean denotational learning curves for English robotic corpus.

From here on these values of d_n are constant, i.e. for $n \geq 3$, $d_n(please) = d_n(now) = 1$, $d_n(forward) = 0$. (This computation holds in both Models I and II.) What is also interesting about this example is that if we define denoting and nondenoting internally just in terms of the values of d_n for various words, with a word a_i denoting whenever $d_n(a_i) \geq 1 - \epsilon$ and nondenoting whenever $d_n(a_i) \leq \epsilon$, then in this example $m^* = 3$ for any $\epsilon > 0$.

Asymptotic behaviour. We can study analytically the asymptotic behaviour of various denotational learning models only for trivial languages. Even small fragments of several hundred utterances of a natural language can only be studied numerically from the standpoint of denotational learning. Nevertheless, analysis of trivial examples gives some insight into how the models behave.

We first examine the asymptotic behaviour of Learning Model I for language L_3. Figure 5.1 shows the initial part of the tree. We can focus our analysis entirely on the right-most branch of the tree, for all other branches have after a finite number of trials the correct association.

For simplicity of computation, let

$$d_1(now) = d_1(please) = d_1(forward) = 1. \tag{5.3}$$

Then it is easy to see that the probability p_r of the right-most branch is the infinite product:

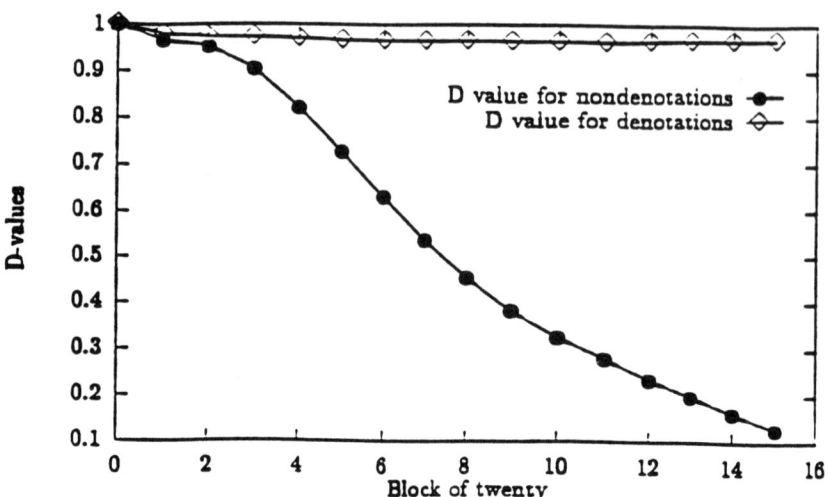

Figure 5.3: Mean denotational learning curves for 60 physics word problems in English.

$$p_r = \frac{1}{2} \cdot \frac{1}{1+(1-\theta)} \cdot \frac{(1-\theta)+\theta}{(1-\theta)+\theta+(1-\theta)^2} \cdot \frac{(1-\theta)+\theta}{(1-\theta)+\theta+(1-\theta)^3} \cdots$$
$$= \frac{1}{2} \cdot \frac{1}{1+(1-\theta)} \cdot \frac{1}{1+(1-\theta)^2} \cdot \frac{1}{1+(1-\theta)^3} \cdots \quad (5.4)$$

Let $\alpha = (1-\theta)$. It is most direct to study the behaviour of $1/p_r$.

$$\frac{1}{p_r} = 2 \cdot (1+\alpha) \cdot (1+\alpha^2) \cdot (1+\alpha^3) \cdots \quad (5.5)$$

For $0 < \alpha < 1$, $\sum \alpha^n$ converges to $\frac{1}{1-\alpha} = \frac{1}{\theta}$.
But this convergence is a necessary and sufficient condition for the convergence of (5.4) to a positive real number, say, $c > 0$. So $p_r = \frac{1}{c} > 0$.

Without giving a detailed analysis, we remark that quite similar results are obtained from a different, but familiar class of learning models, ones with commuting operators. Here is a generic example:

$$d_{n+1}(w) = \begin{cases} \alpha\, d_n(w) & \text{if word } w \text{ occurs on trial } n \text{ and is not associated} \\ & \text{with any internal symbol,} \\ \alpha\, d_n(w) & \text{if } w \text{ does not occur on trial } n \text{ but its association} \\ & \text{is broken,} \\ \beta\, d_n(w) & \text{if } w \text{ occurs on trial } n \text{ and is associated with an} \\ & \text{internal symbol,} \end{cases}$$

where $\alpha < 1$ and $\beta \geq 1$. We do not pursue these models here.

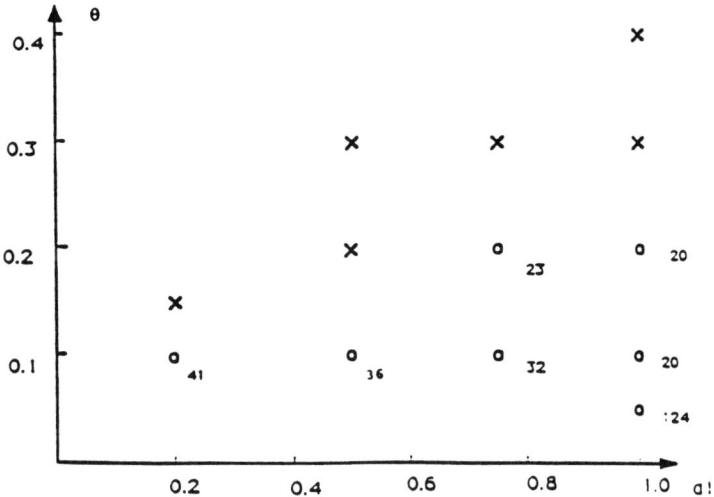

Figure 5.4: Rate of denotational learning with different values of $d_1(a_i)$ and θ.

5.6 Some Empirical Results

In Figure 5.2, we show the mean denotational learning curves of Model II for the English robotic corpus of approximately 400 commands. The only two nondenoting words in the corpus were the definite and indefinite articles *the* and *a*. The bottom curve is for these nondenoting words, and the upper curve for the denoting words. The parameters were set at $d_1(a_i) = 1$ and $\theta = 0.03$. Using the same parameters the results were similar, but with slower learning, for the corresponding Chinese robotic corpus which had eight nondenoting words. (For more linguistic details, see [4].

In Figure 5.3, we show the corresponding mean curves for 60 physics word problems in English, with again the only two nondenoting words being *the* and *a*. The internal language is completely different from the robotic case, but we will not attempt to describe it here, except to say that it is essentially a pure equational language for physical quantities. We emphasize, however, that the learning axioms were the same for the robotic commands and the physics word problems.

Finally, in Figure 5.4 we show the effects of choosing different initial values $d_1(a_i)$ and simultaneously different values of θ in Model II for the robotic English corpus. The criterion of learning was that the ratio (5.2) be such that $\epsilon < 0.01$. The number of trials to achieve this result is printed next to each data point.

5.7 Conclusion

In spite of the problems still to be solved in our theory of machine learning of natural language, prospects for use are not only a distant hope. We list three.

First, in the relatively near future real applications to well-defined domains of activity and their relevant sublanguages will appear. Some of the earliest successful technical examples are likely to be in medicine, from the emergency room to the office dictation of medical records.

Second, it is likely, even if not anything like certain, that within the next decade oral communication with computers, as with people, will be the most important and most used form of communication. If so, single words and phrases will not be good enough. A rich natural sublanguage will be used, and computers will need to learn it. No doubt, the progress on comprehension may be faster than on production in the early years.

Third, important applications early in the next century at least will be in two domains, reflected in the early work reported here. There will be robots that talk and, above all, listen to instructions, and immobile computer-tutors that also talk and listen, and in the process teach the way a good tutor should.

References

[1] Suppes P., Liang L. and Böettner M. (1992) Complexity Issues in Robotic Machine Learning of Natural Language. In Lam L. and Naroditsky V. (eds) *Modelling Complex Phenomena*, Springer-Verlag, New York, 102–127.

[2] Suppes P., Böettner M. and Liang L. (1995) Comprehension Grammars Generated from Machine Learning of Natural Languages. *Machine Learning* 19: 133–152.

[3] Suppes P., Böettner M., Liang L. and Raymond R. (1995) Machine Learning of Natural Language: Problems and Prospects. *Proceedings of the Second World Conference on the Fundamentals of Artificial Intelligence*, France, 3–7 July 1995, 511–525.

[4] Suppes P., Böettner M. and Liang, L. Machine Learning Comprehension Grammars for Ten Languages. To appear.

[5] Chomsky, N. (1959) Review of B. F. Skinner *Verbal Behavior*. *Language* 35, 26-58.

[6] Skinner, B. F. (1959) *Verbal Behavior*. New York: Appleton.

Part II

Causation and Model Selection

6

Causation, Action, and Counterfactuals

J. Pearl

6.1 Introduction

The central aim of many empirical studies in the physical, behavioural, social, and biological sciences is the elucidation of cause-effect relationships among variables. It is through cause-effect relationships that we obtain a sense of a "deep understanding" of a given phenomenon, and it is through such relationships that we obtain a sense of being "in control," namely, that we are able to shape the course of events by deliberate actions or policies. It is for these two reasons, understanding and control, that causal thinking is so pervasive, popping up in everything from everyday activities to high-level decision-making. For example, every car owner wonders why an engine won't start; a cigarette smoker would like to know, given his/her specific characteristics, to what degree his/her health would be affected by refraining from further smoking; a policy maker would like to know to what degree anti-smoking advertising would reduce the costs of health care; and so on. Although a plethora of data has been collected on cars and on smoking and health, the appropriate methodology for extracting answers to such questions from the data has been fiercely debated, partly because some fundamental questions of causality have not been given fully satisfactory answers.

The two fundamental questions of causality are:

1. What empirical evidence is required for legitimate inference of cause-effect relationships?

2. Given that we are willing to accept causal information about a certain

phenomenon, what inferences can we draw from such information, and how?

The primary difficulty is that we do not have a clear empirical semantics for causality; statistics teaches us that causation cannot be defined in terms of statistical associations, while any philosophical analysis of causation in terms of deliberate control quickly reaches metaphysical deadends over the meaning of free will. Indeed, Bertrand Russell noted that causation plays no role in physics proper and offered to purge the word from the language of science.

Philosophical difficulties notwithstanding, scientific disciplines that must depend upon causal thinking have developed paradigms and methodologies that successfully bypass the unsettled questions of causation and that provide acceptable answers to pressing problems of experimentation and inference. Social scientists, for example, have adopted path analysis and structural equations, and programs such as LISREL have become common tools in social science research. Econometricians, likewise, have settled for stochastic simultaneous equations models as carriers of causal information and have focused most of their efforts on developing statistical techniques for estimating the parameters of these models. Statisticians, in contrast, have adopted Fisher's randomized experiment as the ruling paradigm for causal inference, with occasional excursions into its precursor — the Neyman-Rubin model of potential response.

None of these paradigms and methodologies can serve as an adequate substitute for a comprehensive theory of causation, suitable for AI purposes. The structural equations model is based largely on informal modeling assumptions and has hardly been applied beyond the boundaries of linear equations with Gaussian noise. The statisticians' paradigm, on the other hand, is too restrictive, as it does not allow for the integration of large body of substantive knowledge with statistical data. And philosophers have essentially abandoned the quest for the empirical basis of causation, focusing instead on semantical analysis of counterfactual and subjunctive conditionals, with little attention to issues of representation.

A new perspective on the problem of causation has recently emerged from a rather unexpected direction — computer science. When encoding and processing causal relationships on digital machines became necessary, the problems and assumptions that other disciplines could keep dormant and implicit had to be explicated in great detail, so as to meet the levels of precision necessary in programming.

The need to explicate cause-effect relationships arose in several areas of AI: automated diagnosis, robot planning, qualitative physics, and database updates. In the area of robotics, for example, the two fundamental problems of causation got translated into concrete and practical questions:

1. How should a robot acquire causal information through interaction with its environment?

2. How should a robot process the causal information it receives from its creator-programmer?

Attempts to gloss over previous difficulties with causation quickly resulted in a programmer's nightmare. For example, when given the information: "If the grass is wet, then the sprinkler must have been on" and "If I break this bottle, the grass will get wet", the computer will conclude: "If I break this bottle, the sprinkler must have been on". The swiftness and concreteness with which such bugs surface has enabled computer scientists to pinpoint loosely stated assumptions and then to assemble new and more coherent theories of actions, causation, and change.

The purpose of this chapter is to summarize recent advances in causal reasoning, especially those that use causal graphs. Section 6.2 will survey some of the difficulties connected with the formalization of actions in AI. Building on the paradigm that actions are a form of surgeries performed on theories of mechanisms, we will describe the emergence of the causal relation as an abbreviation for the surgery process. Finally, we connect actions to Lewis' theory of counterfactuals using imaging to replace conditioning.

Section 6.3 will provide the formal underpinning for the discussion of Section 6.2 and will demonstrate how causal graphs can resolve many of the confusions and contradictions that have prevented a workable theory of actions from evolving. Specifically, in Section 6.3.1 it will be shown that if qualitative causal information is encoded in the form of a graph, then it is very simple to assess, from non-experimental data, both the strength with which causal influences operate among variables and how probabilities will change as a result of external interventions. Using graph encoding, it is also possible to specify conditions under which manipulative experiments are not necessary and where passive observations suffice. The graphs can also be queried to produce mathematical expressions for causal effects, or to suggest additional observations or auxiliary experiments from which the desired inferences can be obtained.

Finally, Section 6.3.2 will present a symbolic machinery that admits both probabilistic and causal information about a given domain, and produces probabilistic statements about the effect of actions and the impact of observations. The calculus admits two types of conditioning operators: ordinary Bayes conditioning, $P(y|X = x)$, which represents the observation $X = x$, and causal conditioning, $P(y|do(X = x))$, read: the probability of $Y = y$ conditioned on holding X constant (at x) by deliberate action. Given a mixture of such observational and causal sentences, together with the topology of the causal graph, the calculus derives new conditional probabilities of both types, thus enabling one to quantify the effects of actions and observations.

6.2 Action as a Local Surgery

What gives us the audacity to expect that actions should have neat and compact representations? Why did the authors of STRIPS [8] and BURIDAN [19] believe they could get away with such short specification for actions?

Whether we take the probabilistic paradigm that actions are transformations from probability distributions to probability distributions, or the deterministic paradigm that actions are transformations from states to states, such transformations could in principle be infinitely complex. Yet, in practice, people teach each other rather quickly what actions normally do to the world, people predict the consequences of any given action without much hustle, and AI researchers are writing languages for actions as if it is a God given truth that action representation should be compact, elegant and meaningful. Why?

The paradigm I wish to explore in this chapter is that these expectations are not only justified but, mainly, that once we understand the justification, we will be in better shape to craft effective representations for actions.

6.2.1 Mechanisms and Surgeries

Why are the expectations justified? Because the actions we normally invoke in common reasoning tasks are *local surgeries.* The world consists of a huge number of autonomous and invariant linkages or mechanisms (to use Simon's word), each corresponding to a physical process that constrains the behavior of a relatively small groups of variables. In principle, then, the formalization of actions should not be difficult. If we understand how the linkages interact with each other, usually they simply share variables, we should also be able to understand what the effect of an action would be: Simply re-specify those few mechanisms that are perturbed by the action, then let the modified population of mechanisms interact with one another, and see what state will evolve at equilibrium. If the new specification is complete, a single state will evolve. If the specification is probabilistic, a new probability distribution will emerge and, if the specification is logical (possibly incomplete) a new, mutilated logical theory will then be created, capable of answering queries about post-action states of affair.

If this sounds so easy, why did AI ever get into trouble in the arena of action representation? The first answer I wish to explore is that what is local in one space may not be local in another. A speck of dust, for example, appears extremely diffused in the frequency (or Fourier) representation and, vice versa, a pure musical tone requires a long stretch of time to be appreciated. It is important therefore to emphasize that actions are local in the space of mechanisms and not in the space of variables or sentences or time slots. For example, tipping the left-most object in an array of domino tiles does not appear "local" in the spatial representation, because, in the tradition of domino theories, every tile might be affected by such action. Yet the action is quite local in the mechanism domain: Only one mechanism gets perturbed, the gravitational restoring force which normally keeps the left-most tile in a stable erect position. It takes no more than a second to describe this action on the phone, without enumerating all its ramifications. The listener, assuming she shares our understanding of domino physics, can figure out for herself the ramifications of this action, or any action of the type: "tip the ith domino tile to the right". By representing the domain in the form of an assembly of stable mechanisms, we have in fact created an oracle capable of answering queries about the effects of a huge set of actions and action combinations, without us having to explicate those effects.

6.2.2 Laws vs. Facts

This surgical procedure still sounds easy and does not explain why AI got into trouble with action representation. The trouble begins with the realization that in order to implement surgical procedures in mechanism space, we need a language in which some sentences are given different status than others; sentences describing mechanisms should be treated differently than those describing other facts of life, such as observations, assumption and conclusions, because the former are presumed stable, while the latter are transitory. Indeed the mechanism which couples the state of the $(i + 1)th$ domino tile to that of the ith domino tile remains unaltered (unless we set them apart by some action) whereas the states of the tiles themselves are free to vary with circumstances.

Admitting the need for this distinction has been a difficult cultural transition in the logical approach to actions, perhaps because much of the power of classical logic stems from its representational uniformity and syntactic invariance, where no sentence commands special status. Probabilists were much less reluctant to embrace the distinction between laws and facts, because this distinction has already been programmed into probability language by Reverend Bayes in 1763: Facts are expressed as ordinary propositions, hence they can obtain probability values and they can be conditioned on; laws, on the other hand, are expressed as conditional-probability sentences (*e.g.*, $P(accident|careless\text{-}driving) = high$), hence they should not be assigned probabilities and cannot be conditioned on. It is due to this tradition that probabilists have always attributed nonpropositional character to conditional sentences (*e.g.* birds fly); refusing to allow nested conditionals [22], and insisting on interpreting one's confidence in a conditional sentence as a conditional probability judgment [1] (see also [21]. Remarkably, these constraints, which some philosophers view as limitations, are precisely the safeguards that have kept probabilists from confusing laws and facts, and have protected them from some of the traps that have lured logical approaches[1]

---

[1] The distinction between laws and facts has been proposed by Poole (1985) and Geffner (1992) as a fundamental principle for nonmonotonic reasoning. It seems to be gaining

6.2.3 Mechanisms and Causal Relationships

The next issue worth discussing is how causality enters into this surgical representation of actions. To understand the role of causality, we should note that most mechanisms do not have names in common everyday language. In the domino example above I had to struggle hard to name the mechanism which would be perturbed by the action "tip the left-most tile to the right". And there is really no need for the struggle; instead of telling you the name of the mechanism to be perturbed by the action, I might as well gloss over the details of the perturbation process and summarize its net result in the form of an *event*, *e.g.* "the left-most tile is tipped to right", which yields equivalent consequences as the perturbation summarized. After all, if you and I share the same understanding of physics, you should be able to figure out for yourself which mechanism it is that must be perturbed in order to realize the specified new event, and this should enable you to predict the rest of the scenario.

This linguistic abbreviation defines a new relation among events, a relation we normally call "causation": Event A causes B, if the perturbation needed for realizing A entails the realization of B.[2] Causal abbreviations of this sort are used very effectively for specifying domain knowledge. Complex descriptions of what relationships are stable and how mechanisms interact with one another are rarely communicated explicitly. Rather, they are communicated in terms of cause-effect relationships between events or variables. We say, for example: "If tile i is tipped to the right, it causes tile $i + 1$ to tip to the right as well"; we do not communicate such knowledge in terms of the tendencies of each domino tile to maintain its physical shape, to respond to gravitational pull and to obey Newtonian mechanics.

A formulation of action as a local surgery on causal theories has been developed in a number of recent papers [15, 25, 26, 6, 27, 14]. Section 6.3.1 provides a brief summary of this formulation, together with a simple example that illustrates how the surgery semantics generalizes to nonprobabilistic formalisms.

6.2.4 Causal Ordering

Our ability to talk directly in terms of one event causing another, (rather than an action altering a mechanism and the alteration, in turn, having an effect) is computationally very useful, but, at the same time it requires that the assembly of mechanisms in our domain satisfy certain conditions. Some of these conditions are structural, nicely formulated in Simon's "causal ordering" [33], and others are substantive — invoking relative magnitudes of forces and powers.

---

broader support recently as a necessary requirement for formulating actions.

[2] The word "needed" connotes minimality and can be translated to: "...if every minimal perturbation realizing A, entails B".

CAUSATION, ACTION, AND COUNTERFACTUALS 109

The structural requirement is that there be a one-to-one correspondence between mechanisms and variables – a unique variable in each mechanism is designated as the output (or effect), and the other variables, as inputs (or causes). Indeed, the formal definition of causal theories given in Section 6.3.1 assumes that each equation is associated with a unique variable, situated on its left-hand side. In general, a mechanism may be specified as a function

$$G_i(X_1, ..., X_n; U_1, ..., U_m) = 0$$

without identifying any so called "dependent" variable X_i. Simon's causal ordering provides a procedure for deciding whether a collection of such G functions has a unique preferred way of associating variables with mechanisms, based on the requirement that we should be able to solve for the ith variable without solving for its successors in the ordering.

In certain structures, called *webs* [5, 7], Simon's causal ordering determines a unique one-to-one correspondence, but in others, such as those involving feedback, the correspondence is not unique. Yet in examining feedback circuits, for example, people can assert categorically that the flow of causation goes clockwise, rather than counterclockwise. They make such assertions on the basis of relative magnitudes of forces; for example, it takes very little energy to make an input of a gate change its output, but no force applied to the output can influence the input. When such considerations are available, causal directionality can be determined by appealing again to the notion of hypothetical intervention and asking whether an external control over one variable in the mechanism necessarily affects the others. The variable which does not affect any of the others is the dependent variable. This then constitutes the operational semantics for identifying the dependent variables X_i in nonrecursive causal theories (see Section 6.3.1).

6.2.5 Imaging vs. Conditioning

If action is a transformation from one probability function to another, one may ask whether every transformation corresponds to an action, or are there some constraints that are peculiar to exactly those transformations that originate from actions. Lewis' (1976) formulation of counterfactuals indeed identifies such constraints: the transformation must be an *imaging* operator (Imaging is the probabilistic version of Winslett-Katsuno-Mendelzon possible worlds representation of "update").

Whereas Bayes conditioning $P(s|e)$ transfers the entire probability mass from states excluded by e to the remaining states (in proportion to their current $P(s)$), imaging works differently; each excluded state s transfers its mass individually to a select set of states $S^*(s)$, which are considered "closest" to s. The reason why imaging is a more adequate representation of transformations associated with actions can be seen more clearly through a representation theorem due to Gardenfors (1988, Theorem 5.2, pp.113) (strangely, the connection to actions never appears in Gardenfors' analysis). Gardenfors' theorem states that a probability update operator $P(s) \to P_A(s)$ is an imaging operator iff it preserves mixtures, *i.e.*

$$[\alpha P(s) + (1-\alpha)P'(s)]_A = \alpha P_A(s) + (1-\alpha)P'_A(s) \tag{6.1}$$

for all constants $1 > \alpha > 0$, all propositions A, and all probability functions P and P'. In other words, the update of any mixture is the mixture of the updates[3].

This property, called homomorphism, is what permits us to specify actions in terms of *transition probabilities*, as it is usually done in stochastic control and Markov decision process. Denoting by $P_A(s|s')$ the probability resulting from acting A on a known state s', homomorphism (6.1) dictates:

$$P_A(s) = \sum_{s'} P_A(s|s') P(s') \tag{6.2}$$

saying that, whenever s' is not known with certainty, $P_A(s)$ is given by a weighted sum of $P_A(s|s')$ over s', with the weight being the current probability function $P(s')$.

This characterization, however, is too permissive; while it requires any action-based transformation to be describable in terms of transition probabilities, it also accepts any transition probability specification, however whimsical as a descriptor of some action. The valuable information that actions are defined as *local* surgeries, is totally ignored in this characterization. For example, the transition probability associated with the atomic action $A_i = do(X_i = x_i)$ originates from the deletion of just one mechanism in the assembly. Hence, one would expect that the transition probabilities associated with the set of atomic actions would not be totally arbitrary but would constrain one another.

We are currently exploring axiomatic characterizations of such constraints which we hope to use as a logic for sentences of the type: "X affects Y when we hold Z fixed". With the help of such logic we hope to be able to derive, refute or confirm sentences such as "If X has no effect on Y and Z affects Y, then Z will continue to affect Y when we fix X". The reader might find some challenge proving or refuting the sentence above, that is, testing whether it holds in every causal theory, when "affecting" and "fixing" are interpreted by the local-surgery semantics described in this chapter.

---

[3] Assumption (1) is reflected in the (U8) postulate of [17]:
$(K_1 \vee K_2) o \mu = (K_1 o \mu) \vee (K_2 o \mu)$, where o is an update operator.

6.3 Formal Underpinning

6.3.1 Causal Theories, Actions, Causal Effect, and Identifiability

Definition 6.1 *A causal theory is a 4-tuple*

$$T = <V, U, P(\mathbf{u}), \{f_i\}>$$

where

(i) $V = \{X_1, \ldots, X_n\}$ *is a set of observed variables*

(ii) $U = \{U_1, \ldots, U_m\}$ *is a set of unobserved variables which represent disturbances, abnormalities or assumptions,*

(iii) $P(\mathbf{u})$ *is a distribution function over* $U_1, \ldots, U_m$, *and*

(iv) $\{f_i\}$ *is a set of n deterministic functions, each of the form*

$$X_i = f_i(X_1, \ldots, X_n, U_1, \ldots, U_m) \quad i = 1, \ldots, n. \qquad (6.3)$$

We will assume that the set of equations in (iv) has a unique solution for $X_i, \ldots, X_n$, *given any value of the disturbances* $U_1, \ldots, U_m$. *Therefore, the distribution* $P(\mathbf{u})$ *induces a unique distribution on the observables, which we denote by* $P_T(\mathbf{v})$.

We will consider concurrent actions of the form $do(X = x)$, where $X \subseteq V$ is a set of variables and x is a set of values from the domain of X. In other words, $do(X = x)$ represents a combination of actions that forces the variables in X to attain the values x.

Definition 6.2 *(Effect of actions) The effect of the action* $do(X = x)$ *on a causal theory* T *is given by a subtheory* T_x *of* T, *where* T_x *obtains, by deleting from* T, *all equations corresponding to variables in* X *and substituting the equations* $X = x$ *instead.*

Definition 6.3 *(causal effect) Given two disjoint subsets of variables,* $X \subseteq V$ *and* $Y \subseteq V$, *the causal effect of* X *on* Y, *denoted* $P_T(y|\hat{x})$, *is a function from the domain of* X *to the space of probability distributions on* Y, *such that*

$$P_T(y|\hat{x}) = P_{T_x}(y) \qquad (6.4)$$

for each realization x *of* X. *In other words, for each* $x \in dom(X)$, *the causal effect* $P_T(y|\hat{x})$ *gives the distribution of* Y *induced by the action* $do(X = x)$.

Note that causal effects are defined relative to a given causal theory T, though the subscript T is often suppressed for brevity.

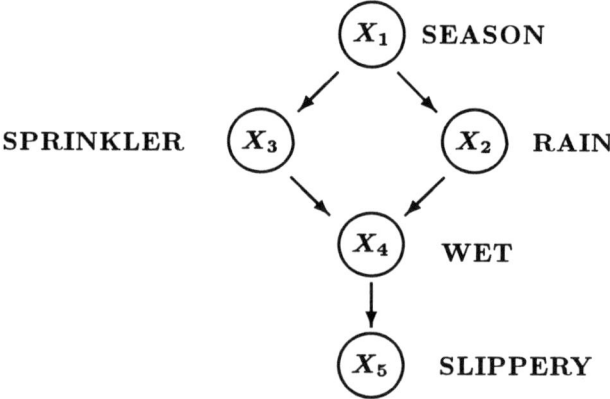

Figure 6.1: A diagram representing a causal theory on five variables.

Definition 6.4 (identifiability) *The causal effect of X on Y is said to be identifiable if the quantity $P(y|\hat{x})$ can be computed uniquely from any positive distribution of the observed variables, that is, if for every pair of theories T_1 an T_2 such that $P_{T_1}(\mathbf{v}) = P_{T_2}(\mathbf{v}) > 0$, we have $P_{T_1}(y|\hat{x}) = P_{T_2}(y|\hat{x})$.*

Identifiability means that $P(y|\hat{x})$ can be estimated consistently from an arbitrarily large sample randomly drawn from the distribution of the observed variables. The notion of identifiability is central to much work in econometrics, where it has become synonymous to the identification of the functions $\{f_i\}$ or some of their parameters [18], mostly under conditions of additive noise. Definition 6.4, which does not assume any parametric represenatation of the functions $\{f_i\}$, extends the notion of identifiabiity to any quantity that can be computed from a given theory. In particular, it replaces the identification of structural parameters with the nonparametric identification of causal effects, $P(y|\hat{x})$, which constitute the ultimate goal of computing structural parameters in policy analysis applications.

Figure 6.1 illustrates a simple causal theory in the form of a diagram. It describes the causal relationships among the season of the year (X_1), whether rain falls (X_2), whether the sprinkler is on (X_3), whether the pavement would get wet (X_4), and whether the pavement would be slippery (X_5). All variables in this figure are binary, taking a value of either true or false, except the root variable X_1 which can take one of four values: Spring, Summer, Fall, or Winter. Here, the absence of a direct link between X_1 and X_5, for example, captures our understanding that the influence of seasonal variations on the slipperiness of the pavement is mediated by other conditions (*e.g.* the wetness of the pavement).

The theory corresponding to Figure 6.1 consists of five functions, each representing an autonomous mechanism:

$$X_1 = U_1$$
$$X_2 = f_2(X_1, U_2)$$

$$X_3 = f_3(X_1, U_3)$$
$$X_4 = f_4(X_3, X_2, U_4)$$
$$X_5 = f_5(X_4, U_5). \tag{6.5}$$

To represent the action "turning the sprinkler ON", $do(X_3 = \text{ON})$, we delete the equation $X_3 = f_3(x_1, u_3)$ from the theory of equation (6.5), and replace it with $X_3 = \text{ON}$. The resulting subtheory, $T_{X_3=\text{ON}}$, contains all the information needed for computing the effect of the action on other variables. For example, it is easy to see from this subtheory that the only variables affected by the action are X_4 and X_5, that is, the descendant of the manipulated variable X_3.

The probabilistic analysis of causal theories becomes particularly simple when two conditions are satisfied:

1. The theory is recursive, *i.e.* there exists an ordering of the variables $V = \{X_1, \ldots, X_n\}$ such that each X_i is a function of a subset $\mathbf{pa}_i$ of its predecessors

$$X_i = f_i(\mathbf{pa}_i, U_i), \quad \mathbf{pa}_i \subseteq \{X_1, \ldots, X_{i-1}\}. \tag{6.6}$$

2. The disturbances $U_1, \ldots, U_n$ are mutually independent, which also implies (from the exogeneity of the U_i's)

$$U_i \parallel \{X_1, \ldots, X_{i-1}\}. \tag{6.7}$$

These two conditions, also called Markovian, are the basis of Bayesian networks [24] and they enable us to compute causal effects directly from the conditional probabilities $P(x_i|\mathbf{pa}_i)$, without specifying the functional form of the functions f_i, or the distributions $P(u_i)$ of the disturbances. This is seen immediately from the following observations.

The distribution induced by any Markovian theory T is given by the product

$$P_T(x_1, \ldots, x_n) = \prod_i P(x_i|\mathbf{pa}_i), \tag{6.8}$$

where $\mathbf{pa}_i$ are the direct predecessors (called *parents*) of X_i in the diagram. On the other hand, the subtheory $T_{x'_j}$, representing the action $do(X_j = x'_j)$ is also Markovian, hence it also induces a product-decomposable distribution

$$P_{T_{x'_j}}(x_1, \ldots, x_n) = \begin{cases} \prod_{i \neq j} P(x_i|\mathbf{pa}_i) = \frac{P(x_1, \ldots, x_n)}{P(x_j|\mathbf{pa}_j)} & \text{if } x_j = x'_j \\ 0 & \text{if } x_j \neq x'_j \end{cases} \tag{6.9}$$

where the partial product reflects the surgical removal of the

$$X_j = f_j(\mathbf{pa}_j, U_j)$$

from the theory of equation (6.6). Given the post-action distribution, the effect of the action $do(X_j = x'_j)$ on any variable (or groups of variables) in the system can readily be computed.

In the example of Figure 6.1, the pre-action distribution is given by the product

$$P_T(x_1, x_2, x_3, x_4, x_5) = P(x_1)P(x_2|x_1)P(x_3|x_1)P(x_4|x_2, x_3)P(x_5|x_4) \quad (6.10)$$

while the surgery corresponding to the action $do(X_3 = \text{ON})$ amounts to deleting the link $X_1 \rightarrow X_3$ from the graph and fixing the value of X_3 to ON, yielding the post-action distribution:

$$P_T(x_1, x_2, x_4, x_5|do(X_3 = \text{ON})) \qquad (6.11)$$
$$= P(x_1) \ P(x_2|x_1) \ P(x_4|x_2, X_3 = \text{ON}) \ P(x_5|x_4).$$

Note the difference between the action $do(X_3 = \text{ON})$ and the observation $X_3 = \text{ON}$. The latter is encoded by ordinary Bayesian conditioning, while the former by conditioning a mutilated graph, with the link $X_1 \rightarrow X_3$ removed. This indeed mirrors the difference between seeing and doing: after observing that the sprinkler is ON, we wish to infer that the season is dry, that it probably did not rain, and so on; no such inferences should be drawn in evaluating the effects of the deliberate action "turning the sprinkler ON". The amputation of $X_3 = f_3(X_1, U_3)$ from (6.5) ensures the suppression of any abductive inferences from any of the action's consequences.

Note also that equations (6.8) through (6.12) are independent of T, in other words, the pre- and post-action distributions depend only upon observed conditional probabilities but are independent of the particular functional form of $\{f_i\}$ or the distribution $P(\mathbf{u})$ which generate those probabilities. This is the essence of identifiability as given in Definition 6.4, which stems from the Markovian assumptions (6.6) and (6.7). Section 6.3.2 will demonstrate that certain causal effects, though not all, are identifiable even when the Markovian property is destroyed by introducing dependencies among the disturbance terms.

Generalization to multiple actions and conditional actions are straightforward. Multiple actions $do(X = x)$, where X is a compound variable result in a distribution similar to (6.9), except that all factors corresponding to the variables in X are removed from the product in (6.8). Stochastic conditional strategies of the form

$$do(X_j = x_j) \text{ with probability } P^*(x_j|\mathbf{pa}_j^*)$$

where $\mathbf{pa}_j^*$ is the support of the decision strategy, also result in a product decomposition similar to (6.9), except that each factor $P(x_j|\mathbf{pa}_j)$ is replaced with $P^*(x_j|\mathbf{pa}_j^*)$. The identification of conditional and multiple actions is treated in [28, 30].

The surgical procedure described above is not limited to probabilistic analysis. The causal knowledge represented in Figure 6.1 can be captured by logical theories as well, for example,

$$x_2 \iff [(X_1 = \text{Winter}) \vee (X_1 = \text{Fall}) \vee ab_2] \wedge \neg ab'_2$$
$$x_3 \iff [(X_1 = \text{Summer}) \vee (X_1 = \text{Spring}) \vee ab_3] \wedge \neg ab'_3$$
$$x_4 \iff (x_2 \vee x_3 \vee ab_4) \wedge \neg ab'_4$$
$$x_5 \iff (x_4 \vee ab_5) \wedge \neg ab'_5$$

where x_i stands for $X_i = true$, and ab_i and ab'_i stand, respectively, for trigerring and inhibiting abnormalities. The double arrows represent the assumption that the events on the r.h.s. of each equation are the *only* causes for the l.h.s.

It should be emphasized, though, that the models of a causal theory are not made up merely of truth value assignments which satisfy the equations in the theory. Since each equation represents an autonomous process, the scope of each individual equation must be specified in any model of the theory, and this can be encoded using either the graph (as in Figure 6.1) or the generic description of the theory, as in (6.5). Alternatively, we can view a model of a causal theory to consist of a mutually consistent set of submodels, with each submodel being a standard model of a single equation in the theory.

6.3.2 Action Calculus

The identifiability of causal effects demonstrated in Section 6.3.1 relies critically on the Markovian assumptions (6.6) and (6.7). If a variable that has two descendants in the graph is unobserved, the disturbances in the two equations are no longer independent, the Markovian property (6.6) is violated and identifiability may be destroyed. This can be seen easily from equation (6.9); if any parent of the manipulated variable X_j is unobserved, one cannot estimate the conditional probability $P(x_j|\mathbf{pa}_j)$, and the effect of the action $do(X_j = x_j)$ may not be predictable from the observed distribution $P(x_1, \ldots, x_n)$. Fortunately, certain causal effects are identifiable even in situations where members of $\mathbf{pa}_j$ are unobservable and, moreover, polynomial tests are now available for deciding when $P(x_i|\hat{x}_j)$ is identifiable, and for deriving closed-form expressions for $P(x_i|\hat{x}_j)$ in terms of observed quantities [29, 10].

These tests and derivations are based on a symbolic calculus to be described in the sequel, in which interventions, side by side with observations, are given explicit notation, and are permitted to transform probability expressions. The transformation rules of this calculus reflect the understanding that interventions perform "local surgeries" as described in Definition 6.2, *i.e.* they overrule equations that tie the manipulated variables to their pre-intervention causes.

Let X, Y and Z be arbitrary disjoint sets of nodes in a DAG G. We say that X and Y are independent given Z in G, denoted $(X \perp\!\!\!\perp Y|Z)_G$, if the set Z d-separates [24] X from Y in G. We denote by $G_{\overline{X}}$ the graph obtained by deleting from G all arrows pointing to nodes in X. Likewise, we denote by $G_{\underline{X}}$ the graph obtained by deleting from G all arrows emerging from nodes in X. To represent the deletion of both incoming and outgoing arrows, we use the notation $G_{\overline{X}\underline{Z}}$. Finally, the expression $P(y|\hat{x}, z) \triangleq P(y, z|\hat{x})/P(z|\hat{x})$ stands for the probability of $Y = y$ given that $Z = z$ is observed and X is held constant at x.

Theorem 6.1 *Let G be the directed acyclic graph associated with a Markovian causal theory, and let $P(\cdot)$ stand for the probability distribution induced by that theory. For any disjoint subsets of variables X, Y, Z and W we have:*

Rule 1 *Insertion/deletion of observations*

$$P(y|\hat{x}, z, w) = P(y|\hat{x}, w) \quad if \quad (Y \perp\!\!\!\perp Z|X, W)_{G_{\overline{X}}} \qquad (6.12)$$

Rule 2 *Action/observation exchange*

$$P(y|\hat{x}, \hat{z}, w) = P(y|\hat{x}, z, w) \quad if \quad (Y \perp\!\!\!\perp Z|X, W)_{G_{\overline{X}\underline{Z}}} \qquad (6.13)$$

Rule 3 *Insertion/deletion of actions*

$$P(y|\hat{x}, \hat{z}, w) = P(y|\hat{x}, w) \quad if \quad (Y \perp\!\!\!\perp Z|X, W)_{G_{\overline{X}, \overline{Z(W)}}} \qquad (6.14)$$

where $Z(W)$ is the set of Z-nodes that are not ancestors of any W-node in $G_{\overline{X}}$.

Each of the inference rules above follows from the basic interpretation of the "$\hat{x}$" operator as a replacement of the causal mechanism that connects X to its pre-action parents by a new mechanism $X = x$ introduced by the intervening force. The result is a submodel characterized by the subgraph $G_{\overline{X}}$ (named "manipulated graph" in Spirtes et al. (1993)) which supports all three rules.

Rule 1 reaffirms d-separation as a valid test for conditional independence in the distribution resulting from the intervention $do(X = x)$, hence the graph $G_{\overline{X}}$. This rule follows from the fact that deleting equations from the system does not introduce any dependencies among the remaining disturbance terms.

Rule 2 provides a condition for an external intervention $do(Z = z)$ to have the same effect on Y as the passive observation $Z = z$. The condition amounts to $\{X \cup W\}$ blocking all back-door paths from Z to Y (in $G_{\overline{X}}$), since $G_{\overline{X}\underline{Z}}$

retains all (and only) such paths.[4]

Rule 3 provides conditions for introducing (or deleting) an external intervention $do(Z = z)$ without affecting the probability of $Y = y$. The validity of this rule stems, again, from simulating the intervention $do(Z = z)$ by the deletion of all equations corresponding to the variables in Z (hence the graph $G_{\overline{XZ}}$).

Corollary 6.1 *A causal effect q: $P(y_1, ..., y_k | \hat{x}_1, ..., \hat{x}_m)$ is identifiable in a model characterized by a graph G if there exists a finite sequence of transformations, each conforming to one of the inference rules in Theorem 6.1, which reduces q into a standard (i.e. hat-free) probability expression involving observed quantities.* □

Although Theorem 6.1 and Corollary 6.1 require the Markovian property, they do not require all variables to be observable and, hence, they can be applied to non Markovian, recursive theories as well. To demonstrate, assume that variable X_1 in Figure 6.1 is unobserved, rendering the disturbances U_3 and U_2 dependent, since these terms now include the common influence of X_1. Theorem 6.1 tells us that the causal effect $P(x_4 | \hat{x}_3)$ is identifiable, because:

$$P(x_4|\hat{x}_3) = \sum_{x_2} P(x_4|\hat{x}_3, x_2) P(x_2|\hat{x}_3)$$

Rule 3 permits the deletion

$$P(x_2|\hat{x}_3) = P(x_2), \text{ because } (X_2 \perp\!\!\!\perp X_3)_{G_{\overline{X_3}}},$$

while Rule 2 permits the exchange

$$P(x_4|\hat{x}_3, x_2) = P(x_4|x_3, x_2), \text{ because } (X_4 \perp\!\!\!\perp X_3|X_2)_{G_{\underline{X_3}}}.$$

This gives

$$P(x_4|\hat{x}_3) = \sum_{x_2} P(x_4|x_3, x_2) P(x_2)$$

which is a "hat-free" expression, involving only observed quantities.

In general, it can be shown [29] that:

1. The effect of interventions can often be identified (from non-experimental data) without resorting to parametric models,

2. The conditions under which such non-parametric identification is possible can be determined by simple graphical criteria, and,

3. When the effect of interventions is not identifiable, the causal graph may suggest non-trivial experiments which, if performed, would render the effect identifiable.

---

[4] A path from X to Y is said to be a back-door path if it ends with arrows into both X and Y [26].

The ability to assess the effect of interventions from non-experimental data has immediate applications in the medical and social sciences, since subjects who undergo certain treatments often are not representative of the population as a whole. Such assessments are also important in AI applications where an agent needs to predict the effect of the next action on the basis of past performance records, and where that action has never been enacted out of free will, but in response to environmental needs or to other agent's requests.

6.3.3 Historical Background

An explicit translation of interventions to "wiping out" equations from linear econometric models was first proposed by Strotz & Wold [36] and later used in Fisher [9] and Sobel [34]. Extensions to action representation in non-monotonic systems were reported in [15, 25]. Graphical ramifications of this translation were explicated first in Spirtes *et al.* [35] and later in Pearl [26]. A related formulation of causal effects, based on event trees and counterfactual analysis was developed by Robins [32, pp. 1422–25]. Calculi for actions and counterfactuals based on this interpretation are developed in [28] and [2], respectively.

6.4 Counterfactuals

A counterfactual sentence has the form

If A were true, then C would have been true

where A, the counterfactual antecedent, specifies an event that is contrary to one's real-world observations, and C, the counterfactual consequent, specifies a result that is expected to hold in the alternative world where the antecedent is true. A typical example is "If Oswald were not to have shot Kennedy, then Kennedy would still be alive" which presumes the factual knowledge of Oswald's shooting Kennedy, contrary to the antecedent of the sentence.

The majority of the philosophers who have examined the semantics of counterfactual sentences have resorted to some version of Lewis' "closest world" approach; "C if it were A" is true, if C is true in worlds that are "closest" to the real world yet consistent with the counterfactuals antecedent A [20]. Ginsberg [13], followed a similar strategy. The drawback of the "closest world" approach is that it leaves the precise specification of the closeness measure almost unconstrained. In the domino example cited in Section 6.1, the manifestly closest world consistent with the antecedent "tile i is tipped to the right" would be a world in which just tile i is tipped, while all the others remain erect. Yet, we all accept the counterfactual sentence "Had tile i been tipped over to the right, tile $i + 1$ would be tipped as well" as plausible and valid. Thus, we see that distances among worlds are not determined merely by surface similarities but require a delicate balance between disturbed and naturally occurring mechanisms. The local surgery paradigm expounded in Section 6.1 in fact offers a concrete explication of the closest world approach. A world w_1 is "closer" to w than a world w_2 is, if the the set of atomic surgeries needed for transforming w into w_1 is a subset of those needed for transforming w into w_2. In the domino example, finding tile i tipped and $i+1$ erect requires the breakdown of two mechanism (*e.g.* by two external actions) compared with one mechanism for the world in which all j-tiles, $j > i$ are tipped. This paradigm conforms to our perception of causal influences and lends itself to economical machine representation.

Counterfactual thinking is important to fault diagnosis, planning, determination of liability, and policy analysis, and it dominates most reasoning in political science and economics. We say, for example, "If Germany were not punished so severely at the end of World War I, Hitler would not have come to power", or, "If Reagan did not lower taxes, our deficit would be much lower today". The messages conveyed by such thought experiments emphasize an understanding of generic laws in the domain, and are aimed to guide future policy making, for example, that defeated countries should not be humiliated, or that lowering taxes (contrary to Reaganomics theory) tend to increase national debts.

Strangely, there is very little formal work on counterfactual reasoning, or even policy analysis in the general literature. An examination of econometric journals and textbooks, for example, reveals a strange imbalance: while an enormous mathematical machinery is centred on problems of estimation and prediction, policy analysis (the ultimate goal of economic theories) receives almost no formal treatment. Currently, the primary methods of economic policy making are based on the so called *reduced-form* analysis [16, 23] which boils down to the following: To find the impact of a policy involving decision variables X on outcome variables Y, one examines past data and estimates the conditional expectation $E(Y|X=x)$, where x is the particular instantiation of X under the policy studied.

The assumption underlying this method is that the data were generated by a process in which the decision variables X act as exogenous variables. Unfortunately, almost every realistic policy making (*e.g.* taxation) involves some endogenous variables, that is, variables whose values are determined by other variable in the analysis. Taking taxation policies as an example, economic data are generated by a process in which Government is reacting to various indicators and various pressures, hence, taxation is endogenous in the estimation phase of the study. Taxation becomes exogenous when we wish to predict the impact of a specific decision to raise or lower taxes. Thus, the reduced-form method is valid only when past decisions are nonresponsive to other variables in the system, and this, unfortunately, rules out almost all interesting control variables (*e.g.* tax, interest-rates, quotas) from the analysis.

This problem is not unique to economics or social policy making, but appears whenever one wishes to evaluate the merit of a plan on the basis of past performance of other agents. Even when the motivations behind past actions of those agents are known with certainty, a systematic method must be devised of selectively ignoring the influence of those motivations from the evaluation process. In fact the very essence of *evaluation* rest on having the freedom to imagine and compare trajectories in various counterfactual worlds, each created by a hypothetical policy that is free of the pressures that compelled the implementation of such policies in the past.

A connection between counterfactuals and policy making was formulated in [2] using the network representation of causal theories. The probability of a counterfactual sentence is computed by the following steps: First, the available observations are used to update the joint probability of the exogenous U variables (viewed as constant, but unknown boundary conditions, see equation (6.3)). Second, the counterfactual antecedent is interpreted as an external intervention that forces that antecedent to hold true. In the network representation, such an intervention is simulated by severing all causal edges that lead into the antecedent variables and setting their values to those specified in the antecedent. Finally, the probability of the counterfactual consequent is evaluated using the mutilated network as in (6.9). It turns out that, unlike causal effect queries, counterfactual queries are not identifiable even in Markovian theories, but require that the functional-form of $\{f_i\}$ be specified. In [3] a method is devised for computing bounds on counterfactual probabilities which, under certain circumstances may collapse to point estimates. This method has been applied to the evaluation of causal effects in studied involving noncompliance, to the determination of legal liability, and to the evaluation of economic policies in non-recursive linear systems [4].

Acknowledgements

The research was partially supported by Air Force grant #F49620-94-1-0173, NSF grant #IRI-9420306 and Northrop/Rockwell Micro grant #94-100.

References

[1] Adams E. (1975) *The Logic of Conditionals,* Chapter 2, D. Reidel, Dordrecht, Netherlands.

[2] Balke A. and Pearl J. (1994) Probabilistic evaluation of counterfactual queries. In *Proceedings of the Twelfth National Conference on Artificial Intelligence (AAAI-94)* Seattle, WA, (I) 230–237, July 31 – August 4.

[3] Balke A. and Pearl J. (1994) Counterfactual probabilities: Computational methods, bounds, and applications. In R. Lopez de Mantaras and

D. Poole (eds), *Proceedings of the Conference on Uncertainty in Artificial Intelligence (UAI-94)* Morgan Kaufmann, San Mateo, CA, 46–54, July 29–31.

[4] Balke A. and Pearl J. (1995) Counterfactuals and Policy Analysis in Structural Models. In P. Besnard and S. Hanks (eds), *Uncertainty in Artificial Intelligence 11*, Morgan Kaufmann, San Francisco, CA, 11–18.

[5] Dalkey N. (1994) "Webs," UCLA Cognitive Systems Laboratory. Technical Report (R-166), Computer Science Department, University of California, Los Angeles, March.

[6] Darwiche A. and Pearl, J. (1994) Symbolic causal networks for planning under uncertainty. In *Symposium Notes of the 1994 AAAI Spring Symposium on Decision-Theoretic Planning*, Stanford, CA, 41–47, March 21-23.

[7] Dechter R. and Pearl J. (1991) Directed constraint networks: A relational framework for Causal Modeling. In *Proceedings, 12th International Joint Conference of Artificial Intelligence (IJCAI-91)*, Sydney, Australia, 1164–1170, August 24–30.

[8] Fikes R.E. and Nilsson N.J. (1971) STIRPS: A new approach to the application of theorem proving to problem solving. *Artificial Intelligence* (2)3/4: 189–208.

[9] Fisher F.M. (1970) A correspondence principle for simultaneous equation models. *Econometrica* 38: 73–92.

[10] Galles D. and Pearl J. (1995) Testing Identifiability of Causal Effects. In P. Besnard and S. Hanks (eds), *Uncertainty in Artificial Intelligence 11* Morgan Kaufmann, San Francisco, CA, 185-195.

[11] Gardenfors P. (1988) *Knowledge in Flux: Modeling the Dynamics of Epistemic States.* MIT Press, Cambridge, MA.

[12] Geffner H.A. (1992) *Default Reasoning: Causal and Conditional Theories.* MIT Press, Cambridge, MA.

[13] Ginsberg M.L. (1986) Counterfactuals. *Artificial Intelligence* 30: 35–79.

[14] Goldszmidt M. and Darwiche A. (1994) Action networks: A framework for reasoning about actions and change under uncertainty. In R. Lopez de Mantaras and D. Poole (eds), *Proceedings of the Tenth Conference on Uncertainty in Artificial Intelligence (UAI-94)*, Morgan Kaufmann, San Mateo, CA, 136–144.

[15] Goldszmidt M. and Pearl, J. (1992) Rank-based systems: A simple approach to belief revision, belief update, and reasoning about evidence and actions. In B. Nebel, C. Rich and W. Swartout (eds), *Proceedings of the Third International Conference on Knowledge Representation and Reasoning*, Morgan Kaufmann, San Mateo, CA, 661-672, October.

[16] *Econometric Models, Techniques, and Applications*. Prentice Hall, Englewood Cliffs, NJ.

[17] Katsuno H. and Mendelzon A. (1991) On the difference between updating a knowledge base and revising it. In *Principles of Knowledge Representation and Reasoning: Proceedings of the Second International Conference*, Boston, MA, 387-394.

[18] Koopman T.C. and Reiersol O. (1950) The identification of structural characteristics. *Annals of Mathematical Statistics* 21: 165-181.

[19] Kushmerick N., Hanks S. and Weld D. (1995) An algorithm for probabilistic planning. *Artificial Intelligence* 76: 239-286.

[20] Lewis D.K. (1973) *Counterfactuals*. Basil Blackwell, Oxford, UK.

[21] Lewis D. (1976) Probabilities of conditionals and conditional probabilities. *Philosophical Review* 85: 297-315.

[22] Levi I. (1988) Iteration of conditionals and the Ramsey test. *Synthese* 76: 49-81.

[23] Maddala G.S. (1992) *Introduction to Econometrics*. Macmillan, NY.

[24] Pearl J. (1988) *Probabilistic Reasoning in Intelligence Systems*. Morgan Kaufmann, San Mateo, CA.

[25] Pearl J. (1993) From Conditional Oughts to Qualitative Decision Theory. In D. Heckerman and A. Mamdani (eds), *Proceedings of the Ninth Conference on Uncertainty in Artificial Intelligence*, Washington, D.C., Morgan Kaufmann, San Mateo, CA, 12-20, July.

[26] Pearl J. (1993) Graphical models, causality, and intervention. *Statistical Science* (8) 3: 266-273.

[27] Pearl J. (1994) From Adams' conditionals to default expressions, causal conditionals, and counterfactuals. In E. Eells and B. Skyrms (eds), *Probability and Conditionals*, Cambridge University Press, New York, NY, 47-74.

[28] Pearl J. (1994) A probabilistic calculus of actions. In R. Lopez de Mantaras and D. Poole (eds), *Proceedings of the Tenth Conference on Uncertainty in Artificial Intelligence (UAI-94)*, Morgan Kaufmann, San Mateo, CA, 454-462.

[29] Pearl J. (1995) Causal diagrams for experimental research (with discussion). *Biometrika*, (82) 4: 669–709.

[30] Pearl J. and Robins J. (1995) Probabilistic evaluation of sequential plans from causal models with hidden variables. In P. Besnard and S. Hanks (eds), *Uncertainty in Artificial Intelligence 11*, Morgan Kaufmann, San Francisco, CA, 444–453.

[31] Poole D. (1985) On the comparison of theories: Preferring the most specific explanations. In *Proceedings of International Conference on Artificial Intelligence (IJCAI-85)*, Los Angeles, CA, 144–147.

[32] Robins J. (1986) A new approach to causal inference in mortality studies with a sustained exposure period – applications to control of the healthy workers survivor effect. *Mathematical Modelling* 7: 1393–1512.

[33] Simon H. (1953) Causal ordering and identifiability. In W.C. Hood and T.C. Koopmans (eds), *Studies in Econometric Method*, New York, NY, Chapter 3.

[34] Sobel M.E. (1990) Effect analysis and causation in linear structural equation models. *Psychometrika* (55) 3: 495–515.

[35] Spirtes P., Glymour, C. and Schienes, R. (1993) *Causation, Prediction, and Search*. Springer-Verlag, New York.

[36] Strotz R.H. and Wold H.O.A. (1960) Recursive versus nonrecursive systems: An attempt at synthesis. *Econometrica* 28: 417–427.

7

Another Semantics for Pearl's Action Calculus

V.G. Vovk[1]

7.1 Introduction

According to Pearl's ([6], Subsection 2.2; [7], Subsection 2.1) picture in its simplest form, the world consists of invariant stochastic mechanisms, each mechanism determining the value of a variable; an action is described as a "local surgery" which replaces the mechanism corresponding to each variable in a set D of variables by a trivial mechanism that fixes the value of that variable. Our purpose is to identify the effect of the action on some other variables B from the sample data when some of the variables (outside D and B) are not observable. The problem is that the sample data are collected while the "natural" mechanisms for the variables in D are working, and we are interested in the "post-intervention" probabilities corresponding to the fixed values of the variables in D. Pearl [6, 7] developed powerful techniques for solving this problem.

A weak side of Pearl's picture is that:

- each of the mechanisms is stable at the stage of collecting the sample data (so that we can estimate the distribution of the observable variables);

---

[1] The research described in this publication was made possible in part by Grant No. MRS300 from the International Science Foundation and Russian Government. It was finished while the author was a Fellow at the Center for Advanced Study in the Behavioral Sciences. I am grateful for financial support provided by National Science Foundation (SES-9022192).

COMPUTATIONAL LEARNING AND PROBABILISTIC REASONING
Editor Alex Gammerman ©1996 John Wiley & Sons Ltd.

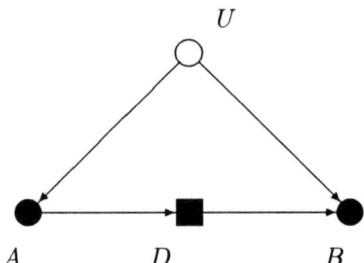

Figure 7.1: Diagram representing the causal relations between the result A of a test, the doctor's decision D on whether to prescribe a drug, the outcome B of the treatment, and the unobservable state U of the patient.

- some mechanisms can be manipulated by the actor (namely, can be replaced by a trivial mechanism).

If the actor is able to manipulate some variables, why he does not do it when the sample data are collected? It seems that Pearl's "actions" should be understood as very serious interventions: in the usual conditions the actor never applies his power to change D. (A good example of such an intervention is given in Section 4 of Balke and Pearl [2]: in that example the government fixes coffee prices for the first time, perhaps just after the Socialist Revolution.)

Let us consider two of Pearl's examples. In the first example (see Figure 7.1), the doctor has an option of prescribing some drug; however, the drug is not very efficient (some even believe that it does more harm than good). The doctor's decision D on whether to prescribe the drug is usually based on the result A of some test. The variable B says whether the patient recovered or died. Both A and B depend upon the patient's state U, which we cannot observe directly.

We are given a huge record

$$a_1 d_1 b_1 a_2 d_2 b_2 \ldots \tag{7.1}$$

of what has happened to the patients so far (a_n, d_n, b_n are the values of A, D, B, respectively, for the nth patient), and our goal is to give some recommendations to the doctor. If the doctor works in regular conditions and can perform the test, our recommendation will depend upon the result of the test; if he works in extremal conditions and cannot perform the test, our recommendation will simply specify whether to prescribe the drug.

In this example, it is clear that, whereas the assumption that U, A and B are determined by stable mechanisms is natural (these mechanisms are dictated by people's biology, the technology of performing the test, and the composition of the drug), the analogous assumption about D is very restrictive. Different patients may be treated by different doctors with different

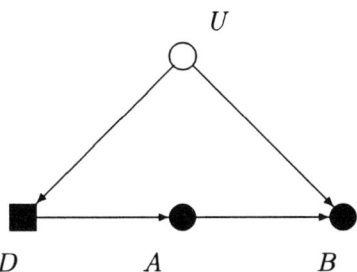

Figure 7.2: Diagram representing the causal relations between smoking D, tar deposits A, lung cancer B, and the unobservable genotype U which may be carcinogenic and may involve inborn craving for nicotine.

beliefs about the effect of the drug (doctor i, $i = 1, 2, \ldots$, taking decisions $d_{n_i}, d_{n_i+1}, d_{n_{i+1}-1}$ in (7.1), with $n_1 := 1$); even the same doctor constantly obtains new information (both provided by new cases, and new recommendations, like the one we are preparing) and so can change his practices.

A simple example of an "unstable" strategy for selecting D is to treat the nth patient if and only if n is odd (it is unstable in the sense that $d_1 d_2 \ldots$ are not the outcomes of independent and identical trials). More realistic examples are provided by the strategies described in the literature on the multi-armed bandit problem (e.g. [5]); in general, any "learning while acting" strategy will be unstable. (Notice that the strategies described in literature are usually in conflict with medical ethics: to find out how to select the best decision, we have to experiment, and so sometimes to select a decision which we do not believe to be the best for this particular patient.)

In the other example (Figure 7.2) we are interested in the effect of smoking D on lung cancer B ([6], Subsection 3.3). Here the assumption of stability of the mechanism corresponding to D means that given U the probabilities for different levels of smoking are constant. This assumption also does not seem quite convincing: these probabilities are influenced by changes in the habits prevailing in the society, anti-smoking campaigns, etc.

In this chapter we will remove Pearl's assumption about the invariance of the mechanism corresponding to D; we will allow the actor to choose the value of D taking into account the past during the collection of sample data. A nice feature of our framework is that all of Pearl's mathematics remains intact, at least when D contains only one variable (as in our examples and the examples of Figure 6 in Pearl [6]); this result will be formulated in the next section.

7.2 Main Result

In this section we first define our semantics for identifying causal effects, then briefly describe Pearl's semantics, and finally state our main result, which says that, provided a single variable is manipulated, a causal effect is identifiable in our semantics if and only if it is identifiable in Pearl's semantics.

A *structural model* is a triple $(G, \text{LAT}, \text{DEC})$, where G is a directed acyclic graph, and LAT and DEC are disjoint sets of G's nodes:

$$\text{LAT} \subseteq \text{VAR}, \ \text{DEC} \subseteq \text{VAR}, \ \text{LAT} \cap \text{DEC} = \emptyset$$

(VAR stands for the set of all nodes of G). The nodes of G are also called *variables*; variables in LAT and DEC are called *latent* variables and *decision* variables, respectively. We put

$$\text{OBS} := \text{VAR} \setminus \text{LAT}, \ \text{RAN} := \text{VAR} \setminus \text{DEC};$$

the variables in OBS and RAN are called *observable* and *random*, respectively. Therefore, the nodes of G are partitioned into three classes: the latent variables LAT; the decision variables DEC; the observable random variables OBS ∩ RAN.

A structural model $(G, \text{LAT}, \text{DEC})$ is *comprehensive* if $\text{DEC} = \emptyset$ (all variables are random). It will be convenient for us to identify a comprehensive structural model with the pair (G, LAT).

Intuitively, a variable is random if its value is determined given the values of its parents by a stable stochastic mechanism. No stability will be assumed for decision variables.

For each variable $X \in \text{VAR}$ we let par X stand for the set of X's parents in G. We call X *exogenous* if par $X = \emptyset$; otherwise, X is *endogenous*.

A *causal theory* T of a structural model $(G, \text{LAT}, \text{DEC})$ specifies the following:

- the *frame* Θ_X (*i.e.* finite nonempty set of possible values) of each variable $X \in \text{VAR}$;

- the probability
$$\text{prob}^T(X = x) \tag{7.2}$$
that X will take value x, for each possible value $x \in \Theta_X$ of each exogenous random variable X;

- the probability
$$\text{prob}^T(X = x \mid \text{par } X = a) \tag{7.3}$$
that X will take value x given par $X = a$, for each possible value $x \in \Theta_X$ of each endogenous random variable X and each possible configuration $a \in \Theta_{\text{par } X}$ of X's parents par X (when A is a set of variables, Θ_A is the Cartesian product of Θ_Y, $Y \in A$). It is required that (7.2) and (7.3) be nonnegative and sum over $x \in \Theta_X$ to 1.

Notice that we drop T from our notation for the frames; this will cause no ambiguity. For simplicity we will usually consider *strictly positive* causal theories, for which the basic probabilities (7.2) and (7.3) are positive.[2] (This assumption concerns the random variables; later we will also make an analogous assumption about the decision variables.) We will use notation (7.3) both for exogenous and for endogenous variables; in the case of exogenous variables the part "$|\operatorname{par} X = a$" of (7.3) should be ignored.

A causal theory of a comprehensive structural model is also called comprehensive. In the case where T is a comprehensive theory, we can extend prob^T to a probability distribution in Θ_{VAR}: for every $a \in \Theta_{\text{VAR}}$, we put

$$\operatorname{prob}^T(\text{VAR} = a) := \prod_{X \in \text{VAR}} \operatorname{prob}^T(X = a(X) \mid \operatorname{par} X = a(\operatorname{par} X)) \quad (7.4)$$

(if A is a set of variables and $a \in \Theta_A$, we let $a(X)$ stand for the value assigned by a to a variable $X \in A$ and also let $a(B)$ stand for the restriction of a to a subset $B \subseteq A$). Also, for every $A \subseteq \text{VAR}$ and $a \in \Theta_A$, we put

$$\operatorname{prob}^T(A = a) := \sum_{b \in \Theta_{\text{VAR} \setminus A}} \operatorname{prob}^T(A = a, \text{VAR} \setminus A = b).$$

In the case of an arbitrary causal theory T, we obtain such a probability distribution fixing the values of all decision variables. Formally, for each configuration $d \in \Theta_{\text{DEC}}$ we obtain a comprehensive causal theory $T_{\text{DEC}:=d}$ adding the probabilities

$$\operatorname{prob}^{T_{\text{DEC}:=d}}(X = x \mid \operatorname{par} X = a) := \begin{cases} 1, & \text{if } x = d(X), \\ 0, & \text{otherwise,} \end{cases}$$

for all $X \in \text{DEC}$, to probabilities (7.3):

$$\operatorname{prob}^{T_{\text{DEC}:=d}}(X = x \mid \operatorname{par} X = a) := \operatorname{prob}^T(X = x \mid \operatorname{par} X = a), \ X \in \text{RAN}.$$

The corresponding probability distribution in Θ_{VAR} (defined by (7.4)) will be denoted by

$$\operatorname{prob}^T_{\text{DEC}:=d} := \operatorname{prob}^{T_{\text{DEC}:=d}};$$

therefore, for $a \in \Theta_{\text{VAR}}$,

$$\operatorname{prob}^T_{\text{DEC}:=d}(\text{VAR} = a)$$

$$= \begin{cases} \prod_{X \in \text{RAN}} \operatorname{prob}^T(X = a(X) \mid \operatorname{par} X = a(\operatorname{par} X)), & \text{if } a(\text{DEC}) = d, \\ 0, & \text{otherwise.} \end{cases}$$

As usual, we define the conditional probabilities

$$\operatorname{prob}^T_{\text{DEC}:=d}(B = b \mid A = a) := \frac{\operatorname{prob}^T_{\text{DEC}:=d}(B = b, A = a)}{\operatorname{prob}^T_{\text{DEC}:=d}(A = a)}, \quad (7.5)$$

---

[2] This is slightly stronger than Pearl's [6, 7] requirement.

where A and B are disjoint sets of random variables; notice that this is well defined if T is strictly positive.

We are interested in probabilities (7.5) when $A, B \subseteq \text{OBS}$. These probabilities are "causal", in the sense that they determine unambiguously the effect of every decision on the observed variables; notice that they do not depend upon the arbitrary actions of people who manipulate the decision variables. We would like to extract these probabilities from the sample data.

A *data sequence* is an infinite[3] sequence

$$\omega = \omega_1 \omega_2 \ldots \in \Theta_{\text{VAR}}^\infty \qquad (7.6)$$

of configurations $\omega_n \in \Theta_{\text{VAR}}$. Our problem is to extract probabilities (7.5) from the "observed components"

$$\omega(\text{OBS}) := \omega_1(\text{OBS})\omega_2(\text{OBS})\ldots$$

of the realized data sequence (7.6).

Let T be a causal theory of a structural model $(G, \text{LAT}, \text{DEC})$. Theory T describes the stable mechanisms that govern the random variables; now we will introduce the unstable mechanisms that will govern the decision variables. A *decision strategy* Δ for T specifies, for every finite sequence $\omega_1 \ldots \omega_{N-1}$ ($N \geq 1$) of configurations $\omega_n \in \Theta_{\text{VAR}}$, every $X \in \text{DEC}$, every $x \in \Theta_X$, and every $a \in \Theta_{\text{par} X}$, the probability

$$\text{prob}_{\omega_1 \ldots \omega_{N-1}}^\Delta (X = x \mid \text{par } X = a) \qquad (7.7)$$

that the decision $X = x$ will be selected at trial N based on knowing the current values a of X's parents and the past configurations[4] $\omega_1 \ldots \omega_{N-1}$; it is required that (7.7) be non-negative and sum to 1 over $x \in \Theta_X$. Causal theory T and decision strategy Δ together determine the stochastic mechanisms for all variables and all trials; therefore, they determine a probability distribution $\text{Prob}^{T,\Delta}$ in $\Theta_{\text{VAR}}^\infty$. Formally, $\text{Prob}^{T,\Delta}$-probability that the first N configurations will be $\omega_1 \ldots \omega_N$ is

$$\prod_{n=1}^{N} \Bigg(\prod_{X \in \text{RAN}} \text{prob}^T(X = \omega_n(X) \mid \text{par } X = \omega_n(\text{par } X))$$

$$\prod_{X \in \text{DEC}} \text{prob}_{\omega_1 \ldots \omega_{n-1}}^\Delta (X = \omega_n(X) \mid \text{par } X = \omega_n(\text{par } X)) \Bigg).$$

We say that Δ is *strictly positive* if there exists $\epsilon > 0$ such that all probabilities (7.7) are greater than ϵ. Now we can give our definition of identifiability of causal effects.

---

[3] For simplicity we consider infinite data sequences. They make mathematics simpler (though do not have direct empirical meaning).

[4] It is easy to generalize all our definitions and results to the case where prob^Δ are allowed to depend, besides the past configurations $\omega_1 \ldots \omega_{N-1}$, also on some outside information (see Section 7.5).

Let A and B be disjoint sets of observable random variables in a structural model $(G, \text{LAT}, \text{DEC})$. We say that the formula $A \to B$ (which reads: the causal effect of DEC on B given A is identifiable; cf. Pearl [6, 7], Definition 4) is *S-true*[5] if, for any family $\{\Theta_X \mid X \in \text{OBS}\}$ of frames for the observable variables, there exists a function

$$F: \Theta_{A,B,\text{DEC}} \times \Theta_{\text{OBS}}^\infty \to [0,1]$$

such that, for any strictly positive causal theory T of $(G, \text{LAT}, \text{DEC})$ with the frames Θ_X for all $X \in \text{OBS}$ and any strictly positive decision strategy Δ for T,

$$\forall a \in \Theta_A, b \in \Theta_B, d \in \Theta_{\text{DEC}}:$$
$$\text{prob}^T_{\text{DEC}:=d}(B = b \mid A = a) = F(a, b, d, \omega(\text{OBS})) \qquad (7.8)$$

for $\text{Prob}^{T,\Delta}$-almost all $\omega \in \Theta_{\text{VAR}}^\infty$. The intuition behind $A \to B$ being S-true is that we can extract the "causal probabilities" $\text{prob}^T_{\text{DEC}:=d}(B = b \mid A = a)$ from the sample data $\omega(\text{OBS})$. (Of course, we cannot hope to be able to do it for all ω; at best we can do it for "typical," or almost all, ω.)

We are interested in finding as many S-true formulas as possible. In the next section we will discuss Pearl's action calculus, which will allow the deduction of surprisingly many S-true formulas.

The notion of S-truth is the most important for us. However, we will also need the notion of truth in the sense of Pearl; besides enabling us to compare our semantics with Pearl's, it will serve as a powerful technical tool.

Pearl considers formulas of the form $\hat{D}A \to B$, where D A, and B are disjoint sets of observable variables in a comprehensive structural model (G, LAT). Our semantics for such a formula is as follows: $\hat{D}A \to B$ is S-true in (G, LAT) if $A \to B$ is S-true in the structural model (G, LAT, D) with $\text{DEC} = D$. Now we define Pearl's semantics.

Let (G, LAT) be a comprehensive structural model and T be a comprehensive causal theory of (G, LAT). For any subset $D \subseteq \text{OBS}$ and any $d \in \Theta_D$, we can formally (without attaching any intuitive meaning to it) define Pearl's post-intervention probability distribution

$$\text{prob}^T_{D:=d} := \text{prob}^{T_{D:=d}},$$

where $T_{D:=d}$ is the comprehensive causal theory obtained from T by replacing the probabilities $\text{prob}^T(X = x \mid \text{par}\, X = a)$, where $X \in D$, $x \in \Theta_X$, and $a \in \Theta_{\text{par}\, X}$, by

$$\text{prob}^{T_{D:=d}}(X = x \mid \text{par}\, X = a) := \begin{cases} 1, & \text{if } x = d(X), \\ 0, & \text{otherwise.} \end{cases}$$

(In other words, we replace the original mechanism for determining D by simply fixing $D := d$.) We say that $\hat{D}A \to B$, where $D, A, B \subseteq \text{OBS}$ are

---

[5] I believe that this definition is in the spirit of Shafer [8]; in any case, it was inspired by Shafer's critique of Pearl's philosophy.

disjoint, is *P-true* in (G, LAT) if, for any family $\{\Theta_X \mid X \in \text{OBS}\}$ of frames for the observable variables, there exists a function

$$f: \Theta_{A,B,D} \times \mathcal{P}(\Theta_{\text{OBS}}) \to [0, 1]$$

(where $\mathcal{P}(\Theta_{\text{OBS}})$ is the set of all probability distributions in Θ_{OBS}) such that, for any strictly positive comprehensive causal theory T of (G, LAT) with the frames Θ_X for all $X \in \text{OBS}$,

$$\forall a \in \Theta_A, b \in \Theta_B, d \in \Theta_D:$$
$$\text{prob}^T_{D:=d}(B = b \mid A = a) = f\left(a, b, d, \text{prob}^T\Big|_{\text{OBS}}\right) \quad (7.9)$$

(where $\text{prob}^T\big|_{\text{OBS}}$ is the probability distribution in Θ_{OBS} defined by

$$\text{prob}^T\Big|_{\text{OBS}} (\text{OBS} = a) := \text{prob}^T(\text{OBS} = a)).$$

Remark The notion of P-truth, unlike S-truth, does not involve time and data sequences explicitly (though Pearl's intuition behind P-truth involves them: see Definition 4 of [6, 7]).

We let $\#M$ stand for the cardinality of set M. Our main result, which will be proven in Section 7.4, is:

Theorem 7.1 *Let D, A, B be disjoint sets of observable variables in a comprehensive structural model and $\#D = 1$. The formula $\hat{D}A \to B$ is S-true if and only if it is P-true.*

7.3 Formal Action Calculus

In this section we will consider an effective tool for deducing P-true formulas $\hat{D}A \to B$: Pearl's action calculus [7] (or calculus of intervention [6]). Unlike Pearl, we will present this calculus in a formal way (see, for example, Shoenfield [10], Section 1.2); to apply such a formal calculus we need not remember the probabilistic meaning of our formulas.

Let A, B and C be disjoint sets of variables. We write $(A \perp\!\!\!\perp B \mid C)_G$ to mean that C d-separates A from B in G (see Definition 1 in Pearl [6]). We denote by $G_{\overline{A}\underline{B}}$ the graph obtained by deleting from G all arrows pointing to A-nodes (*i.e.* nodes in A) and all arrows emerging from B-nodes; $G_{\overline{A}}$ is the same as $G_{\overline{A}\underline{\emptyset}}$. We let AB (or A, B) stand for the union of A and B.

Let (G, LAT) be a comprehensive structural model. The action calculus associated with (G, LAT) is defined as follows. The *formulas* of the calculus are $\hat{D}A \to B$, where D, A and B are disjoint sets of observable variables in (G, LAT). The *axiom* is the formula $\to \text{OBS}$ (which is the same as $\hat{\emptyset}\emptyset \to \text{OBS}$). The *inference rules* are:

Insertion/deletion of observations

$$\frac{\hat{D}CA \to B}{\hat{D}A \to B} \text{ and } \frac{\hat{D}A \to B}{\hat{D}CA \to B} \text{ if } (C \perp\!\!\!\perp B \mid DA)_{G_{\overline{D}}}; \tag{7.10}$$

Action/observation exchange

$$\frac{\widehat{DC}A \to B}{\hat{D}CA \to B} \text{ and } \frac{\hat{D}CA \to B}{\widehat{DC}A \to B} \text{ if } (C \perp\!\!\!\perp B \mid DA)_{G_{\overline{DC}}}; \tag{7.11}$$

Insertion/deletion of actions

$$\frac{\widehat{DC}A \to B}{\hat{D}A \to B} \text{ and } \frac{\hat{D}A \to B}{\widehat{DC}A \to B} \text{ if } (C \perp\!\!\!\perp B \mid DA)_{G_{\overline{D,C(A)}}}, \tag{7.12}$$

where $C(A)$ is the set of C-nodes that are not ancestors of any A-node in $G_{\overline{D}}$;

Marginalization

$$\frac{\hat{D}A \to BC}{\hat{D}A \to B}; \tag{7.13}$$

Conditioning

$$\frac{\hat{D}A \to BC}{\hat{D}AB \to C}; \tag{7.14}$$

Composition

$$\frac{\hat{D}A \to B, \hat{D}AB \to C}{\hat{D}A \to BC} \tag{7.15}$$

((7.15) is obtained by turning upside down and merging (7.13) and (7.14)).

The *theorems* of the action calculus are defined inductively as follows:

- The axiom is a theorem.
- If $\frac{F_1,...,F_k}{F}$ is an inference rule ($k = 2$ in the case of composition; otherwise, $k = 1$) and $F_1, \ldots, F_k$ are theorems, then F is a theorem.

(In other words, the set of theorems is the least set satisfying these two conditions.)

The action calculus is *P-sound*, in the sense that all its theorems are P-true (Pearl [6, 7]). It is obvious that the axiom $\to$ OBS is P-true, so to show that the action calculus is P-sound it is sufficient to show that the inference rules are P-valid ($\frac{F_1,...,F_k}{F}$ is *valid*, or truth-preserving, if F is always true when $F_1, \ldots, F_k$ are true). The P-validity of inference rules (7.13)–(7.15) follows from the formulas

$$\text{prob}^T_{D:=d}(B = b \mid A = a) = \sum_{c \in \Theta_C} \text{prob}^T_{D:=d}(B = b, C = c \mid A = a) \tag{7.16}$$

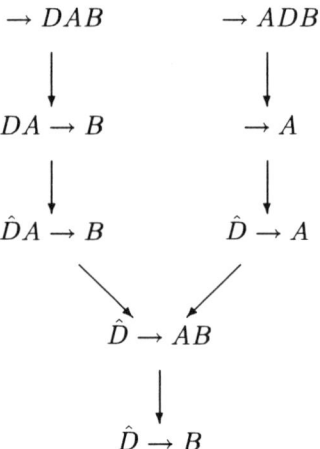

Figure 7.3: Inference tree for $\hat{D}A \to B$ and $\hat{D} \to B$ in the example of Figure 7.1

(marginalization),

$$\text{prob}_{D:=d}^T(C = c \mid A = a, B = b)$$
$$= \frac{\text{prob}_{D:=d}^T(B = b, C = c \mid A = a)}{\sum_{c \in \Theta_C} \text{prob}_{D:=d}^T(B = b, C = c \mid A = a)} \qquad (7.17)$$

(conditioning), and

$$\text{prob}_{D:=d}^T(B = b, C = c \mid A = a)$$
$$= \text{prob}_{D:=d}^T(B = b \mid A = a) \text{prob}_{D:=d}^T(C = c \mid A = a, B = b) \qquad (7.18)$$

(composition), where T is a strictly positive comprehensive theory. The P-validity of rules (7.10)–(7.12) is proven in Pearl [6]. It is an open question (cf. [6], Subsection 4.2) whether the action calculus is *complete*, i.e. whether every P-true formula is a theorem of the calculus.

Theorem 7.1 shows that the action calculus is S-sound in a weak sense: every theorem $\hat{D}A \to B$ with $\#D = 1$ is S-true.

Let us check that the causal effects in the examples of Figures 7.1 and 7.2 are identifiable. Figures 7.3 and 7.4 present inference trees for the corresponding formulas in the action calculus. (Each exogenous node of an inference tree is an axiom and each endogenous node is obtained from its parents by applying some inference rule.) Therefore, the formulas $\hat{D}A \to B$ and $\hat{D} \to B$ in the example of Figure 7.1 and the formula $\hat{D} \to B$ in the example of Figure 7.2 are S-true. The proofs of Figures 7.3 and 7.4 follow Pearl's [6] Subsections 3.1 and 4.3, respectively.

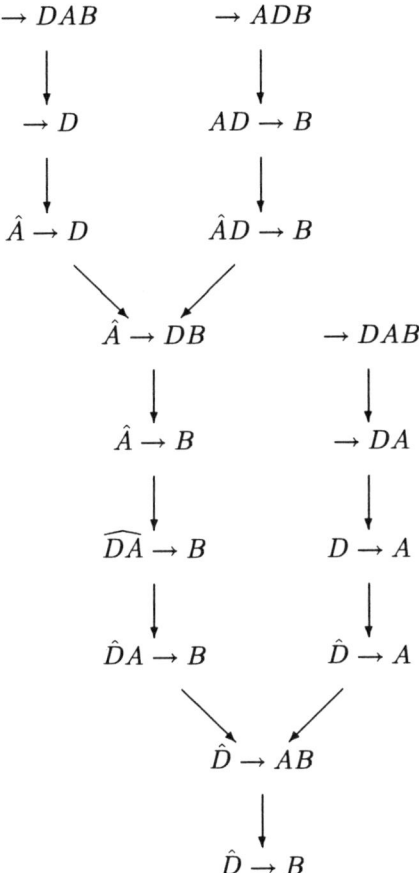

Figure 7.4: Inference tree for $\hat{D} \to B$ in the example of Figure 7.2

Notice that in inference trees we need not specify the inference rules we apply: what rule was applied to obtain an endogenous node is easily seen from that node and its parents. Also notice that it is not difficult to explicitly write down the function f (see (7.9)) corresponding to any formula in an inference tree: rules (7.10)–(7.12) "do not change probabilities" (*e.g.* when rule (7.11) can be applied,

$$\mathrm{prob}^T_{D:=d,C:=c}(B=b\mid A=a) = \mathrm{prob}^T_{D:=d}(B=b\mid C=c, A=a), \ \forall a,b,c,d)$$

and rules (7.13)–(7.15) change probabilities in accordance with (7.16)–(7.18), respectively. For example, for the root $\hat{D} \to B$ of Figure 7.3 we obtain

$$\mathrm{prob}^T_{D:=d}(B=b) = \sum_{a\in\Theta_A} \mathrm{prob}^T(B=b\mid D=d, A=a)\,\mathrm{prob}^T(A=a)$$

$$= \sum_{a\in\Theta_A} \left(\frac{\mathrm{prob}^T(B=b, D=d, A=a)}{\sum_{b\in\Theta_B}\mathrm{prob}^T(B=b, D=d, A=a)} \sum_{b\in\Theta_B, d\in\Theta_D} \mathrm{prob}^T(B=b, D=d, A=a) \right).$$

(7.19)

In Section 7.4 we will see that a similar formula explicitly defines the function F in the definition of S-truth (see (7.8)); we only need to replace prob^T by $Q^{\omega(\mathrm{OBS})}$ in (7.19), where $Q^{\omega(\mathrm{OBS})}$ is a data-dependent probability distribution (essentially, the empirical distribution in Θ_{OBS}): for almost all ω,

$$\mathrm{prob}^T_{D:=d}(B=b) = \sum_{a\in\Theta_A} Q^{\omega(\mathrm{OBS})}(B=b\mid D=d, A=a) Q^{\omega(\mathrm{OBS})}(A=a).$$

(7.20)

Having identified the probabilities $\mathrm{prob}^T_{D:=d}(B=b)$ by (7.20) from the sample data, we can give a recommendation to the doctor who cannot perform the test: it is natural to recommend the decision

$$\arg\max_{d\in\Theta_D} \mathrm{prob}^T_{D:=d}(B = \text{``recovery''})$$

which makes recovery most likely. Analogously, having identified the probabilities $\mathrm{prob}^T_{D:=d}(B=b|A=a)$ (their identifiability can be seen from Figure 7.3), we can recommend the decision

$$\arg\max_{d\in\Theta_D} \mathrm{prob}^T_{D:=d}(B = \text{``recovery''} \mid A=a)$$

to the doctor who has just performed the test and obtained result a.

The situation is more complicated in the case of the example of Figure 7.2. Figure 7.4 tells us that the probabilities $\mathrm{prob}^T_{D:=d}(B=b)$ are identifiable. Let us suppose, for simplicity, that $\#\Theta_D = 2$ and $\#\Theta_B = 2$, namely

$$\Theta_D = \{\text{``smoke,''} \text{``not smoke''}\},$$
$$\Theta_B = \{\text{``health problems,''} \text{``no health problems''}\}.$$

Suppose
$$\text{prob}^T_{D:=\text{"smoke"}}(B = \text{"health problems"}) \qquad (7.21)$$
$$> \text{prob}^T_{D:=\text{"not smoke"}}(B = \text{"health problems"}).$$

Does this inequality justify recommending people to abstain from smoking? It is clear that the answer is negative: (7.21) is not enough. It is true that, on average, it is better for people not to smoke; however, the decision on whether to smoke also depends upon the unobservable genotype, and it is not excluded that people subconsciously feel whether smoking is beneficial or harmful for them. It is possible that the people who smoke are exactly the people for whom smoking is beneficial (inborn craving for nicotine being a means to moderate the effects of a carcinogenic genotype), and (7.21) merely reflects the fact that for most people smoking is harmful. In this case, our recommendation to give up smoking, if anyone follows it, will make people's health problems worse. A promising approach to this kind of problems is described in Balke and Pearl [1] (however, that approach assumes that the decision variables are governed by stable mechanisms).

This decision-making side of the story gives rise to very interesting questions; however, the topic of this chapter is different and we will not go beyond the above obvious remarks.

Remark Of course, to prove theorems in the action calculus it is not necessary to actually deduce formulas from the axiom using inference rules (7.10)–(7.15): as in mathematical logic (see, for example, Shoenfield [10], Chapter 3), we can use various "derivative rules." For example, the inference trees in Figures 7.3 and 7.4 would simplify if we were allowed to start not from the axiom but from the theorems $A \to B$, where $A, B \subseteq \text{OBS}$, and use the truth-preserving rule of cut

$$\frac{\hat{D}A \to B, \hat{D}AB \to C}{\hat{D}A \to C}$$

(this rule obtains from combining the rules of composition and marginalization).

7.4 Proof of Theorem 7.1

We begin by proving a simple "strong law of large numbers" for causal theories (Theorem 7.2). With each data sequence $\omega = \omega_1 \omega_2 \ldots \in \Theta^\infty_{\text{VAR}}$ we associate the sequence P^ω_N, $N = 1, 2, \ldots$, of empirical distributions in Θ_{VAR}:

$$P^\omega_N(\text{VAR} = a) := \frac{1}{N} \sum_{n=1}^{N} \mathbf{I}(\omega_n = a), \; a \in \Theta_{\text{VAR}}$$

($\mathbf{I}(E) = 1$ if property E holds and $\mathbf{I}(E) = 0$ otherwise). If P and Q are probability distributions in Θ_{VAR}, we measure the distance between them by

$$\|P - Q\| := \sum_{a \in \Theta_{\text{VAR}}} |P(\text{VAR} = a) - Q(\text{VAR} = a)|$$

(the *variation distance*; in our finite case it matters little which distance to use). Let $\mathcal{P}^\omega$ be the set of limit points (in the sense of the variation distance) of the sequence P_N^ω, $N \to \infty$; $\mathcal{P}^\omega \neq \emptyset$ because the set $\mathcal{P}(\Theta_{\text{VAR}})$ of all probability distributions in Θ_{VAR} is compact.

Let T be a causal theory of $(G, \text{LAT}, \text{DEC})$. We say that a probability distribution P in Θ_{VAR} is *G-decomposable* if $P = \text{prob}^{T'}$ for some comprehensive causal theory T' of (G, LAT). We say that such P *agrees* with T if we can take as T' an extension of T. We say that P is *positive* if $P(\text{VAR} = a) > 0$, for all $a \in \Theta_{\text{VAR}}$.

The next theorem shows how the properties of the causal theory are reflected in the sample data.

Theorem 7.2 *Let T be a strictly positive causal theory of a structural model $(G, \text{LAT}, \text{DEC})$ with $\#\text{DEC} = 1$ and let Δ be a strictly positive decision strategy for T. For $\text{Prob}^{T,\Delta}$-almost all $\omega = \omega_1 \omega_2 \ldots \in \Theta_{\text{VAR}}^\infty$, all $P \in \mathcal{P}^\omega$ are positive, are G-decomposable, and agree with T.*

Proof By *almost surely* (a.s.) we will mean "for $\text{Prob}^{T,\Delta}$-almost all ω." We begin the proof by noting that

$$\forall a \in \Theta_{\text{VAR}}: \liminf_{N \to \infty} P_N^\omega(\text{VAR} = a) > 0 \text{ a.s.} \qquad (7.22)$$

This is obvious from the strict positivity of both T and Δ (formally, we can reduce (7.22) to Borel's strong law of large numbers, or directly apply a standard martingale strong law of large numbers such as Theorem VII.5.4 of Shiryayev [9]).

We will need not only the empirical distributions P_N^ω but also their "modifications" $\overline{P_N^\omega}$ which already satisfy our desiderata: are G-decomposable and agree with T. We define $\overline{P_N^\omega}$ to be the only G-decomposable probability distribution in Θ_{VAR} such that, for all $X \in \text{VAR}$, $x \in \Theta_X$, and $a \in \Theta_{\text{par} X}$,

$$X \in \text{RAN} \implies \overline{P_N^\omega}(X = x \mid \text{par } X = a) = \text{prob}^T(X = x \mid \text{par } X = a),$$
$$X \in \text{DEC} \implies \overline{P_N^\omega}(X = x \mid \text{par } X = a) = P_N^\omega(X = x \mid \text{par } X = a)$$

(almost surely, $\overline{P_N^\omega}$ is defined from some N on). We will prove that

$$\|P_N^\omega - \overline{P_N^\omega}\| \to 0 \; (N \to \infty) \text{ a.s.} \qquad (7.23)$$

Fix a construction ordering $X_1 \ldots X_{\#\text{VAR}}$ of the variables in G (it is required that every variable go before its parents). Let $\text{pred } X_k$ stand for the set $\{X_1, \ldots, X_{k-1}\}$ of X_k's predecessors in this ordering. To prove (7.23), it suffices (in view of (7.22)) to prove

$$\lim_{N \to \infty} \left(P_N^\omega(X = x \mid \text{pred } X = a) - \overline{P_N^\omega}(X = x \mid \text{pred } X = a) \right) = 0 \text{ a.s., } (7.24)$$

where $X \in \text{VAR}$, $x \in \Theta_X$, and $a \in \Theta_{\text{pred } X}$.

In the case $X \in \text{RAN}$, (7.24) reduces to

$$\lim_{N \to \infty} P_N^\omega(X = x \mid \text{pred } X = a) = \text{prob}^T(X = x \mid \text{par } X = a(\text{par } X)) \text{ a.s.} \quad (7.25)$$

Let N_k be the kth value of N for which $\omega_N(\text{pred } X) = a$, $k = 1, 2, \ldots$. (Formally, N_k are the stopping times defined inductively by

$$N_k := \min \{N > N_{k-1} \mid \omega_N(\text{pred } X) = a\},$$

with $N_0 := 0$.) Define ξ_k to be 1 if $\omega_{N_k}(X) = x$ and 0 otherwise. Since ξ_k are independent Bernoulli random variables (notice that each ξ_k is defined almost surely) with probability of success $\text{prob}^T(X = x \mid \text{par } X = a(\text{par } X))$, Borel's strong law of large numbers gives

$$\lim_{K \to \infty} \frac{1}{K} \sum_{k=1}^{K} \xi_k = \text{prob}^T(X = x \mid \text{par } X = a(\text{par } X)) \text{ a.s.,}$$

which is equivalent to (7.25).

In the case $X \in \text{DEC}$, (7.24) reduces to

$$\lim_{N \to \infty} \left(P_N^\omega(X = x \mid \text{pred } X = a) - P_N^\omega(X = x \mid \text{par } X = a(\text{par } X)) \right) = 0 \text{ a.s.}$$

This asymptotic conditional independence condition is equivalent, in view of (7.22), to

$$\lim_{N \to \infty} \Big(P_N^\omega((\text{pred } X \setminus \text{par } X) = c \mid \text{par } X = b, X = x)$$
$$- P_N^\omega((\text{pred } X \setminus \text{par } X) = c \mid \text{par } X = b) \Big) = 0 \text{ a.s.,}$$

where $b := a(\text{par } X)$, $c := a(\text{pred } X \setminus \text{par } X)$. The last relation will follow from

$$\lim_{N \to \infty} P_N^\omega((\text{pred } X \setminus \text{par } X) = c \mid \text{par } X = b, X = x)$$
$$= \text{prob}^T_{\text{DEC} := d}((\text{pred } X \setminus \text{par } X) = c \mid \text{par } X = b) \text{ a.s.} \quad (7.26)$$

and

$$\lim_{N \to \infty} P_N^\omega((\text{pred } X \setminus \text{par } X) = c \mid \text{par } X = b)$$
$$= \text{prob}^T_{\text{DEC} := d}((\text{pred } X \setminus \text{par } X) = c \mid \text{par } X = b) \text{ a.s.} \quad (7.27)$$

(the right-hand sides do not depend upon the choice of d: recall that $\text{DEC} = \{X\}$). To prove (7.26) (resp. (7.27)) define N_k to be the kth value of N for which $\omega_N(\operatorname{par} X) = b$ and $\omega_N(X) = x$ (resp. $\omega_N(\operatorname{par} X) = b$). Define ξ_k to be 1 if $\omega_{N_k}(\operatorname{pred} X \setminus \operatorname{par} X) = c$ and 0 otherwise. Then ξ_k are independent Bernoulli with probability of success given by the right-hand sides of (7.26) and (7.27), and so (7.26) and (7.27) follow from Borel's law.

Now we can easily deduce the conclusion of the theorem from (7.23). The positivity of all $P \in \mathcal{P}^\omega$, for almost all ω, follows from (7.22). Since the set Σ of all G-decomposable probability distributions in Θ_{VAR} that agree with T is closed and (7.23) implies that the distance between P_N^ω and Σ tends to 0 a.s., we have $\mathcal{P}^\omega \subseteq \Sigma$ a.s. □

Now we are ready to prove Theorem 7.1: a formula $\hat{D}A \to B$, where $A, B, D \subseteq \text{OBS}$ are disjoint and $\#D = 1$, is S-true if and only if it is P-true.

Proof of Theorem 7.1 Let us first prove part "only if" (without assuming $\#D = 1$). Arguing indirectly, we suppose that $\hat{D}A \to B$ is S-true but not P-true in a comprehensive structural model (G, LAT). Therefore, there are two strictly positive comprehensive causal theories T_1 and T_2 of (G, LAT) (with the same frames Θ_{OBS} for the observable variables) and configurations $a \in \Theta_A$, $b \in \Theta_B$, and $d \in \Theta_D$ such that

$$\operatorname{prob}^{T_1}\big|_{\text{OBS}} = \operatorname{prob}^{T_2}\big|_{\text{OBS}} \tag{7.28}$$

but

$$\operatorname{prob}^{T_1}_{D:=d}(B = b \mid A = a) \ne \operatorname{prob}^{T_2}_{D:=d}(B = b \mid A = a). \tag{7.29}$$

Let T'_k be the causal theory of (G, LAT, D) obtained from T_k by deleting all probabilities $\operatorname{prob}^{T_k}(X = x \mid \operatorname{par} X = a)$ for $X \in D$, and let Δ_k be the decision strategy

$$\operatorname{prob}^{\Delta_k}_{\omega_1\ldots\omega_{N-1}}(X = x \mid \operatorname{par} X = a) := \operatorname{prob}^{T_k}(X = x \mid \operatorname{par} X = a),$$
$$N \ge 1, X \in D,$$

$k = 1, 2$. (In other words, we have split T_k into T'_k and Δ_k.) Then, for $\operatorname{Prob}^{T'_k, \Delta_k}$-almost all $\omega \in \Theta_k^\infty$ (Θ_k being the frame Θ_{VAR} for the theory T_k),

$$\operatorname{prob}^{T'_k}_{D:=d}(B = b \mid A = a) = F(a, b, d, \omega(\text{OBS})) \tag{7.30}$$

($k = 1, 2$), where F is the function from the definition of S-truth. Of course, we can replace T'_k by T_k in (7.30). Since $\operatorname{Prob}^{T'_k, \Delta_k} = \left(\operatorname{prob}^{T_k}\right)^\infty$, we have, for $\left(\operatorname{prob}^{T_k}\right)^\infty$-almost all $\omega \in \Theta_k^\infty$,

$$\operatorname{prob}^{T_k}_{D:=d}(B = b \mid A = a) = F(a, b, d, \omega(\text{OBS})),$$

$k = 1, 2$. Let R be the probability distribution (7.28) in Θ_{OBS}. Then

$$\operatorname{prob}^{T_k}_{D:=d}(B = b \mid A = a) = F(a, b, d, \phi)$$

for R^∞-almost all $\phi \in \Theta_{\text{OBS}}^\infty$, $k = 1, 2$, which contradicts (7.29).

Now we prove part "if". Let $\hat{D}A \to B$ be P-true, $\#D = 1$, and Θ_{OBS} be the frames for the observable variables. We know that there exists a function f such that, for any strictly positive comprehensive causal theory T' of (G, LAT) with the frames Θ_{OBS} for the observable variables,

$$\forall a, b, d: \text{prob}_{D:=d}^{T'}(B = b \mid A = a) = f\left(a, b, d, \text{prob}^{T'}\big|_{\text{OBS}}\right). \tag{7.31}$$

For each $\phi = \phi_1 \phi_2 \ldots \in \Theta_{\text{OBS}}^{\infty}$ fix an arbitrary limit point $Q^{\phi} \in \mathcal{P}(\Theta_{\text{OBS}})$ of the sequence Q_N^{ϕ},

$$Q_N^{\phi}(\text{OBS} = a) := \frac{1}{N} \sum_{n=1}^{N} \mathbf{I}(\phi_n = a), \ a \in \Theta_{\text{OBS}},$$

of the observable empirical distributions, as $N \to \infty$. We define the function F (see (7.8)) by

$$F(a, b, d, \phi) := f\left(a, b, d, Q^{\phi}\right). \tag{7.32}$$

We will need the following simple assertion: for each $\omega \in \Theta_{\text{VAR}}^{\infty}$, $Q^{\omega(\text{OBS})} = P\big|_{\text{OBS}}$ for some limit point $P \in \mathcal{P}^{\omega}$ of P_N^{ω}. To see this, extract from $Q_N^{\omega(\text{OBS})}$ a subsequence such that $Q_{N_k}^{\omega(\text{OBS})} \to Q^{\omega(\text{OBS})}$ ($k \to \infty$), notice that $Q_{N_k}^{\omega(\text{OBS})} = P_{N_k}^{\omega}\big|_{\text{OBS}}$, and use the compactness of $\mathcal{P}(\Theta_{\text{VAR}})$ to extract from $P_{N_k}^{\omega}$ a convergent subsequence.

Now we can prove that F satisfies the definition of S-truth. Let T be a strictly positive causal theory of (G, LAT, D) with the frames Θ_{OBS} for the observable variables and let Δ be a strictly positive decision strategy for T. We are required to prove

$$\forall a, b, d: \text{prob}_{D:=d}^{T}(B = b \mid A = a) = F(a, b, d, \omega(\text{OBS})) \text{ a.s.}$$

("a.s." meaning "for $\text{Prob}^{T, \Delta}$-almost all $\omega \in \Theta_{\text{VAR}}^{\infty}$"). From (7.31) and (7.32), we obtain

$$F(a, b, d, \omega(\text{OBS})) = f\left(a, b, d, Q^{\omega(\text{OBS})}\right)$$
$$= f\left(a, b, d, P\big|_{\text{OBS}}\right) = f\left(a, b, d, \text{prob}^{T'}\big|_{\text{OBS}}\right) \tag{7.33}$$
$$= \text{prob}_{D:=d}^{T'}(B = b \mid A = a) = \text{prob}_{D:=d}^{T}(B = b \mid A = a) \text{ a.s.},$$

where P is a limit point of P_N^{ω} satisfying $P\big|_{\text{OBS}} = Q^{\omega(\text{OBS})}$ and T' is the strictly positive comprehensive causal theory such that $\text{prob}^{T'} = P$ (see Theorem 7.2); the last equality of (7.33) follows from the fact that P agrees with T (Theorem 7.2). □

7.5 Discussion

In this section we will briefly discuss possible extensions of our result. The most obvious extension would be to discard the assumption that there is only one decision variable. Sometimes P-true formulas $\hat{D} \to B$ with $\#D > 1$ are obviously S-true: e.g. if A satisfies the back-door criterion relative to (D, B) (Pearl [6], Definition 3), where $A, D, B \subseteq \text{OBS}$ are disjoint, we have

$$\text{prob}^T_{D:=d}(B = b) = \sum_{a \in \Theta_A} \text{prob}^T_{D:=d}(A = a) \text{prob}^T_{D:=d}(B = b \mid A = a) \quad (7.34)$$

(T is a strictly positive causal theory of (G, LAT, D)); and since

$$\text{prob}^T_{D:=d}(A = a) = \text{prob}^{T'}(A = a), \quad (7.35)$$

$$\text{prob}^T_{D:=d}(B = b \mid A = a) = \text{prob}^{T'}(B = b \mid A = a, D = d), \quad (7.36)$$

for any strictly positive comprehensive extension T' of T, the left-hand sides of (7.35) and (7.36) and, therefore, the left-hand side of (7.34) can be extracted from the sample data $\omega(\text{OBS})$ without assuming the stable mechanisms for D. (The explicit formulas for extracting probabilities (7.35) and (7.36) are:

$$\text{prob}^T_{D:=d}(A = a) = \lim_{N \to \infty} \frac{1}{N} \sum_{n=1}^{N} \mathbf{I}(\omega_n(A) = a),$$

$$\text{prob}^T_{D:=d}(B = b \mid A = a)$$
$$= \lim_{N \to \infty} \frac{\sum_{n=1}^{N} \mathbf{I}(\omega_n(B) = b, \omega_n(A) = a, \omega_n(D) = d)}{\sum_{n=1}^{N} \mathbf{I}(\omega_n(A) = a, \omega_n(D) = d)}.)$$

In other situations, however, it is not excluded that the presence of several manipulated variables may cause even conceptual difficulties: several decision variables may represent several decision makers, and our notion of S-truth may become not appropriate; it is possible that we will have to, for instance, accurately define what information about the past is disclosed to every decision maker.

Another interesting direction of study is to perform a rigorous analysis of decision problems such as those considered at the end of Section 7.3. Such analysis may involve investigation of long-run properties of various decision rules. A possible next step is to mix the learning phase (where the causal effects are identified) and the working phase (where a decision rule constructed on the basis of what has been learned is applied).

In our definition of S-truth we assumed that the values of the decision variables are chosen with probabilities depending only on the past configurations $\omega_1 \ldots \omega_{N-1}$, $\omega_n \in \Theta_{\text{VAR}}$. The reader familiar with the device of a filtered probability space (see, for example, Shiryayev [9], Chapter VII) will easily "translate" our Theorems 7.1 and 7.2 to the case where the probabilities $\text{prob}^\Delta_N(X = x \mid \text{par} X = a)$ (cf. (7.7)) specified by a decision strategy for trial N, $N = 1, 2, \ldots$, are only required to form a predictable sequence (with the sequence $\omega_1 \omega_2 \ldots$ being adapted). This framework allows the probabilities $\text{prob}^\Delta_N(X = x \mid \text{par} X = a)$ to depend not only upon the past configurations

$\omega_1 \ldots \omega_{N-1}$ but also on some extra information; however, its drawbacks are that (a) it is rather complicated technically (and would make our presentation much less comprehensible), and (b) it presumes a fully probabilized world. Both drawbacks of the framework of a filtered probability space are overcome by the "martingale" (or "prequential") approach to the foundations of probability (see Dawid [3] or, in more detail, Vovk [11]); that approach enables us to define the notion of "almost surely" even in the case where the extra information is unprobabilized. (We have not based our presentation on that approach so as not to create additional problems for the reader who is not familiar with it.) The newest version of the martingale approach (Chapter 12 of Shafer [8] and Vovk [12]) allows considering truly unprobabilized decisions; however, it is still an open problem whether the results of this chapter can be restated in that version.

It is quite possible that there are also deeper connections between Pearl's framework and Dawid's prequential framework (see Dawid [4]).

Acknowledgements

This chapter is a result of my attempt to reconcile Judea Pearl's beautiful mathematics with Glenn Shafer's sober philosophy. Discussions with them in Aalborg during Research Seminar on Probability and Causality (June, 1995) were extremely thought-provoking and inspiring to me. It is my pleasure to thank Judea, Glenn and Steffen Lauritzen, who organized the seminar. Judea's advice was also most helpful on later stages of preparing the manuscript.

References

[1] Balke A. and Pearl J. (1994) Counterfactual probabilities: Computational methods, bounds, and applications. In Lopez de Mantaras R. and Poole D. (eds) *Proceedings of the Conference on Uncertainty in Artificial Intelligence*, 46–54, Morgan Kaufmann, San Mateo, CA.

[2] Balke A. and Pearl J. (1995) Counterfactuals and policy analysis in structural models. In Besnard P. and Hanks S. (eds) *Uncertainty in Artificial Intelligence 11*, 11–18, Morgan Kaufmann, San Francisco, CA.

[3] Dawid A.P. (1985) Calibration-based empirical probability (with discussion). *Ann. Statist.* 13: 1251–1273.

[4] Dawid A.P. Comment on "Causal diagrams for empirical research" by J. Pearl. To appear in *Biometrika*.

[5] Lai T.L. and Robbins H. (1985) Asymptotically efficient adaptive allocation rules. *Adv. Appl. Math.* 6: 4–22.

[6] Pearl J. (1995) Causal diagrams for empirical research. Technical Report R-218-B-L, Cognitive Systems Laboratory, Computer Science Department, University of California, Los Angeles, February 1995. Expanded version of a paper to appear in *Biometrika*.

[7] Pearl J. Causation, action, and counterfactuals. To appear in Gammerman A. (ed) *Computational Learning and Probabilistic Reasoning*. (This volume.) Wiley, New York.

[8] Shafer G. *The Art of Causal Conjecture*. To appear, MIT Press.

[9] Shiryayev A.N. (1989) *Veroyatnost'*, 2nd ed. Nauka, Moscow. English translation of the 1st ed.: (1984) *Probability*. Springer, New York.

[10] Shoenfield J.R. (1967) *Mathematical Logic*. Addison-Wesley, Reading, MA.

[11] Vovk V.G. (1993) A logic of probability, with application to the foundations of statistics (with discussion). *J. R. Statist. Soc.* B 55: 217–251.

[12] Vovk V.G. A purely martingale version of Kolmogorov's strong law of large numbers. To appear in *Theory Probab. Appl.*

8

Efficient Estimation and Model Selection in Large Graphical Models

D. Wedelin

8.1 Introduction

In this chapter we study the problem of determining the interaction structure in a multidimensional binary sample. As an important special case we consider directed causal models, and how to infer the directions.

Problems of this kind are studied in statistics and artificial intelligence, especially in relation to the analysis of large databases, and to automated knowledge acquistion for probabilistic expert systems (see [12, 9, 15]). We emphasize that in contrast to most statistical and neural network approaches, we aim at recovering true underlying structure, rather than just a parameter estimation in arbitrary models. However, serious approaches in this direction have been severely restricted by computational limitations, making it impossible to apply such methods to realistically large data sets with many, say 100, variables. The main contribution of this work is to provide efficient algorithms that make this possible. For other approaches to this problem see [1, 3, 15].

Mathematically, we wish to find a probability distribution of the form

$$p(X) = \prod_k q(X_k) \qquad (8.1)$$

that fits the data as well as possible. Here, $X = (x_1, \ldots, x_n)$, where the x_i are the binary variables and n the number of variables. The problem consists both of finding the best *structure* or *model* of $p(X)$ in terms of the subsets X_k of X, and to estimate the *potentials* $q(X_k)$, which are distinct functions over these subsets. Models of this kind arise naturally when conditional independence and causal relations are present and are known under the names of *graphical models*, *log-linear models* and *Markov random fields* (see [4]).

As an example of what we would like to find, consider the special form

$$p(X) = q_1(x_1, x_2, x_3)\, q_2(x_2, x_4)\, q_3(x_3, x_4, x_5). \tag{8.2}$$

This structure would occur if the variables had been stochastically generated from each other according to the *directed* or *causal* model

$$p(X) = p(x_1)\, p(x_2)\, p(x_3|x_1, x_2)\, p(x_4|x_2)\, p(x_5|x_3, x_4). \tag{8.3}$$

In our approach the main objective is to determine the structure in (8.2), and models of the form (8.3) are considered as a further analysis of (8.2). Some advantages of this approach are that undirected models are more general than directed models, and do not require specification of directions that cannot be reliably inferred. Also, we avoid the algorithmic problem of having to define the directions before the model can be tested. The most critical part of this strategy is to estimate the parameters in the undirected model, which in contrast to the case for directed models is a difficult problem, and a main topic of the chapter.

We will assume that the reader has a general familiarity with the problem considered, and refer to [16] for more background and technical detail. For simplicity, we will typically assume that $p(X)$ is positive and ignore the complications that otherwise may arise (see [10]).

In Section 8.2 we define the orthogonal interaction model on which the algorithm is based, and give a result on independence properties in this model. Section 8.3 gives the algorithm for approximate maximum likelihood estimation in a given model. Section 8.4 outlines the heuristic search algorithm for model selection, and in Section 8.5 the algorithm for determining the directions is described. Section 8.6 gives the computational results.

8.2 An Orthogonal Interaction Model

The model (8.1) is not ideal algorithmically, one reason being that the potentials are usually not uniquely defined, and we will therefore use another model based on orthogonal functions. There are several equivalent ways of defining an orthogonal interaction model, and we will choose a way that is well suited for our purposes. Other interpretations are in terms of the Hadamard transform (multidimensional binary Fourier transform) or by using so-called Walsh functions (see [11]).

Let $Z = (z_1, \ldots, z_n)$ where the z_i are binary. Then for every Z, define 2^n new variables as

$$x_Z = \sum_i^{\oplus} z_i\, x_i. \tag{8.4}$$

This is the scalar product ZX with addition modulo 2. To distinguish between the different kinds of variables, we denote the original n variables x_i as the *basic* variables, and the new variables as the *interaction* variables. It is sometimes useful to view Z as a set of basic variables, and we will also allow the notation $x_{12} = x_1 \oplus x_2$.

The interaction variables can be used just as the basic variables, and for every interaction variable we define its *marginal* as

$$p(x_Z) = \sum_{X, ZX = x_Z} p(X). \tag{8.5}$$

If the marginal for some Z is not uniform we say that there is a *correlation* between the basic variables in Z.

We will now consider factorizations of the form

$$p(X) = \prod_Z q(x_Z), \tag{8.6}$$

where Z ranges over all its 2^n possible values. For $Z = 0$ we receive a proportionality constant which typically will be of no interest to us. For (8.6) the following well known theorem holds:

Theorem 8.1 *Any positive function $p(X)$ can be factorized to the form (8.6), and the potentials are unique up to a proportionality constant.*

The proof is simple and consists of identifying the potentials with the Hadamard components of $\log p(X)$, and we refer to [16] for details. Expanding to the basic variables, for two variables we receive

$$p(X) = q(x_1)\, q(x_2)\, q(x_1 \oplus x_2) \tag{8.7}$$

and for three variables

$$p(X) = q(x_1)\, q(x_2)\, q(x_3)\, q(x_1 \oplus x_2)\, q(x_1 \oplus x_3)\, q(x_2 \oplus x_3)\, q(x_1 \oplus x_2 \oplus x_3). \tag{8.8}$$

Note that there is no simple relation between these potentials and the probabilities $p(x_1)$, $p(x_2)$, $p(x_2|x_1)$ and $p(x_1|x_2)$, except in the special case when x_1 and x_2 are independent and $q(x_1 \oplus x_2)$ is absent from the expansion.

Typically, we will be interested in models where $p(X)$ has a special structure, in that many of the potentials $q(x_Z)$ are uniform. If a potential $q(x_Z)$ is nonuniform we say that there is an *interaction* between the basic variables in Z. We express such an *interaction model* with the *active set* A, so that

$$p(X) = \prod_{Z \in A} q(x_Z). \tag{8.9}$$

We now consider how (8.9) is related to (8.1). Just as $p(X)$ can be factorized according to (8.6), each potential in (8.1) can be further factorized to the form (8.6). For example, the second factor in (8.2) may always be factorized to

$$q_2(x_2, x_4) = q_2(x_2)q_2(x_4)q_2(x_2 \oplus x_4). \tag{8.10}$$

Therefore, for any model of the form (8.1), there is exactly one equivalent model of the form (8.9), consisting of the union of the potentials $q(x_Z)$ obtained by factorizing each $q(X_k)$ individually. This class of models are called *hierarchical*, since the presence of some potential $q(x_{Z_i})$ implies the presence of all $q(x_{Z_j})$, for which $Z_i \supset Z_j$. Conversely, for any hierarchical model of the form (8.9), we can receive an equivalent model of the form (8.1) by choosing X_k corresponding to the interactions which dominate in A, i.e. for which $Z_i \supseteq Z$ for all $Z \in A$.

8.2.1 Independence Properties in the Orthogonal Model

We will now show that it is possible to identify a large number of independence relations between the interaction variables. This result will be used in the approximation algorithm of Section 8.3.2. The idea can be illustrated in the two variable case, where

$$p(X) = q(x_1)\, q(x_2)\, q(x_{12}).$$

If $q(x_{12})$ is absent from (8.9), x_1 and x_2 are independent of each other in the sense that

$$p(X) = p(x_1)p(x_2) = q(x_1)q(x_2).$$

This is the ordinary independence relation in two variables. However, due to the symmetry between x_1, x_2 and x_{12}, we also have

$$p(X) = p(x_1)p(x_{12}) = q(x_1)q(x_{12}),$$

when $q(x_2)$ is absent, and similarly when $q(x_1)$ is absent. For two basic variables, we therefore have *three* independence relations, and not just one. In general, it is convenient to define independence in the following way:

Definition 8.1 *A variable x_Z is independent with respect to the structure A if*

$$p(x_Z) = q(x_Z) \tag{8.11}$$

for any numerical choice of the potentials.

If x_Z is not independent it will generally be true that $p(x_Z) \neq q(x_Z)$ and we will say that this marginal is *disturbed*. We now have the following theorem (see [16] for the proof):

Theorem 8.2 *The variable x_Z is independent if and only if Z cannot be written as a linear combination of the interactions in $A \setminus Z$.*

We note that if only one potential is present, then clearly no marginals are disturbed. If two potentials are present, exactly one marginal is disturbed. If n potentials are present and their names are linearly independent, all $2^n - 1$ marginals are disturbed.

8.3 Efficient Parameter Estimation in a Given Undirected Model

We now turn to estimation of $p(X)$ in a given model of the form (8.9). The input data will be described by the relative frequency function $\hat{p}(X)$ (of which most entries will typically be 0), and we let $\hat{p}(x_Z)$ be the marginals of $\hat{p}(X)$. The desired output will be the potentials of $p(X)$.

We first describe some well known results for this problem. The following theorem is a variant of a similar theorem for hierarchical models (see [16] for details).

Theorem 8.3 *The maximum likelihood (ML) estimate of $p(X)$, given the structure A, is the unique distribution of the form (8.9) satisfying*

$$p(x_Z) = \hat{p}(x_Z) \tag{8.12}$$

for all $Z \in A$.

We see that no other information is needed from $\hat{p}(X)$ other than the marginals $\hat{p}(x_Z)$ for $Z \in A$. We also note that the theorem implies that the marginals $p(x_Z)$ for $Z \in A$ are always sufficent to determine $p(X)$ uniquely, given that it is of the form (8.9).

We will now show how to compute $p(X)$ from Theorem 8.1 by using the well known IPS algorithm (Iterative Proportional Scaling, also known as the Deming-Stephan algorithm, see [9, 16]). In general, this algorithm can be used to adjust any initial probability distribution to satisfy given marginals, and is known to converge to a unique solution except in degenerate cases. The algorithm is based on the simple observation that if a single potential of $p(X)$ is changed, the corresponding marginal will change accordingly.

The following version of the IPS algorithm will converge to the unique estimate $p(X)$ according to Theorem 8.1. Note that the assignments in the algorithm are not scalar values but probability distributions.

$p(X) :=$ uniform

FOR ALL $Z \in A$, $q(x_Z) :=$ uniform

REPEAT

FOR ALL $Z \in A$

$$p(x_Z) := \sum_{X, ZX = x_Z} p(X)$$
$$\Delta q(x_Z) := \frac{\hat{p}(x_Z)}{p(x_Z)}$$
$$p(X) := p(X)\Delta q(x_Z)$$
$$q(x_Z) := q(x_Z)\Delta q(x_Z)$$

UNTIL convergence.

We see that in this version of the IPS algorithm, $p(X)$ is represented explicitly and also implicitly through its potentials $q(x_Z)$. If we represent $p(X)$ by its potentials only, we can rewrite this algorithm into the expression

$$q(x_Z) := q(x_Z)\frac{\hat{p}(x_Z)}{p(x_Z)} \tag{8.13}$$

where $p(x_Z)$ is computed from (8.5) using $p(X)$ from (8.9). The potentials are initialized to uniform values and the equation is repeated iteratively over and over again for all $Z \in A$, until the $q(x_Z)$ no longer change.

Unfortunately, this algorithm is intractable for large n, due to the exponential complexity of calculating $p(x_Z)$. This is an instance of the basic problem of computing marginals of a probability distribution given in factorized form. For special structures decomposition techniques can be applied (see [12, 9, 16]), but these methods are very sensitive to global properties of the structure, and frequently fail for large problems.

8.3.1 The Reverse Algorithm

We will now try to develop approximations to (8.13) that can be computed efficiently. The approximations are presented in two steps and the final algorithm is described in Section 8.3.2.

We first note that (8.13) gives the ML estimate even if there is no similarity between the the given structure and the true underlying distribution from which the sample is taken. This is nice but perhaps unnecessary, especially since we intend to use this algorithm as a component in an active search for the best structure. Instead, to enable us to find an efficient algorithm, we will reduce our ambition to receiving good estimates when $\hat{p}(X)$ is sampled from a distribution of the form (8.9). As a guideline in our approximations we will require that they give the correct ML estimate $p(X) = \hat{p}(X)$ when $\hat{p}(X)$ is exactly of the form (8.9).

Our new algorithm will be based on the following approximation:

$$\frac{\hat{p}(x_Z)}{p(x_Z)} = \frac{\sum_{X, ZX = x_Z} \hat{p}(X)}{\sum_{X, ZX = x_Z} p(X)} \approx \sum_{X, ZX = x_Z} \frac{\hat{p}(X)}{p(X)}. \tag{8.14}$$

EFFICIENT ESTIMATION AND MODEL SELECTION

Observe that while we are unable to compute the exact expression, we can efficiently compute the approximation since $\hat{p}(X)$ is sparse. It contains mostly zeros, so that X only has to range over the nonzero values of $\hat{p}(X)$. We also see that if $p(X) = \hat{p}(X)$, both expressions are the same.

Our first efficient approximation will therefore be to replace (8.13) by

$$q(x_Z) := q(x_Z) \sum_{X, ZX = x_Z} \frac{\hat{p}(X)}{p(X)} \qquad (8.15)$$

which by using (8.9) can also be written as

$$q(x_Z) := \sum_{X, ZX = x_Z} \frac{\hat{p}(X)}{\prod_{Y \in A \setminus Z} q(x_Y)}. \qquad (8.16)$$

As before, the potentials should be initialized to uniform values, and (8.16) is repeated iteratively for all $Z \in A$ until the potentials converge.

When $\hat{p}(X)$ is of the form (8.9), the algorithm will converge to the correct answer $q(x_Z) = \hat{q}(x_Z)$, where $\hat{q}(x_Z)$ are the potentials of $\hat{p}(X)$. We also note that the new algorithm can be interpreted as another instance of the IPS algorithm, where an initial distribution of $\hat{p}(X)$ is adjusted to uniform marginals. Applying the IPS algorithm in this way is essentially the reverse to the former case, in which the uniform distribution was adjusted to the marginals of $\hat{p}(X)$, and will therefore be called the *reverse* algorithm.

Unfortunately, it turns out that even when the given structure is correct, the estimates from this approximation are significantly inferior to those of the ML method, and the fact that both methods give the same result asymptotically with increasing sample size is of little practical value. The main reason for this can be seen directly from (8.16) by noting that in $\hat{p}(X)/p(X)$ relative frequencies with a very low relative accuracy can be considerably amplified due to small values in the denominator, giving a correspondingly low accuracy in the estimated potentials.

8.3.2 An Improved Reverse Algorithm

We now use the result of Section 8.2.1 to maximally reduce the errors of the reverse algorithm. For this purpose we develop the notion of a *cancellation set*:

Definition 8.2 *A cancellation set* $\mathrm{CS}(Z)$ *for a variable* Z *is a set of interaction variables such that if their potentials are removed from the structure, x_Z is independent of the remaining structure of $p(X)$.*

A simple way to find a cancellation set, though not the only way, is to choose a basic *cancellation variable* $x_i \in Z$, and let the cancellation set be all other variables x_{Z_j}, for which it holds that $x_i \in Z_j$. As an example, consider the following structure (in terms of (8.1)),

$$p(X) = q(x_1, x_2) \, q(x_2, x_3) \, q(x_3, x_4) \, q(x_4, x_5) \, q(x_5, x_6) \, q(x_6, x_1). \qquad (8.17)$$

In terms of (8.6), factors for all basic variables x_i, and interaction variables for all pairs of basic variables $x_{ij} = x_i \oplus x_j$, will be present. Here, the cancellation set for x_2 becomes $\{x_{12}, x_{23}\}$, and for x_{23} either $\{x_2, x_{12}\}$ or $\{x_3, x_{34}\}$.

We now observe that to preserve equilibrium for $q(x_Z) = \hat{q}(x_Z)$ when $\hat{p}(X)$ is of the form (8.9), it is sufficient to cancel out from $\hat{p}(X)$ only so many potentials that x_Z is independent with respect to the *remaining* structure of $\hat{p}(X)$. Our final approximation then becomes

$$q(x_Z) := \sum_{X, ZX = x_Z} \frac{\hat{p}(X)}{\prod_{Y \in \mathrm{CS}(Z)} q(x_Y)}, \qquad (8.18)$$

which we will refer to as the *improved reverse* algorithm. The improvement is that we have only a few potentials in the denominator, so when $\hat{p}(X)$ is based on a true sample the amplification of relative frequencies with a low relative accuracy is much lower, and the problems with (8.16) are significantly reduced. We note, however, that for (8.18) the form of the equations explicitly depend upon the assumption that the structure of $\hat{p}(X)$ should be reasonably close to that of $p(X)$.

In Table 8.1 we show a numerical comparison of the three estimation algorithms for a problem of the form (8.17). We see that the results of our best approximation are still not identical to the optimal results of (8.13), but in practice they appear to give values that are very close to the ML estimates.

8.4 Efficient Model Selection

When the structure is not known, the ML criterion is insufficient, and we choose instead to minimize the description length (MDL, see [14]). For this we need an encoding of the model, and we choose to encode each individual interaction with its identity Z and its parameter value to optimal precision. Typically, the description length for low order interactions will be around 10–30 bits.

We develop a heuristic algorithm where the structure is determined incrementally, which can be expected to work well when the model is hierarchical. Candidate potentials are tested and possibly accepted as part of the structure in a bottom-up fashion, assuming no structure at all at the beginning, and searching successively higher order interactions between the basic variables. This is implemented by maintaining an active set of interactions whose values are calculated from (8.18), assuming the structure found so far. The active set is intitialized to all basic interactions. All first order interactions (between

Table 8.1: A comparison between the different estimation algorithms.

Z	correct	ML	reverse	improved
100000	0.36000	0.37464	0.30711	0.37512
010000	0.88033	0.88515	0.89354	0.88550
001000	0.96291	0.95950	0.98289	0.95861
000100	0.98182	0.97833	0.99257	0.97782
000010	0.79412	0.78178	0.91290	0.78220
000001	0.96644	0.96682	0.99073	0.96681
110000	0.55501	0.56208	0.65483	0.56221
011000	0.55501	0.57438	0.55168	0.57539
001100	0.55501	0.51330	0.69764	0.50647
000110	0.55501	0.54828	0.72904	0.54868
000011	0.55501	0.55554	0.53993	0.55580
100001	0.55501	0.56432	0.47770	0.56475
$\log p(D|H)$	-29773.35	-29766.55	-30578.61	-29766.57

two basic variables) are then tested. Second order interactions (between three variables) are tested selectively, based on the first order interactions that were found, and higher order interactions are generated in the same way, until no more candidates are generated. This corresponds to one *pass* of the algorithm. Two passes are usually sufficient for reasonable convergence, but more passes may lead to small additional changes. We note that if we were to test all interactions, the algorithm would be asymptotically correct in the trivial sense that it would give the correct structure when the number of cases tend to infintity.

A potential is accepted into the active set by a hypothesis test that checks if its acceptance reduces the total description length, and can be rejected at any time if this is no longer the case. We note that this is the only place in the algorithm where the MDL-criterion is used, and the hypothesis test could in principle be based on some other overall criterion than MDL.

We give an outline of this search algorithm:

ActiveSet = all basic interactions;

Iterate(ActiveSet);

FOR pass=1 TO maxpass

    FOR order=1 TO $n-1$

        Candidates=GenerateCandidates(ActiveSet,order);

        IF Candidates=∅ exit;

        FOR ALL $Z \in$ Candidates

ActiveSet=ActiveSet $\cup Z$;
LocalIterate(Z);
Test(Z);
Iterate(ActiveSet);
TestAll(ActiveSet);

FindDirections(ActiveSet);

In this outline, the GenerateCandidates function generates a set of candidates of the current order, based on the interactions of lower order in the active set. The procedure Test does an evaluation of Z and deletes it from the active set if it is not significant. Iterate iterates all active potentials with (8.18), and TestAll throws out interactions that are no longer significant. The LocalIterate procedure implements a fast iteration of Z and its neighbours, to improve execution speed. As the active set changes, the procedures for iterating the potentials must at all times maintain updated cancellation sets and marginal tables. It should be pointed out that as long as the active structure remains reasonable in size, and does not grow so much that the potentials become degenerate, it does not matter much if irrelevant potentials are included temporarily. As a postprocessing step we add a procedure that tries to find the causal directions for all maximal interactions, described in the next section. For more details on the generation of candidates and the MDL calculations see [16].

8.5 Directed Models

We will now outline how a directed model can be reconstructed when we know $p(X)$, or an estimate of $p(X)$. The algorithm uses local properties in the structure, and cannot determine directions when only two variables are involved. The main steps are:

1. Factorize $p(X)$ to determine its interactions. If only sampled data is available the algorithms of the previous sections can be used in this step;

2. Translate into a model of the form (8.1). The sets X_k in (8.1) are obtained by identifying the dominating interactions in (8.6);

3. For each set X_k with three or more variables test each possible direction in turn to see if it is consistent with $p(X)$. If the test eliminates all but one of the possible directions this must be the correct direction.

We now describe the test. In Section 8.2 we noted that the individual potentials in (8.1) can be further factorized into interactions. Clearly, an active interaction in $p(X)$ means that this interaction is present in at least one of the potentials $q(X_k)$. In many cases, it is possible to inspect the structure and trace an interaction in $p(X)$ to a single contributing potential in (8.1). The value of the interaction in $p(X)$ must then be the same as the value of this interaction for the corresponding conditional probability. For example, in (8.3) the interactions $q(x_{12})$, $q(x_{13})$, $q(x_{23})$ and $q(x_{123})$ of $p(x_3|x_1, x_2)$ must have the same values as the corresponding interactions in $p(X)$. These observations can be used in a test in the following way:

1. Assume a direction by choosing a variable x_i, i.e. so that the assumed conditional probability is $p(x_i|X_k \setminus x_i)$;

2. Compute the conditional probablity directly from $p(X)$ with

$$p(x_i|X_k \setminus x_i) = \frac{p(X_k)}{p(X_k \setminus x_i)} \qquad (8.19)$$

where $p(X_k)$ is the marginal of $p(X)$ with respect to the variables in X_k, and similarly for $p(X_k \setminus x_i)$. If only sampled data is available use $\hat{p}(X)$ instead;

3. Factorize this conditional probability into interactions and check all interactions that should match with those already computed for $p(X)$. If they are not equal, the assumed direction must be incorrect.

When $p(X)$ is estimated from a sample we cannot expect a perfect numerical match for the interactions. In our implementation we therefore use a hypothesis test based on the MDL-criterion to accept a direction only if it reduces the description length. Note that this method does not assume a priori that a causal relation is present. However, it is quite possible for the algorithm to determine a best guess for the directions, if we know that they should exist.

8.6 Computational Results

In these tests we show the results graphically using a special graphical representation, where both directed and undirected arcs may occur. First order interactions are always undirected, but in all cases when a second or higher order interaction has been found, we use the directed formalism using the causal direction recommended by the program. We note that if the model is to be used for subsequent inference, the undirected model can be used directly, and it is not necessary to specify the directions if they cannot be reliably inferred.

We begin by considering a problem known as the Alarm problem (see[2]), which has been tested also by others (see [3, 15]). In our test we use only the structure of this problem, and have assigned random weights to the conditional probability tables. This has been done since the original problem contains variables with up to four values, and therefore cannot be handled directly by

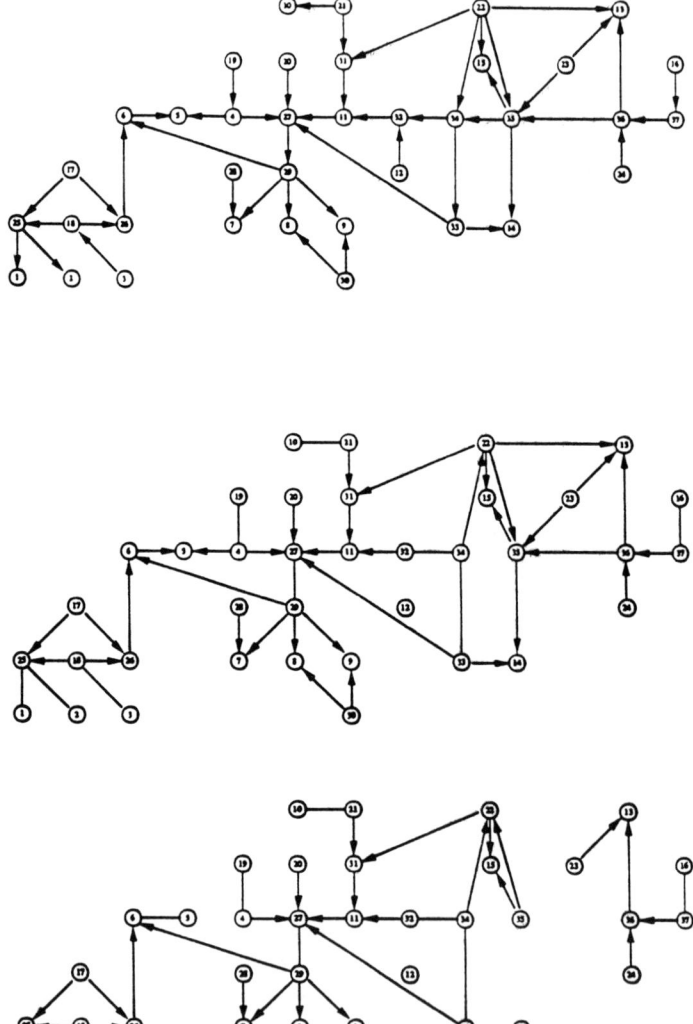

Figure 8.1: (a) The Alarm structure. (b) Results for 10000 and 2000 cases.

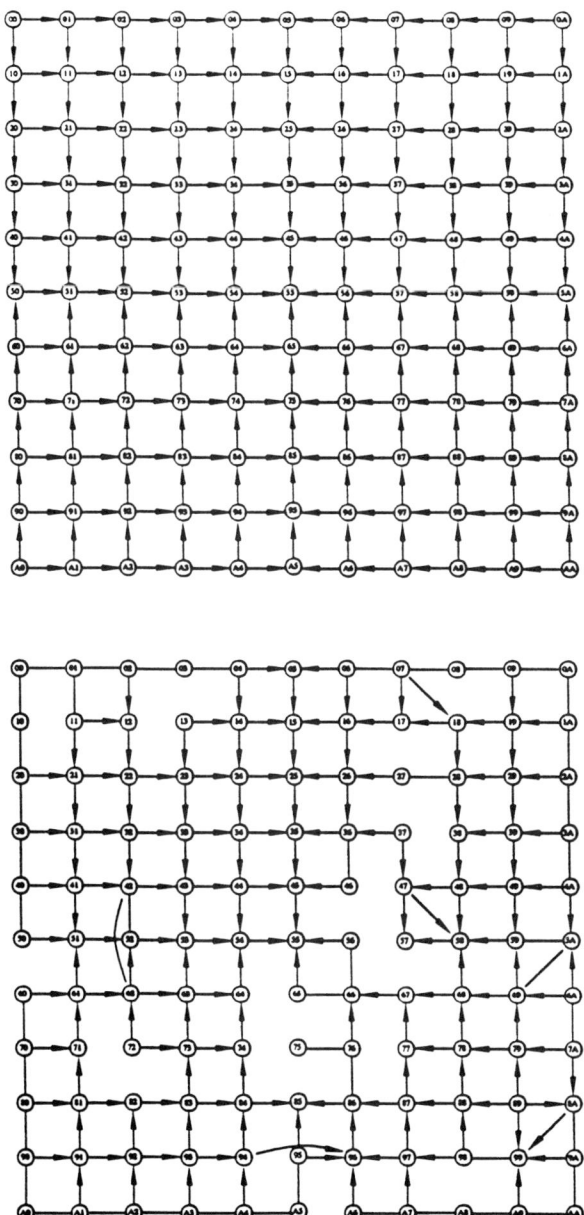

Figure 8.2: (a) The big121 problem. (b) Result for 10000 cases.

our algorithm. The structure is shown in Figure 8.1a. From this structure, we simulated 10000 cases which were used as input to the program. Figure 8.1b shows the result graphically for 10000 and 2000 cases. With 10000 cases almost all of the structure is correctly reproduced, with a few exceptions. With 2000 cases most of the structure is still intact, but some of the weaker interactions have disappeared. The running time for three passes was about 20 minutes for 10000 cases and 5 minutes for 2000 cases (Macintosh SE/30).

Our second test is a larger example, which we call the big121 problem, with 121 variables. The structure is shown in Figure 8.2. It has been constructed with a regular structure for easy evaluation, and has a structure which is considerably denser than the previous problem. The apparent regularity of the structure is of no help to the algorithm, which has to test all first order interactions. The results for 10000 cases is shown in Figure 8.2b. The execution times for three passes on a Macintosh SE/30 were approximately six hours for 10000 cases and one hour and 40 minutes for 2000 cases. About 45% of the time was spent in the third pass, during which only small changes took place.

8.7 Conclusion

We have developed a set of algorithms which can efficiently identify the interaction structure between given variables from samples only. The algorithms provide the parameters of the undirected structure, and can also reliably determine many of the causal directions if such exist. The algorithms have been designed to efficiently handle large problems, and we have shown that they can solve problems with more than 100 variables. The estimation algorithm is not based on decomposition methods, and is therefore sensitive only to the local connectivity of the structure, and not on global properties of the structure.

References

[1] Badsberg J.H. (1991) A Guide to CoCo. Technical Report 91–43, Institute for Electronic Systems, University of Aalborg.

[2] Beinlich I., Suermondt H., Chavez R. and Cooper G. (1989) The ALARM monitoring system: A case study with two probabilistic inference techniques for belief networks. Technical Report KSL-88-84, Knowledge Systems Laboratory, Stanford University.

[3] Cooper G. and Herskovits E. (1991) A Bayesian Method for the Induction of Probabilistic Networks from Data. Technical Report KSL-91-02, Knowledge Systems Laboratory, Stanford University.

[4] Darroch J.N., Lauritzen S. L. and Speed T. P. (1980) Markov fields and log-linear models for contingency tables. *Ann. Statist.* 8: 522–539.

[5] Goodman L.A. (1970) The multivariate analysis of qualitative data: Interaction among multiple classifications. *J. Amer. Statist. Ass.* 65: 226–256.

[6] Haberman S.J. (1974) *The Analysis of Frequency Data*. The University of Chicago Press. Midway Reprint 1977.

[7] Herskovits E. and Cooper G. (1990) Kutato: An Entropy-Driven System for Construction of Probabilistic Expert Systems from Databases. Technical Report KSL-90-22, Knowledge Systems Laboratory, Stanford University.

[8] Lauritzen S.L. (1982) *Lectures on Contingency Tables*. Aalborg University Press.

[9] Lauritzen S.L. and Spiegelhalter D.S. (1988) Local computations with probabilities on graphical structures and their application to expert systems. *J. R. Statist. Soc.* (B 50).

[10] Moussouris J. (1974) Gibbs and Markov Random Systems with Constraints. *J. of Statistical Physics* 10: 11–33.

[11] Ott. J. and Kronmal R.A. (1976) Some Classification Procedures for Multivariate Binary Data Using Orthogonal Functions. *J. of the Am. Stat. Ass.* 71, 391–399.

[12] Pearl J. (1988) *Probabilistic reasoning in intelligent systems.* Morgan Kaufmann, San Mateo, CA.

[13] Plackett R.L. (1974) *The Analysis of Categorical Data.* London: Griffin & Co.

[14] Rissanen J. (1983) A universal prior for integers and estimation by minimum description length. *Ann. Statist.* (11) 2: 416–431.

[15] Spirtes P., Glymour C. and Scheines R. (1992) Causation, Prediction and Search. Book manuscript. Dept. of Philosophy, Carnegie Mellon Univ.

[16] Wedelin D. (1993) Efficient Algorithms for Probabilistic Inference, Combinatorial Optimization and the Discovery of Causal Structure from Data, Ph.D. Thesis, Dept. of Computer Science, Chalmers Univ. of Technology.

9

T-Normal Distribution on the Bayesian Belief Networks

Yu.N. Blagoveschensky

9.1 Introduction

The difficulties which multivariate statistical analysis always comes across when probabilities are not normal or when the set of unknown parameters is very big are well known. Graphical models on the set of components of a random vector can help such tasks, but in practice they have problems connected with constructing real graphs. These two directions can stimulate the development of each other. In this chapter, we consider some approaches to the subject[1].

It should be observed that Bayesian Belief Networks (BBNs) is a methodology rather than a formal theory. So here we shall discuss acyclic directed Markov graphs (or M-graphs for short) as a specimen of a BBN. In [4] base knowledge about the subject has been discussed in detail.

In the first part of this chapter, necessary definitions and designations are given, and in the second part we discuss some of the results and new problems connected with them.

---

[1] The investigation is supported by **INTAS**, project INTAS-93-0725

9.2 Necessary Definitions and Denotations

Graphs

Acyclic directed graph $G = (V, E)$ is determinated by a finite set V of its nodes and a set E of ordered pairs (v, w) or arrows $v \to w$ from $v \in V$ to $w \in V$. Here we shall use following subsets of V for each $v \in V$:

pa(v) — the *parent* set of v or, formally, $w \in$ pa(v) if $(w \to v) \in E$;

tr(v) — the *tree* set of v or, formally, $w \in$ tr(v) if there exists a sequence $(v_1, \ldots, v_t)$ of nodes for which $v_1 = v$, $w \in (v_1, \ldots, v_t)$ and $(v_k \to v_{k+1}) \in E$, $1 \leq k \leq t-1$ (for $t > 1$);

nt(v) — the *non-tree* set of v or, formally, $w \in$ nt(v) if $w \notin$ tr(v), thus evidently there is the equality nt$(v) \cup$ tr$(v) = V$.

Random variables

We shall use the denotation $X^\top = (X_1, X_2, \ldots, X_r)$, where "⊤" is the sign of transposing, for random vector X. Then let X_i, $\mathbf{X}_i$ and x_i be accordingly denotations of the i'th component of X, the value space for X_i and a point of $\mathbf{X}_i$, $i = 1, 2, \ldots, r$. Simplifying our situation, we assume $\mathbf{X}_i$ can be either a finite set of any symbols or a connected subset of real numbers $\mathbf{R}^1 = (-\infty, +\infty)$; and let the space have accordingly the measure as the number of elements, in the first variant, and the Lebesgue measure, in the second one. We suppose also that X has a limited positive probability density $f(x)$, and we use the denotation $f(x_i \mid x_J)$ for the conditional density X_i given $X_J = x_J$, where J is any subset of $(1, 2, \ldots, r)$, $X_J = \{X_i, i \in J\}$ and $x_J = \{x_i, i \in J\}$. Analogous denotations will be used for other cases of the index of a set of X.

We shall say that a random vector X has *T-normal distribution* if its components are continuous random variables and if there exist monotonous transformations[2] from X_i to $Z_i = h_i(X_i), i \in (1, 2, \ldots, r)$, such that $Z^\top = (Z_1, Z_2, \ldots, Z_r)$ is jointly Gaussian.

Markov property

Let the nodes in the graph G = (V,E) represent a numbered set of random variables $X_i, i \in V = (1, 2, \ldots, r)$. We shall write $A \perp B \mid C$ if A,B and C are subsets of V and if X_C-conditionally *set* variables $X_A = \{X_i, i \in A\}$ and X_B are independent. Now we shall say that the triplet (V, E, X) is an *M-graph* if there is the following:

---

[2] We have to say that normalizing marginal distributions is well known, but it is a useful procedure if and only if the normalized vector

$$Z = \{\Phi^{(-1)}(F_i(X_i)), \ i = 1, 2, \ldots, r\},$$

where $\Phi(t)$ is the function of Gaussian standard distribution and $F_i(t)$ is the marginal distribution of X_i normally distributed in collection.

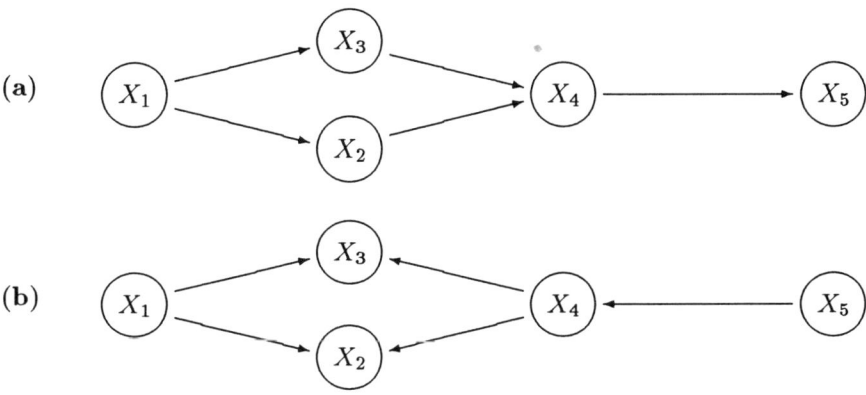

Figure 9.1: M-graph given formally (a) and expertly (b).

(M) The property $\star$ $i\perp\mathrm{nt}(i)\mid \mathrm{pa}(i)$ $\star$ is true for all $i \in V$.

Example

As an example we shall consider two M-graphs, shown in Figure 9.1. It is important to note that both graphs describe a single situation equally well; the common "agricultural" sense of X_i ($i = 1, 2, 3, 4, 5$) in these graphs can be determinated in the following way: **(1)** X_1 — the symbol of *herbicide spraying* or the absence of herbicide spraying; **(2)** X_2 — the yield of crop (actual); **(3)** X_3 — the biomass of the weed per area unit at the moment of crop flowering; **(4)** X_4 — the difference between the sum of rainfall during May-August in the year studied and an average for the same period over many years; **(5)** X_5 — the soil moisture content at the moment of herbicide spraying.

It also is importance that *both* graphs correspond to a *single* probability situation. Let Z_1 be any Bernoulli random variable and $Z_i, i = 2, 3, 4, 5$, be independent standard normal variables on $\mathbf{R}^1$; then we take $X_1 = g_1(Z_1)$, $X_4 = g_4(Z_4)$, $X_5 = g_5(Z_4, Z_5)$ and $X_i = g_i(Z_i, Z_1, Z_4)$ for $i = 2, 3$, where functions $g_i(\ldots)$ are some transformations into R^1. As a result, for our situation there are two models of M-graphs which are represented in Figure 9.1 (it should be observed that X_1 is discrete and X_i with $i > 1$ are continuous variables).

9.3 Results and some New Problems

A. Here we represent results which are proved accurately, but we don't give the proofs because it needs considerable space.

• It is proved that there is the following property for any M-graphs:

$$f(x) = \prod_{i=1}^{r} f(x_i \mid x_{\mathrm{pa}(i)}), \tag{9.1}$$

where $f(x)$ is an arbitrary density function. In detail the multiplication property is considered by Lauritzen and Wermuth [5] for the so-called Conditional Gaussian distribution, or CG-distributon.

Definition. A random vector X has CG-distribution if its continuous components are jointly Gaussian variables when its discrete components are given.

• The example is constructed so that there are two M-graphs ((**a**) and (**b**) in Figure 9.1) which give different formulae (1) for single random vectors $X = (X_1, \ldots, X_5)$. The important fact points to "hidden problem" connected with bringing to light arrows E in the expert way.

Note that Markov distribution in [4] (Theorem 2.6) is unique when the set of arrows E is given. In contrast to the problem here, the base point is a random vector X and we choose E in order to get a Markov property as a result.

• We considered the new family of distributions and showed that it gives constructive procedures for statistical analysis when there are sample observations of the random vector X.

Namely, we say that the distribution of X belongs to the TCG-family (Transformed to Conditional Gaussian) if there exist arbitrary smooth monotonous functions $h_i(\ldots)$, $i = 1, \ldots, r$, on $\mathbf{R}_1$ such that the random vector Y with components $Y_i = h_i(X_i)$, $i = 1, \ldots, r$, have CG-distribution.

Here we do not consider all these facts in detail. Only a word is said here about the statistical problems of models with TCG-distributions.

Let $M = \{j = 1, 2, \ldots, m\}$ be the index set of all observations; thus $X^{(j)}$ is the j-th observation of X, and so some components can be missing besides the different observed vectors that can have different missing components.

Let $j \in M_{ik}$ if $j \in M$ and if the observation $X^{(j)}$ contains sampled values of $X_i^{(j)}$ and $X_k^{(j)}$; also, $j \in M_i$ if the component $X_i^{(j)}$ is really observed (certainly, the quality of estimating depends upon the sizes of sets M_i, M_{ik}, $i, k \in M$, but not of set M directly).

In addition, we note that in the case of TCG-distributions, base estimators are rank correlation coefficients (by Van der Waerden B.L.) and order statistics for components of vector X.

B. We'll now look at some problems connected with that often we can know a set of latent factors, which cause the structure of links among our variables to be measured.

In practice, as a rule these latent factors are quite clear. For our example they can be described by the following patterns: Z_1 — herbicide, Z_2 — average crop yield in case of favourable weather, Z_3 — seed bank of weed, Z_4 — rainfall of the concrete year, and Z_5 — soil struciure.

Table 9.1:

factors $\Rightarrow$	Z_1	Z_2	Z_3	Z_4	Z_5
variables:	(Influences or not)				
X_1	Yes	No	No	No	No
X_2	Yes	Yes	No	Yes	No
X_3	Yes	No	Yes	Yes	No
X_4	No	No	No	Yes	No
X_5	No	No	No	Yes	Yes

Data about their properties (about "soil structure" in this instance) can be gradations, real numbers, finite sets of real numbers, graphical curves or complex information, but with confidence we can assume that any characteristics $Z_i^{\bullet}, i = 1, \ldots, 5$, over factors $Z_1, \ldots, Z_5$ are independent as random variables.

In our case the real interrogation of experts in soil science determined that variables $X_i, i = 1, \ldots, 5$, are functions of subsets $\mathbf{Z_i} \in \{Z_1, \ldots, Z_5\}$, $i = 1, \ldots, 5$. The result of the interrogation can be given by Table 9.1. Thus $X_i = H_i(\mathbf{Z_i}), i = 1, \ldots, 5$ and both M-graphs (a) and (b) in Figure 9.1 can be used as graphical models for our situation.

In the more general case, we suppose that there is a collection of random factors which are variables of arbitrary sorts, and which are jointly independent. If $Z_i, i = 1, \ldots, N$, are these factors, then let $X_j = T_j(Z_{A(j)})$, $A(j) \in \{1, \ldots, N\}$, where $T_j(\ldots)$ are transformations of the subcollections $Z_{A(j)} = \{Z_i, i \in A(j)\}$ to real-valued random variables $X_j, j = 1, \ldots, r$.

In contrast to Pearl's causal theory [6], which considers unobserved factors $Z_i, i = 1, \ldots, N$, within the Markov graphical model given beforehand, we propose[3] to construct M-graphs from the fundamental information about subsets $A(j), j = 1, \ldots r$.

It's natural that the first question which requires an answer is whether there exists a non-trivial M-graph on the collection $X = (X_1, \ldots, X_r)$ when subsets $A(j), j = 1, \ldots r$ are given. Unfortunately, at the present time we have no real way to answer this in the general case, but we hope that models with latent factors have good prospects.

---

[3]The differences between models are more serious, namely, Pearl assumes that there exists an ordering of the variables $X = (X_1, \ldots, X_r)$ such that Z_i is independent of $X_1, \ldots, X_{i-1}$ for $i > 1$.

9.4 Conclusions

Reduction of dimensionality is our particular concern when we have to deal with multivariate data. It seems to us that models with latent factors are a progressive approach to bring to light links among observed variables, and to make for a reduction of dimensionality using expert information.

Concrete results are connected with an expansion of distribution families (jointly normal after monotonous transformations of axes), and we note that this chapter contains only a heuristic scheme of the statistical research, and not the formulae for estimating, because of space restrictions.

References

[1] Aitken C.G. and Gammerman A. (1990) An illustrative example of the use of probabilistic reasoning in agricultural forecasting. Technical appendix to progress report: DOSES Project B6, Likely Phase 2, 3–11.

[2] Blagoveschensky Yu.N. (1995) T-normal distribution on the Bayesian belief networks. *Applied decision technologies (ADT) 1995, Stream 1: Computational learning & Probabilistic reasoning*, 3–5 April 1995, Brunel Conference Centre, London, 131–134.

[3] Blagoveschensky Yu.N. and Meshalkin L.D. (1986) Multidimensional T-normal distributions. *1st World Bernoulli Congr. on Prob. Theory and Math. Stat., Theses* (1), Nauka, Moscow, 79.

[4] Dawid A.P. and Lauritzen S.L. (1993) Hiper Markov laws in the statistical analysis of decomposable graphical models. *The Annals of Statistics* (21) 3: 1272–1317.

[5] Lauritzen S.L. and Wermuth N. (1988) Graphical models for associations between variables, some of which are qualitative and some quantitative. *The Annals of Statistics* (17) 1: 31–57.

[6] Pearl J. (1995) Causation, Action and Counterfacuals. *Applied decision technologies (ADT) 1995, Stream 1: Computational learning & Probabilistic reasoning*, 3-5 April 1995, Brunel Conference Centre, London, 1–17.

[7] Spiegelhalter D.J., Dawid A.P., Lauritzen S.L. and Cowell R.G. (1992) *Bayesian analysis in expert systems*. Reprint, University College, London UK, 1–48.

Part III

Bayesian Belief Networks and Hybrid Systems

# 10

Bayesian Belief Networks with an Application in Specific Case Analysis

C.G.G. Aitken, A. Gammerman, G. Zhang,
T. Connolly, D. Bailey, R. Gordon, R. Oldfield

10.1 Introduction

Specific case analysis or "offender profiling" is the name given to a collection of scientific and psychological theories and techniques that seek to relate offender, victim and crime characteristics together in order to assist the police with crime investigation. This has long been considered the domain of the behavioural sciences. The research work in this area, exemplified by Canter [3] and Canter and Heritage [4], has led to some remarkable successes in helping police with their enquiries.

Recently the statistical techniques have started to be applied to offender profiling. The use of these techniques were previously thought infeasible in such application which has insufficient number of past cases for analysis. For a small sample base, each case would be too distinct to provide assistance in the identification of the criminal. However, for a well constructed set of past samples, statistical techniques still produce positive suggestions. A review of earlier work, in particular in relation to epidemiological methods applied to criminology, is given in Aitken *et al.* [2]. The role of control groups and the construction of the data set used for the analysis are also discussed. A brief review of the construction is given here with further details in Aitken *et al.* [2].

In this chapter, we describe the construction of a dataset for offender profiling and an approach using Bayesian belief networks (BBN) on this dataset. The results produced from this approach are compared with those produced by police detectives and prove that the use of statistical techniques is both plausible and useful.

10.2 The CATCHEM Project

The CATCHEM project started in 1986 to construct a dataset containing details of all sexually motivated child murders and abductions which had been reported in Great Britain since 1960. The dataset, now containing more than 400 cases, cover the details of characteristics of the offender, where known, of the victim, and of the crime scene in coded form. These details were obtained by constructing a *pro forma* listing all the details which were considered relevant by the police. Each case record was then studied and the details transferred to the *pro forma*. These data have been used to assist with police investigation of similar cases in a manual manner. This led to certain categorizations being used which may not be appropriate in other cases or other types of crime.

10.2.1 Offender Variables

The following details regarding the offender were used in the analysis:

- Age: though this is a continuous variable, in practice only a crude classification is of interest; for example, the offender may be classified as infant (0–7), pre-puberty (8–12) or post-puberty (13 or older);

- Marital status: this was recorded as one of four states (married; single; single and divorced; single and cohabiting) but here only a binary classification is considered: living with a partner (married or single and cohabiting) or not living with a partner (single or single and divorced);

- Previous convictions: classified as a binary variable (yes; no);

- Relationship to the victim: classified as a binary variable (known-parent, cohabitee, relative or acquaintance; stranger);

- Proximity: this is the proximity of the point of contact of the offender and the victim to the normal residence of the offender and classified as a binary variable (within 5 miles or not).

The states of the last attribute (proximity) were chosen by detectives because of their particular relevance to the cases under investigation when the database was constructed. In retrospect, the states are too coarse to be of much use. It was found that over 90% of offenders lived within 5 miles of the point of contact. Thus, prediction of this variable will not provide much additional information since an initial prediction to the effect that a criminal lived within five miles of the scene of the crime would be correct 90% of the time. Notice that the sex of the offender is not of interest. For the cases under investigation here (child murders with sexual connotations) the offender was male except for two cases in which the female offenders were accomplices of male offenders.

10.2.2 Victim Variables

For each case over 200 variables were recorded, in binary form, present or absent, about the victim and the scene of the crime. The variables recorded could be conveniently arranged into groups which were: (1) victim age and sex, (2) victim location at point of contact, (3) offender activity, (4) crime locations, (5) activities during offence, (6) distances and proximities, (7) times (*e.g.* day of the week, time of day) and (8) offender peculiarities. Three victim variables, sex, age and abduction (whether the victim was abducted or not) were of particular interest as these were notable features of the cases which led to the formation of the database. In practice, the number of variables considered was reduced considerably. Not all of sex, age and abduction were always considered.

10.2.3 Combination of Offender and Victim Variables

The information about offenders and victims were initially in two separate datasets with linking reference numbers. These two datasets were brought together to form a working dataset. The analyses were done on the subset of the database relating to single offender-single victim cases only. Undetected cases were excluded. Serial offenders and cases with multiple offenders and/or victims were also excluded. This left 320 cases for analysis. It was felt that in the excluded cases there might be different motivations which would complicate inferences drawn from the analyses. Further, with single offender-single victim cases independence of the observations when fitting the models is ensured. Such independence may not be so when considering serial criminals. Various models were investigated. The results for *BBN* are reported here. Aitken *et al.* [2] discuss approaches using logistic regression and, briefly, *BBN*.

10.3 PRESS — a Bayesian Belief Network System

The technique described here involves a graphical representation of the relationships amongst the various characteristics, offender, victim and crime. A graph in this context is a set of nodes and directed arcs (or edges). Each node represents a particular characteristic. A directed arc links two nodes. The

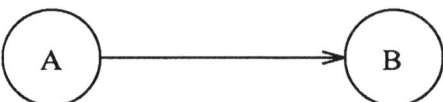

Figure 10.1: Diagrammatic representation that attribute A causes or influences attribute B.

direction of the arc represents a causal or influential relationship. The absence of an arc between two nodes implies that the two attributes associated with these nodes are independent of each other, conditional on knowledge of the values of the other attributes. There is also a restriction that the directed arcs cannot form a closed loop in that it cannot be possible to start from a particular node and follow arcs to return to that node. This particular form of graphical representation is a Bayesian belief network (see [1, 5]).

As an example shown in Figure 10.1, the notation $A \rightarrow B$ is taken to mean attribute A causes or influences attribute B. A and B are called the *parent* and *child* node respectively, *i.e.* A is the parent of B and B is the child of A.

In a criminal case, attribute A may be interpreted as "Offender is living with a partner", and attribute B may be interpreted as "Victim is male". From the database, two conditional probabilities $Pr(B \mid A)$ and $Pr(B \mid \bar{A})$, namely the probability that B (the victim is male) is true if A is true (the offender is living with a partner) and the probability that B (the victim is male) is true if A is false (the offender is not living with a partner) may be determined. The complementary probabilities are the probability the victim is female if the offender is living with a partner and the probability the victim is female if the offender is not living with a partner. These latter probabilities do not need to be specified explicitly as they are complementary to the appropriate one of the first two probabilities given above. The strength of the relationship between A and B may be represented by the magnitude of these probabilities. The particular implication of this two-node network is that the choice of victim may be affected by the marital status of the offender. An offender who is living with a partner may have a different sexual preference for his victim from one who is not living with a partner. In practice, the reverse implication to that given above is required. Namely, it is of interest to know, given the sex of the victim, the probability of the offender living with a partner. It is a highly attractive property of the system that a suitable mathematical theory has been developed so that such implications can be made from the network.

The analysis reported here was carried out using a system known as PRESS, a *BBN* inference system implemented under a UNIX environment. It provides facilities for constructing and editing the Bayesian networks, extracting network conditional probabilities and evaluating the classification results on randomly selected cases. The PRESS user interface supports both interactive and batch processing. The former is used for evaluating individual cases and the latter for evaluation of a test database.

In addition to the prediction of characteristics, PRESS offers a unique feature in suggesting the most influential of the explanatory attributes for a response attribute at any stage of an analysis. This can be done either using prior probabilities or after evidence has been absorbed into the network. For example, in a network with 7 nodes as illustrated in Figure 10.2, the most significant factors to offender's marital status is the victim's sex, followed by the method of killing and victim age, based on information extracted from the past cases. Thus, the victim's sex is more important than the others and requires immediate attention in deciding the offender's marital status. This feature helps to prioritise different lines of investigation and will aid in the allocation of limited resources to the more important issues.

10.3.1 Network Construction

At the moment, the structure of a *BBN* is provided by users, based on their knowledge about the problem domain. This process is aided by the graphical interface of PRESS which provides the editing, printing, display and evaluation facilities. The automatic extraction of such structure from past databases is also possible as demonstrated by Hanson *et al.* [6] when sufficient number of past cases are available. Once a network has been produced, each node may be manually linked to an attribute in a database to allow the automatic extraction of the network conditional probabilities, which may be inspected and edited by users. An alternative procedure allows users to supply these conditional probabilities directly.

10.3.2 System Testing and Performance

PRESS outputs a list of posterior probabilities for each state of an attribute. These are determined from the initial conditional probabilities through the use of Bayes' Theorem. The state with the larger, or largest, probability is chosen as the final outcome. If there is no observed evidence, the probabilities are simply those obtained from the relative frequencies in the database. Given observed evidence of the state of a particular attribute, that state is taken as known and other conditional probabilities are adjusted. For testing purposes, the whole data set is split randomly into two parts, one of which is used as a training set, the other as a test set. The training set is used to estimate probabilities. The performance is measured by the percentage of the states

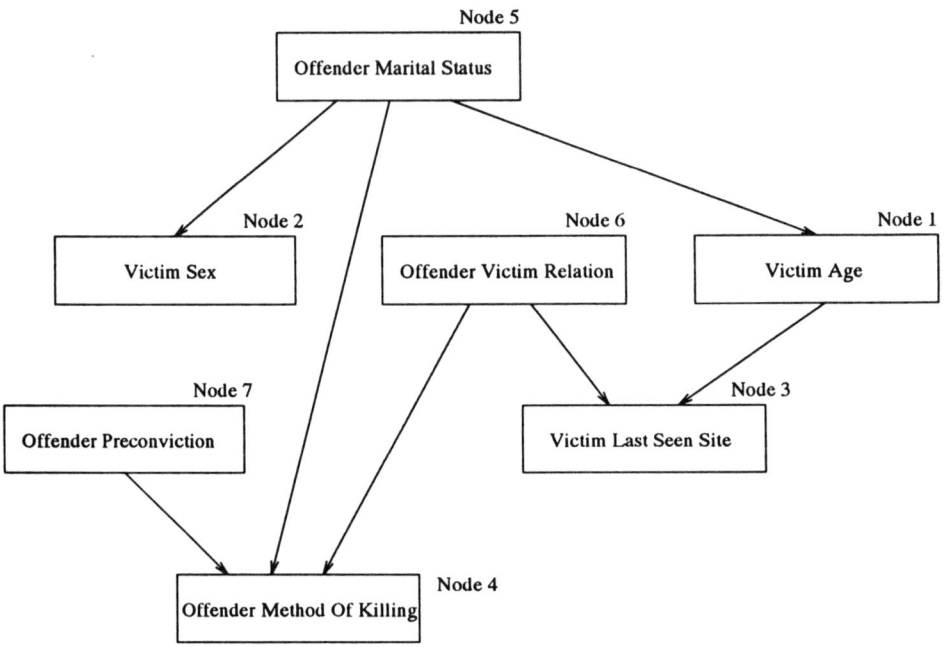

Figure 10.2: Seven node network showing relationship among three offender characteristics and four victim and crime characteristics. The numbering and description of nodes are as in the text.

of the attributes correctly predicted (*i.e.* by comparison of the state of an attribute of a member of the test set as predicted by PRESS with the true state of the attribute).

10.4 Examples of Networks

10.4.1 Seven Node Network

A network consisting of seven nodes was constructed in discussion with a senior detective. There were three stages involved in the construction of a network:

1. the specification of the attributes to be included in the network (choice of nodes);

2. the relationships amongst the attributes (choice of edges);

3. the specification of the conditional probabilities implied by the relationships.

Stages 1 and 2 were completed in discussion and are subjective in nature. For this example, the seven nodes, or attributes, with their possible states (not necessarily only two) and description were as follows:

1. age of victim (0–7, 8–12, 13+), (Victim_age);

2. sex of victim (male, female), (Victim_sex);

3. location of last sighting (home, other), (Last_seen_site);

4. method of killing (strangulation, other), (Method_of_killing);

5. marital status of offender (living_with_partner, other), (Offender_marital);

6. relationship of offender to victim (known, stranger), (Offender_victim_relation);

7. preconviction status of offender (yes, no), (Offender_preconviction).

There are four victim or crime scene attributes (numbers 1 to 4, above) and three offender attributes (numbers 5 to 7) in this network. The network was designed to be used to predict the offender attributes from the victim and crime scene details given. The relationships amongst the attributes, as decided in discussion with the detective, are shown in Figure 10.3. Once this network has been constructed it is apparent which conditional probabilities are required. These probabilities may be obtained from the database or may be representatives of measures of the beliefs of an expert or group of experts.

Two sets of conditional probabilities were used here: one supplied by the senior detective based on his personal experience, the other extracted from the database. In general, these two sets of probabilities were similar and some of these probabilities are given in Table 10.1. Notice that care has to be taken with the determination of the conditional probabilities. For example, the probabilities for Pr(victim male | offender not lwp) for the database and the senior detective are very different. This may well be because of a misunderstanding when eliciting the probability from the detective.

The performance of PRESS is shown in Table 10.2. The figures refer to the percentage of correct classifications amongst the number of cases in the test set. A particular case is classified to a particular state, for a binary attribute, if the probability of belonging to that state is greater than 0.5. For testing, the age and sex of the victim, the location of the last sighting of the victim and the

Table 10.1: Comparison between some of the conditional probabilities provided by a senior detective and those extracted from the database (lwp: living with partner).

	Conditional probability	
Name	Database	Senior detective
Pr(offender with preconvictions)	0.73	0.70
Pr(offender living with partner)	0.30	0.24
Pr(offender known to victim)	0.57	0.60
Pr(victim male | offender lwp)	0.23	0.10
Pr(victim male | offender not lwp)	0.10	0.90
Pr(victim 0–7 | offender lwp)	0.18	0.60
Pr(victim 8–12 | offender lwp)	0.27	0.20
Pr(victim 0–7 | offender not lwp)	0.18	0.20
Pr(victim 8–12 | offender not lwp)	0.30	0.40

Table 10.2: Prediction accuracy as a percentage of the BBN for the seven node network of Figure 10.2; SD – Senior Detective

	Offender profile		
Methods	Preconvictions	Living with partner	Known to victim
BBN (290/30)	77.0	83.0	60.0
BBN (SD/30)	77.0	60.0	53.0
BBN (256/64)	73.4	81.3	46.9
BBN (SD/64)	73.4	81.3	46.9
BBN (320/64)	73.4	81.3	45.3
BBN (320/320)	73.4	70.0	57.8
BBN (SD/320)	55.3	41.6	69.1
Prior probs.	73.4	70.0	57.8

method of killing were entered into the network for each member of the test set. PRESS then determines the probabilities for the other three attributes using the propagation method described in Lauritzen and Spiegelhalter [8].

Consider the first row of Table 10.2. There are 290 cases in the training set and 30 cases in the test set. For 23 of these 30 cases (77%) the offender was correctly predicted as having, or not having, preconvictions. Similarly, in 25 (83%) the offender was correctly predicted as living, or not living, with a partner. For 18 (60%) the offender was correctly predicted as being known, or not known, to the victim. These results are based on 290 cases, selected at random from the original 320 cases to form the training set, and from which the initial conditional probabilities are determined. The victim and crime scene characteristics are those from the 30 cases in the test set.

Table 10.3: The three most important attributes for prediction of particular offender attributes when no information about the victim is available.

Offender	The most important attributes		
attributes	First	Second	Third
Marital status	Sex	Method of killing	Age
Preconviction status	Method of killing	Site last seen	Relationship
Relationship	Method of killing	Site last seen	Preconviction status

The second row shows the results from using the subjective probabilities provided by the detective and the same 30 cases for testing. The third row shows the results from using 256 cases for training and 64 cases for testing. The fourth row shows the results from using the subjective probabilities and the same 64 cases for testing. The fifth row shows the result from using 320 cases for training and 64 cases for testing. The sixth row uses the results from using 320 cases for training and the same 320 cases for testing, an approach which will give biased estimates of the classification rates because the method is being tested on the same set which was used to determine the conditional probabilities. The seventh row shows the results from using the subjective probabilities and 320 cases for testing. The last row lists the relative frequencies provided from the database without consideration of victim and crime scene variables.

It should be noted that the large difference in the prediction accuracy using the two sets of conditional probabilities when the number of test samples is large. This may be caused by the bias in choosing the test cases as for each randomly selected 30 or 64 test cases the accuracy is different. An appropriate test procedure may be to take the average of the results from *all* the test cases. This would result in huge number of possible combinations to be realistic in practice, *e.g.* the number of ways to choose 30 and 64 samples from 320 can be computed by $C_{320}^{30} \approx 7.7 \times 10^{41}$ and $C_{320}^{64} \approx 9.0 \times 10^{66}$ respectively.

PRESS also offers a planning facility when considering the possible yet unknown factors. It suggests the most influential factors from victim and crime scene characteristics based on either the available or hypothesised factors for an offender attribute under consideration. Suppose evidence B is found, the factors which most likely cause B or are caused by B can be obtained using this feature. These factors which are most influential vary according to the current situation regarding the amount of evidence which has been accumulated. Table 10.3 shows the most important attributes to be considered when all that is known is that a child murder with sexual connotations has been committed. Table 10.4 shows the most important attributes to be considered when it is known that the victim is a girl aged between 8 and 12 years old.

When moving from Table 10.3 to Table 10.4, the influence of the observed evidence (age and sex of the victim) is absorbed into the network. The influen-

Table 10.4: The first three most important attributes when a female victim of age 8–12 has been found.

Offender attributes	The most important attributes		
	First	Second	Third
Marital status	Method of killing	Preconvictions	Site last seen
Preconvictions	Method of killing	Relationship	Site last seen
Relationship	Site last seen	Method of killing	Preconvictions

tial attributes in Table 10.4 are different from those in Table 10.3; in particular victim age and sex are known). Based on the network and the results shown in Table 10.4, the method of killing and the site at which the victim was last seen are the most important attributes for identifying the killer. (Note that the preconviction status is ignored because it is a characteristic of the offender.)

It should be noted that not all the important attributes mentioned in Tables 10.3 and 10.4 can be used for prediction as some are obviously the offender's attributes. However, this suggests the possible links between different factors and helps the police to focus on the most relevant factors, which is not really easy for a complicated crime with more factors.

10.4.2 Ten Node Network

A network with ten nodes is shown in Figure 10.3. All five offender characteristics are included. Notice that the variables from the network with seven nodes are not all included. The two networks are for illustrative purposes only. It is not necessary for bigger networks to build on smaller ones though this can be made a condition if desired.

For this example, the description of the ten nodes, or attributes, with their possible states (not necessarily only two) are as follows and the results are presented in Table 10.5:

1. proximity of offender within 5 miles (yes, no), (Offender_proximity);

2. marital status of offender (living with partner, other), (Offender_marital);

3. offender age (0–20, 21+), (Offender_age);

4. preconviction status of offender (yes, no), (Offender_preconviction);

5. relationship of offender to victim (known, stranger), (Offender_victim_relation);

6. abduction (unknown, yes, no), (Victim_abduct);

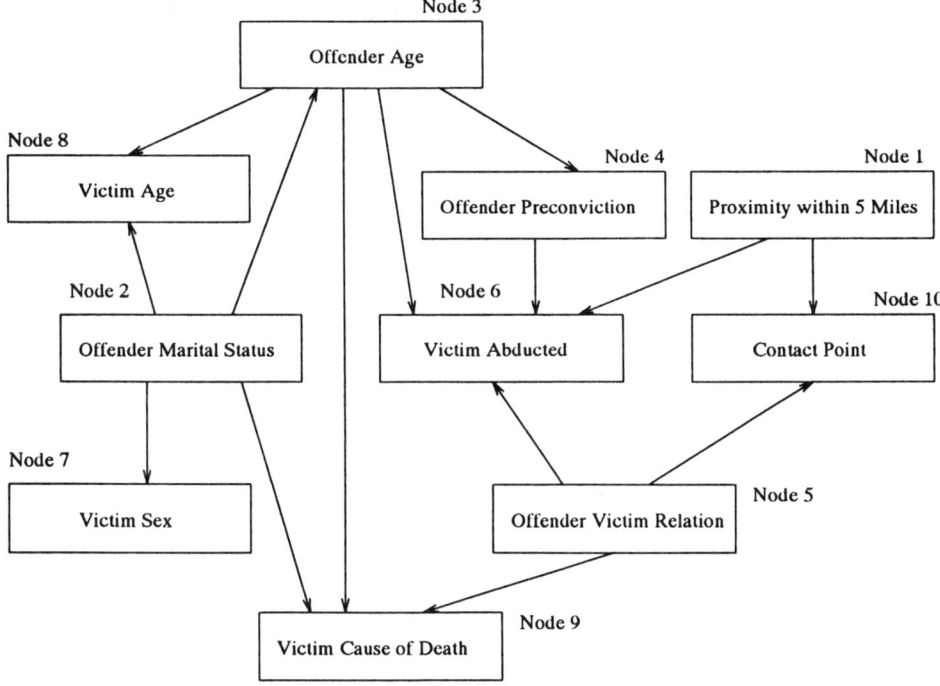

Figure 10.3: Ten node network showing relationship among four offender characteristics and six victim and crime characteristics. The numbering and description of nodes are as in the text.

7. sex of victim (male, female), (Victim_sex);

8. age of victim (0–7, 8–12, 13+), (Victim_age);

9. method of killing (strangulation, drowning, other), (Method_of_killing);

10. contact point (indoor private, indoor public, outdoor public), (Contact_point).

Table 10.5: Prediction accuracy as a percentage of the BBN for the ten node network of Figure 10.3 (lwp: living with partner).

Methods	Offender profile				
	Preconv.	Age	lwp	Known to victim	Live within 5 miles
BBN (256/64)	76.6	46.9	76.6	65.6	93.8
BBN (320/64)	76.6	51.6	76.6	65.6	93.8
BBN (320/320)	73.4	57.5	76.3	66.9	91.6
Prior probs.	73.4	57.2	76.3	57.8	91.6

10.4.3 Networks with a Large Number of Nodes

Networks which contain nodes representing attributes not recorded in the current database may be constructed. There are two possibilities for determination of the conditional probabilities for such networks. First, the states of a node which has not been recorded may be all assumed equally likely (a uniform prior). For example, if such a node has four states then the probability 0.25 would be assigned to each state. Alternatively, these probabilities may be elicited from experienced detectives using subjective judgment. A problem with this latter approach is that maintenance of the system may be difficult if nodes are added to or deleted from an existing network.

10.5 Comments

PRESS can be of considerable assistance in determining and evaluating the strength of relationships amongst variables. In the context of the example here these are as follows.

It permits users (detectives) to hypothesize and reason about a crime using their own expertise and that from other sources not covered by the network. For example, the dataset used was deliberately constructed from the original dataset to omit any rare variables. Detectives may, however, be able to incorporate information from these rare events, if present in a particular case, and use it to make deductions about the offender. PRESS enables the combination of objective information from a database and subjective information from users. However, care has to be exercised. The accuracy of the predictions of the offender profile depends, crucially, on how well the network models the structure of these types of crimes.

Also, the relationships amongst the offender, victim and crime scene details are made explicit. The structure of the graph, in terms of nodes (attributes) and edges (relationships) to be included is decided upon in discussion amongst investigating officers, statisticians and computer scientists. It is a highly interactive process and the nature of the system is such that different models can be tried very quickly.

The graphical nature of the network enables complicated relationships to be represented very clearly. Nodes which are not directly connected by an edge in that one can only move between them along edges by passing through one or more other nodes are assumed to be conditionally independent of each other. This means that large complex relationships can be broken down into small simple relationships amongst only those nodes which are directly connected to each other (Lauritzen and Spiegelhalter, 1988).

The conditional probabilities relating the attributes can be determined automatically from the database. However, they can also be obtained as measures of belief in consultation with police officers. For example, consider offenders with preconvictions. The database gives 73% of offenders as having preconvictions, a senior detective suggested 70%. Consider marital status: the database gives 24% of offenders as living with a partner, the detective also suggested 24%. Consider the relationship between offender and victim: the database gives 57% of offenders as being known to the victim, the detective suggested 60%. The network may be extended to include attributes which are not included in the database and to use measures of belief as probabilities within the system, alongside probabilities estimated from the database for other relationships. The accuracy of these measures of belief as true reflections of the uncertainty will depend to a large extent upon the experience of the people providing them. Careful discussion will be required by the police officers and statisticians involved in their determination.

In theory, it should be possible to construct a network with graph and probability tables for a particular type of crime which may be regarded as a standard network for that type of crime. In this way, the expertise of senior police officers, which would be used to construct the initial network, might be made available to all and might be used in operational conditions.

There are, however, some limitations. The current system was evaluated by dividing the data set into two parts, for example, one consisting of 256 cases, the other of 64. The conditional probabilities were obtained from the 256 cases. The system was then evaluated by comparing the predictions of the other 64 cases with the correct answer, as given by the database. This is not very effective because of the small numbers of cases involved.

In general, the great advantage of a Bayesian belief network is that it gives great flexibility to the users. The relationships amongst the variables are displayed in a way which may be readily understood and which has considerable intuitive appeal. If a particular network provides a relationship with which people are not happy it is straightforward to alter the network and use the altered network for analysis. Alternatively, different networks may be constructed and their results compared. The system has much to recommend it.

10.6 Discussion

The work on the database raised a number of issues, including assessment of performance, modelling in a real case, the interpretation of probabilities and the incorporation of detective expertise.

10.6.1 Measures of Performance

There are a number of ways in which the statistical model can be assessed. In data analytic terms, a numerical measure such as the proportion of correct classifications provides a reasonable overview of the performance. However, an improvement in the percentage of cases detected is not the only measure of success. The contribution of the model to the time taken to solve the crime and the resources saved in its solution may be useful measures also. Further, investigating officers might consider the variables included in the model, to see if they make practical sense or if they go against intuition.

The models can also be continually updated as new cases arise and are solved.

10.6.2 Models in Real Time

A real case consists of successive incidents over time. Profiling could be carried out from the beginning of the case, and the profile could be updated as successive incidents occur. Each incident might have an associated set of profiling requirements. Thus, there could be successive models for each stage of the investigation.

10.6.3 Interpretation of Probabilities

Results from the statistical models would have to be couched in simple terms and with appropriate caveats. A quoted probability or performance measure should always be set in context, otherwise unsatisfactory inferences might be derived. Also, the presentation of the models to police forces has to be made with care. Technical statistical terms can dissuade people from further interest. For example, the following translations could apply in operations. "Selected variables" could be described as "priority paths". "Offender attributes" could be referred to as "suspect acquisitions". The whole procedure could be viewed as a "decision support mechanism".

10.6.4 Incorporation of Detective Expertise

An important consideration is the incorporation of detective expertise, and what part this plays in statistical profiling. The issue of including practical knowledge in statistical inference is considered in the discussion by Holgate [7]: "Resistance to the use of statistics ... is often derived from a feeling that the practical person has knowledge that can be incorporated into the inferential process in an intuitive way, but which is not allowed for in the formal scheme of inference used by a statistician." In the models reported here the detective expertise was concentrated in the construction of the database and, to a lesser extent, in the construction of the belief networks. The resultant statistical profiling was largely a data analysis task, where the detectives provided constructive criticism of the final models. In future, investigating officers may wish to specify which variables (and their effects) should be included in which models. However, greater input of detective expertise at the modelling stage might have an effect on reproducibility, *e.g.* different officers may propose quite different models.

10.7 Conclusion

The application of BBN to a database requires the construction of a network. This process requires an extensive amount of knowledge about the problem domain and it has a major influence on the effectiveness of the network. This introduces subjective factors into the networks since different people may construct different networks. Several networks have been described here and are fairly consistent, However, subjectivity is useful to validate links between nodes obtained from the information extracted from a database. The database may suggest a line between pairs of nodes and hence minimize the subjective components of a network.

The BBN described here have been tested on a database with a reduced number of attribute values. This is because BBN require a full table of conditional probabilities locally for each node. The number of parent nodes and the number of states for each node determine the size of the table of conditional probabilities. For example, for a node with two states and three parent nodes whose number of states are 2, 3 and 3, then the conditional probability table requires $2 \times 2 \times 3 \times 3 = 36$ probabilities (though only 18 need direct evaluation). For these probabilities to be well estimated, the training database has to be of a size which is at least several times the number of required probabilities. If not, there is sparseness and useful predictions cannot be made. In cases where it is difficult to obtain a large database the number of nodes in the network and the number of states for each node has to be reduced.

It should be noted that the evaluation method used, which splits a database into two parts, is very crude. A better method needs to be built into PRESS. Also, the selection of a particular state by choosing the one with the largest probability can be improved, for example, by incorporating some measure of utility to be associated with each decision.

References

[1] Aitken C.G.G. and Gammerman A. (1989) Probabilistic reasoning in evidential assessment. *Journal of the Forensic Science Society* 29: 303–316.

[2] Aitken C.G.G., Connolly T., Gammerman A., Zhang G., Bailey D., Gordon R. and Oldfield R. (1995) Statistical modelling in specific case analysis. *Submitted for publication*

[3] Canter D. (1989) Offender profiles. *The Psychologist* 2: 12–16.

[4] Canter D. and Heritage R. (1990) A multivariate model of sexual offence behaviour: developments in "offender profiling". *The Journal of Forensic Psychiatry* 1: 185–212.

[5] Gammerman A. (1990) A causal probabilistic reasoning system. *Fourth International Symposium on Machine Intelligence* Barcelona, Spain, 45–61.

[6] Hanson R., Stutz J. and Cheeseman P. (1990) Bayesian Classification Theory. Technical Report FIA-90-12-7-01, Artifical Intelligence Research Branch, NASA Ames Research Center.

[7] Holgate P. (1983) Contribution to Discussion of "Statistical Inference of Phylogenies" by Felsenstein J. *Journal of the Royal Statistical Society* A (146)3: 246–272.

[8] Lauritzen S.L. and Spiegelhalter D.J. (1988) Local computations with probabilities on graphical structures and their application to expert systems (with Discussion). *Journal of the Royal Statistical Society* B (50) 157–224.

11

Baysian Belief Networks and Patient Treatment

L.D. Meshalkin, E.K. Tsybulkin

11.1 Introduction

The remote consulting centres are functioning in 35 reanimation units of regional children's hospitals of the former Soviet Union. They use the informational system "DINAR" [18, 4, 16]. The main function of "DINAR" is to assist a physician-consultant to make a decision about tactical and medical treatment. During the system development the principle of a "weak suggestion" was accepted. It consists of a system that helps the user to come to a proper solution instead of giving him the ready solution [5, 12]. Step by step evaluation of the quality of patient treatment is the most important task of any medical decision supporting system. In "DINAR" this evaluation is based on comparison between doctor prognosis made in time, when treatment is prescribed, and actual patient state. To facilitate comparison, we use the sequential (lexicographical) classification of the patient state: at the first level, classification is made according to the severity of the patient state, at the second level according to the dynamics (how a patient come to the state), at the third level according to the intensity of the treatment, and at last level according to the degree of the doctor control of the patient state (the better control — the easier patient). This system is effective for finding the doctors mistakes but seems not very useful for choice's of treatment.

One can hope that being powerful new mathematical technique and productive generalization of Markovian dependencies [2], Bayesian Belief Networks (BBN) will hopefully assist patient treatment further and from more realistic positions [17]. The methodological difficulties one meets in this direction may be due to: (a) changes in the structure of relations between variables as a function of the overall severity of a patient's state, (b) incomplete medical knowledge, (c) surplus of details in the dynamics description, (d) the necessity to reflect the process of development in the graph structure. We will solve problem (a) by the construction of a frame of the scale of the severity of a patient's state and a recommendation of the types of mathematical tools, which correspond better to every gradation of the scale (Section 11.2). Some development of the influence graph technique for solution of problems (b)-(c) will be suggested in Section 11.3. Some technical details of disease dynamics description are given in Section 11.4. The next small section (11.5) is devoted to a statement of the problem of discounting future health events in comparison with today's events. After that, we concretize the general approach using the example of one form of diphtheria (Section 11.6).

11.2 Scale of Severity of a Patient's State and Mathematical Tools

The grade of pathological process generalization is put in the base of a new classification. Three stages are picked out: a) the local process, developing according to its own rules (the first three grades), b) the generalized process, developing according to the rules of organism regulation (the next three grades) and c) the final grade, when the organism is fighting for survival. The classification bears the experience gained by the "DINAR" system and it follows previous international classifications such as CCS (Clinical Classification System [10]), TISS (Therapeutic Intervention Scoring System [9]), MOSF (Multiple Organ System Failure [19]), APACHE (Acute Physiology And Chronic Health Evaluation [11]), and PSI (Physiology Stability Index [15]). The term "typical time" reflects the speed of the patient's state changes, and it corresponds to the time interval between patient observations by a doctor.

Scale and mathematical tools. The relations graph, whose nodes correspond to clinical conditions, patient reaction parameters, and treatment directions, is the most suitable mathematical tool for description of the disease development on the first three gradations (the local process). Two special mathematical methods can be used here for graph simplification. They are the factorizing of conditional probabilities [1] and the statistical measurements of the dynamics of influential factors [14].

For three following gradations (the generalized process) potential mathematical tools are much more varied. One can hope that it is possible to use here, as in the previous case, the relations graph. But the main instruments are the mathematical models of organism regulation in pathological conditions. Physiological models of circulation one can find elsewhere [6, 7]. The corresponding models have to include two types of components: for continuous

Table 11.1: Frame of the scale of patient state severity, therapy and typical time (in order of severity increase).

1. Local process without the necessity of compensation by other systems of the organism. Local therapy. Typical time is determined by the local process;

2. Local process compensated by other systems of the organism, and without any symptoms of decompensation. Local therapy and control. Typical time is determined by the local process;

3. Local process compensated by other systems of organism and accompanied by some symptoms of decompensation or the affection of additional organs or non-vital systems of the organism. Local therapy and substitute therapy. Typical time is mainly determined by the local process;

4. Generalized process with one vital system of the organism affected (The vital systems are the following: respiratory system, central nervous system and cardio-vascular system). Polyorgan treatment and local therapy if necessary. The steady state lasts more than a week;

5. Generalized process with more than one vital system affected. The organism's response is regulated by the central nervous system, so one can say that the organism reacts as a whole. The organism is in a steady state for 6-12 hours. Polyorgan treatment and invasive monitoring with periodically active treatment;

6. Generalized process with more than one vital system affected. The organism's response is regulated by the central nervous system. The state changes every 2 hours. Polyorgan treatment and actions against the change to the state 7. The readiness to it;

7. Acute threat to life. The relations between systems are lost, so the organism could no longer be regarded as a whole. Full replacement of the vital systems functions is necessary. Typical time is minutes or dozens of minutes.

development description one uses dynamical components; and for description of sudden changes in a patient state one uses probabilistic components.

The main statistical inference is made about parameters related the patient's reactions to a treatment. The computer sorting and analysis of possible development versions, even with approximate values of parameters, let us allocate possible dangers and attract the doctor's attention to them. Record-keeping of combinations of parameters leading to adverse reactions with the estimation of corresponding probabilities is the basis of the assistance.

Special models of the function of the basic systems of an organism are necessary for the seventh gradation of severity (disintegration of the reactions).

Local process description. We suggest characterizing the local process as follows:

1. Degree of stress of the organism. In the case of infection, stress can be characterized as:

    (a) quantity of the agent;

    (b) nature and quantity of the mediators of inflammation;

    (c) response of the immune system.

2. Characteristic time of the process, which is inversely proportional to the activity of the process;

3. Degree of disturbance of a function.

The above material about BBN application leads us to a general conclusion that: *in many practical situations mathematical apparatus used for dynamics description could depend upon the state of the object described.* In medicine this dependance is relatively simple: there are the local processes and the disease of a whole organism.

11.3 Mathematical Tools Adjustment

As mentioned above, there seems to be the possibility of BBN use for the prognosis of a patient's state in the future. But the transition from the possibility to the technology is difficult and requires some development of mathematical tools. One needs to overcome the following difficulties:

- Deficiency of medical knowledge. Except for the simplest cases medicine has never before been able to estimate the numerical probabilities of different variants of disease development. Some concepts which are necessary for this have not yet been developed;

- Inconstancy of a disease in time;

- Avalanche of information collected during the course of a disease.

Let us begin with the construction of a scheme which enables us to obtain more precise knowledge during a function.

Hierarchical sequence of graphs (HSG). Let us look at two graphs $G_i = (V(G_i), E(G_i))$ $(i = 1, 2)$, where $V(G_i)$ is a finite set of vertices (nodes) and $E(G_i)$ is a set of edges of graph G_i. Both graphs have no multiple edges and no loops.

We say that graph G_1 is a *contraction* of graph G_2 and write $G_1 \prec G_2$, if (1) there is homeomorphous (several to one) mapping $V_2 \rightarrow V_1$, under which the edges of graph G_2 map into the corresponding edges of G_1, for the except edges, which join nodes mapped into one node of G_1; and (2) in G_1 there is no node or edge which does not have an inverse image in G_2. A finite sequence of graphs $(G_1, G_2, \ldots, G_k)$, such that for all i $(1 \leq i < k)$, $G_i \prec G_{i+1}$ is said to be *an hierarchical sequence of graphs of order k* (HSG(k)). The Graph G_1 is said to be the *initial* of HSG.

HSG helps us to make a bridge between our present knowledge and the more complete knowledge of tomorrow. Working with HSG it is convenient to use an hierarchical system of notation $(v_{i,j,\ldots,l} \in V)$, in which the first index (i) is the number of a node in G_1, the second one (j) is the subnumber of a node, which is the first one that is different from its inverse image, and so on.

Leading graph. For a uniform description of dependencies between variables used during decision making at relatively advanced stages of diseases, we suggest exploiting HSG with a common initial graph, which we call *the leading (in time) graph*.

For the first three grades of the Section 11.2 the leading graph is given in the next section.

11.4 Notation, Time Structure and Leading Graph

We use two systems of notation, one of which is related to a role that a variable has in the treatment process and the other one of which is related to a variable time position in relation to the moment of decision making (t).

The first system. Notation for a local process is L, for a clinical state X, for severity of a patient state $\bar{X}$, for an organism parameter F, and for a treatment U. All variables depend upon time except for anamnesis-vite parameters.

The time classification is in relation to t (in brackets there letters are used for corresponding data notation). The characteristic time of a process is τ, $s = t + \tau$, anamnesis - vite (A), anamnesis - disease (B), present $(L(t), X(t), \bar{X}(t), U(t))$, next-step $(L(s), X(s), \bar{X}(s), U(s))$, future (time from s to clinical recovery) (D, E), future-catamnesis (time after clinical recovery) (D). The letter D here denotes a direction of the possible development of complications, the letter E is a acute treatment complication ("doctor error").

Leading graph for local grades of disease. The leading initial graph is shown at the Figure 11.1, where letters denote sets of variables which have to be developed during HSG construction. Letters with low indexes are corresponding

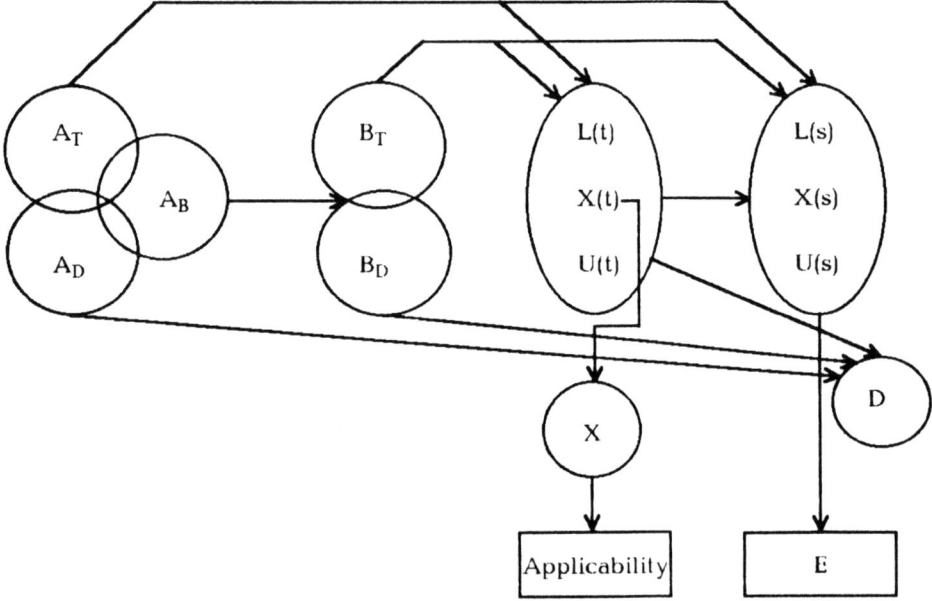

Figure 11.1: The leading graph for lacal grades of disease.

subsets. Two rectangles (below right) mean that the scheme is not more applicable: the first from the left because of a disease transition from a local to a generalized stage; and the second one because of the appearance of a acute treatment complication. The division of A and B into subsets guides a researcher on the elaboration and employment of generalized indicators instead of primary variables. The local character of a disease simplifies the study of possible complications.

We recommend begining construction of HSG (in a limited BBN) for the typical course of a disease. Only after some experience has been obtained can one approach the complete course in all details. It is very important that moving to a BBN should not be uniform on all nodes, as assistance to a patient's treatment is possible even at the graph stage.

11.5 The Choice Between Different Trajectories of Disease Development as a New Problem

BBN and mathematical models in principal provide us with the possibility of a compact description of a disease's process and a prognosis of future events under a given treatment. The next step should be a choice between admissible treatments. It is well known that almost all medical treatments make positive and negative influences. To compare different treatments one ought to know *how to compare different states of health, which occur at different times and have a different duration.* This is not a new problem for practical medicine. Every experienced statistician in this field knows how a doctor can totally ignore the future negative sequences of a current treatment. But it is a new problem for theoretical medicine because only now does the doctor have the option of performing all the calculations necessary to compare similar outputs. We are not intending to discuss this problem here in detail. Probably, some analogy of "discounting" future events in comparison with recent ones can be used here in the same way in which it is used in economics. There is only one serious difference: money almost additive, states of health are not. The first scientist who applied discounting to human population was Fisher [3].

11.6 Example

Diphtheria in children. As an example we examine a course of a combined form of diphtheria in children. It is chosen because of the relative simplicity of versions of development.

In Russia it is accepted to distinguish among the following forms of diphtheria: (a) nontoxic diphtheria of nose and pharynx, (b) nontoxic diphtheria of larynx, (c) combined (nose, pharynx, larynx) diphtheria, (d-f) malignant diphtheria of 1-st, 2-nd and 3-th degrees. Complications of the combined diphtheria with its current frequency characteristics are given in the Table 11.2. This table is constructed on the basis of an expert estimation of one of the authors. The death rate is estimated under the condition of well-timed medical aid.

Local process. In accordance with Section 11.2 we use the following characteristics of the local process of diphtheria:

$L_{1,1}$ – there is not a commonly used characteristic of the quantity of agent for diphtheria;

$L_{1,2}$ – tumor necrosis factor (1 - norma, 2 - increased, 3 - sharply increased);

$L_{1,3}$ – immunological response (1 - weak, 2 - equivalent, 3 - increased);

$L_2 = \tau$ – one day;

Table 11.2: Main current specific complications of combined diphtheria and their numerical characteristics (in %).

Complications of diphtheria	Frequency	Death rate
1. Respiratory insufficiency (2-6 days)	15	0
2. Early myocarditis (its begin approximately after the 4 day)	4	< 0.5
3. Late myocarditis without neurological disorders	5	1
4. Late myocarditis with neurological disorders	15	1
5. Neurological disorders without myocarditis	25	< 1
6. Toxic nephritis	10	< 0.5
7. Liver complications	< 1	< 0.5
8. Noncomplicated curse of disease	75	0
9. Serum disease	< 1	?
10. Complications of treatment	?	?

$L_{3,1}$ – breathing work (norm, increased and effective, increased and ineffective, weakened and ineffective);

$L_{3,2}$ – swallow (norm, pain, mechanical disturbance).

Other characteristics. We use the letter X for the characteristics of a patient's state. A general estimate of the severity of a state is $\bar{X}$. Systems that give the main complications are: respiratory system (X_1: $X_{1,1}$ – provoked by the stenosis of airways; $X_{1,2}$ – provoked by toxicoses), circulatory system (X_2), peripheral nervous system (X_3), cleansing function of kidneys (X_4), liver (X_5 : $X_{5,1}$ – cleansing function of liver; $X_{5,2}$ — barrier function of liver), the immune system (X_6). The following parameters (F) are used: F_0 — inoculated or non-inoculated child, reaction to stress (F_6, the reaction types are hyporeaction, normoreaction, and hyperreaction. Later the dimension of the reaction, probably, should be increased), efficiency of functioning of respiratory system F_1, anomaly of heart development (F_2), perinatal nervous system damage (F_3), kidney weakness in anamnesis (F_4), liver weakness in anamnesis (F_5). The following groups of treatment are considered (the letter U): antibacterial therapy (U_1), antitoxical therapy (U_2: $U_{2,1}$ — initial doses of the serum; $U_{2,2}$ — extra-corporal methods: haemabsorption and others; $U_{2,3}$ — cortical-steroid hormones), the therapy against respiratory insufficiency (U_3: $U_{3,1}$ — the insufficiency provoked by the stenosis; $U_{3,2}$ — the insufficiency provoked by toxicoses), hormonal therapy directed at the regulation of the autoimmune answer (U_4); in the case of complications one can add treatments for heart support (U_5), peripheral nervous system support (U_6), kidney cleansing function support (U_7, diuretics and dialysis), liver support

(U_8: $U_{8,1}$ cleansing liver function support; $U_{8,2}$ barrier liver function support) and also normalization of the blood coagulation system (U_9).

Variables numerical values. The variable groups X have non-negative integer values. They begin from 0 and correspond to sequential gradations of severity. F_6 receives value 2 for the hyperreaction, value 1 for the normo-reaction and value 0,5 for the hyporeaction. F_1 is a continuous number, measured in special units. F_0 has a value 0 for the inoculated child or value 1 for the non- inoculated child. $F_2 - F_5$ have value 0 if there is no corresponding weakness in the anamnesis, or the value 1 if there is. The variable groups U have a value 0 if the corresponding treatment is not applied, and a value 1 for moderate treatment or 2 for intensive treatment.

The influence of parameters $F_2 - F_5$ is obvious at the qualitative level. If the anamnesis is bad, then the probability of complications is higher. The same with the variable F_1: if the parameter is higher, then the difficulties for the organism arise.

Examples of different influences of traditional treatments of diphtheria. The application of antitoxins serum $U_{2,1}$, on the one hand, reduces the effect of toxins, and, on the other hand, increases the autoimmune reaction with the delay of 2–8 weeks, and by this the probability of the complications related to the autoimmune reaction increases. The extra-corporal methods (plasmapheresis, hemocarboperfusion) are usually used in case of malignant forms of diphtheria to reduce heart complications, but if heart complication occurs this treatment becomes dangerous to the patient. The respiratory system support by the elimination of stenosis ($U_{3,1}$) increases the probability of pneumonia.

The absence of a quantitative approach, which puts in a state of balance pluses and minuses of a treatment, does not allow us today to make precise recommendations about the doses to use.

Construction of HSG for the third day of a typical course of local process of combined diphtheria without possible treatment complications. It must be taken into account that the words "third day","local process", "typical course" and "without treatment complications" simplify the situation greatly.

We start with definition of the sets of Figure 11.1.

$A_B = \{F_0, F_6, F_2\}, A_T = \{F_0, F_6, F_2, F_4\}, A_D = \{F_0, F_6, F_1, F_2, F_3, F_4, F_5\},$
$B_S = B_D = \{L(k), X(k), U_1(k), U_{2,1}(k) \quad k = 1, 2\},$
$L(3) = \{L_{1,2}, L_{1,3}, L_{3,1}, L_{3,2}\}$ all values are in limits of local process ,
$U(3) = \{U_1, U_{2,1}, U_{10}\},$
$D = \{X_1, X_2, X_3, X_4, X_5\},$

We will need the following concepts:

$\bar{F}$ is the summarized weakness in anamnesis, which depends upon values of all F_i;

D_i $k = 1, \ldots, 5$ is the direction of the development of the i-th complication (1-st is respiratory system, 2-nd is cardio-vascular system, 3-rd is nervous system, 4-th is kidney, 5-th is liver).

The existing clinical knowledge can be summed up as follows:

1. The frequency of inoculated children is about 10%. Severity of disease in them is much lower than in non-inoculated children. The frequency of complications is 2-3 times lower;

2. Antitoxical serum reduces the frequency and severity of early complications and increases the frequency and severity of later ones;

3. On the fourth day there is the possibility of the development of early myocarditis and toxical nephritis;

4. Frequencies of the weakness of a system in anamnesis are approximately (in %): for the respiratory system < 1, for circulation 1, for the nervous system 10, for kidneys 20, for the liver < 1;

5. According to the power of influence of the weakness of a system in anamnesis on the frequency of the development of corresponding complications, organism systems are ordered as follows (in order of decreasing influence): circulation, nervous system, kidneys, respiratory system, liver;

6. Development of one complication increases the probabilities of the development of others;

7. Under conditions where a process continues to be the local one, peculiarities of a disease do not have any influence on the probabilities of the development of complications. The same is true for local and supporting therapy, but this statement is valid only for the first few days of diphtheria;

8. Antibiotics against corynebacterium diphtheria reduce the duration and severity of the disease. According to current doctrine, they are always administered. The numerical characteristics of their influence on the probabilities of the development of complications are unknown.

A fragment of the graph of relations on the third day of a typical course of a local process of combined diphtheria is shown in Figure 11.2.

The introduction of the new language, which is more powerful in comparison with the traditional medical one, has shown that the current level of knowledge is not sufficient for the construction of complete BBN models for diphtheria. At the same time, the "ignorance" is formulated more precisely, and allows us to make a quantitative estimation. Until the appropriate quantitative relations are established, one can use only qualitative analysis.

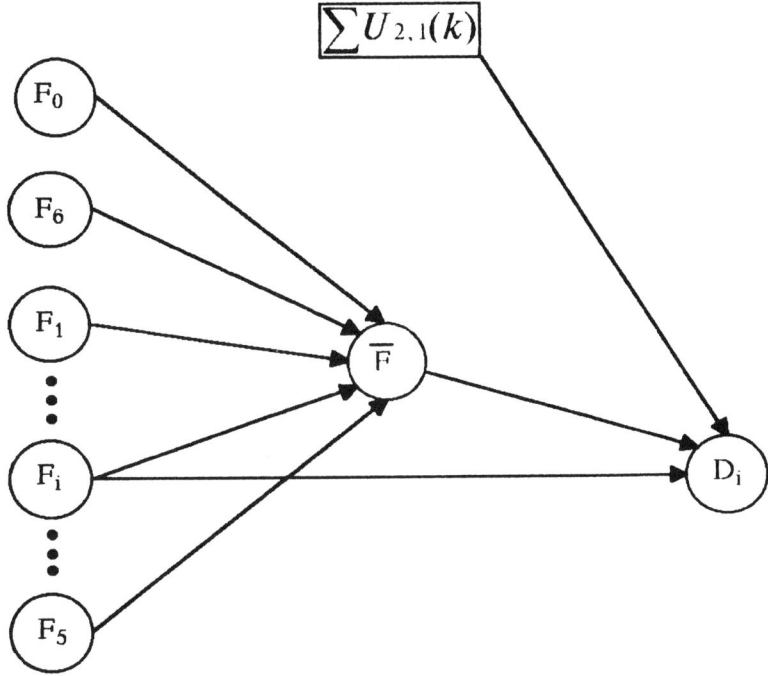

Figure 11.2: The fragment of the graph of relations for the third day of diphtheria.

11.7 Conclusion

Heathfield and Wyatt [8] compared the frequencies of questions, which are interesting to clinicians, and the frequencies of topics, in which the mathematicians were engaged in work. According to the MEDLINE data the clinicians were interested in therapy in 41% of all enquiries and only 60% in diagnostics, while only 19% from all developed (1973–93) medical expert systems were devoted to therapy and 53% to diagnostics. We are beginning to see that the approach proposed can help to reduce this gap. BBN lets us to formulate precisely the frequency laws of disease development, and the influence on it of various factors for a wide range of diseases and patient states. BBN is a convenient language for the theoretical schemes of patient treatment, at least for the local stages of disease. The first attempt to use this quantitative language has shown the numerous gaps in medical knowledge. Nevertheless, it is time to put forward, where possible, *the program of elaboration of a unified BBN — schemes of disease treatment and appropriate intelligent assistance to a doctor.*

Acknowledgements

This chapter has been written within the INTAS-93-725 program, funded by the International Organization for the Promotion of Co-operation with Scientists from the Independent States of the Former Soviet Union.

References

[1] Cox D.R. (1972) Regression models and life tables. *J.R. Statist.Soc.* (B) 2: 187.

[2] Dawid A.P. and Lauritzen S.L. (1993) Hyper markov laws in the statistical analysis of decomposable graphical models. *The Annals of Statistics* (21) 3: 1272–1317.

[3] Fisher R.A. (1958, 1929) *Theory of Natural Selection* 2nd rev. ed. New York: Dover Publications.

[4] Goldberg S.I. et al. (1991) Expert system "DINAR-2" — methodological basis for the pediatric emergency aid organization in a large region. *Medical Informatics Europe* Vienna, Austria.

[5] Goldberg S.I. and Meshalkin L.D. (1992) A New Class of AI Systems (DrWT-systems) *Techn. Kibernetika* 5: 217–223 (in Russian).

[6] Guyton A.C., Coleman T.G. and Granger H.J. (1972) Circulation: overall regulation. *Ann. Rev. Physiol.* 34: 13–46.

[7] Guyton A.C., Montani J.P., Hall J.E. and Manning R.D. (1988) Computer Models for Designing Hypertension Experiments and Studying Concepts. *The American Journal of the Medical Sciences* (31) 4: 320—326.

[8] Heathfield H.A. and Wyatt J. (1993) Philosophies for Design and Development of Clinical Decision-Support System. *Meth. Inform.Med.* 32: 1–8.

[9] Keen R. and Cullen D.S. (1983) Therapeutic Intervention Scoring System. *Crit. Care Med.* 11: 1–3.

[10] Knaus W.A., Wagner D.P., Draper E.A. *et al.* (1981) The range of intensive care services today. *JAMA* 246: 2711-6.

[11] Knaus W.A., Draper E.A., Wagner D.P. and Zimmerman J.E. (1985) APACHE II: A severity of disease classification system. *Crit. Care Med.* 13: 818-29.

[12] Meshalkin L., Goldberg S. *et al.* (1993) Children emergency aid information system. *Proceedings of MIE-93, Eleventh International Congress European Federation for Medical Informatics*, Jerusalem, Israel, April,18-22, Freund P., 195-197.

[13] Meshalkin L.D. and Goldberg S.I. (1994) Intellectual Systems: New Methods Development. *Priroda* 10: 66-75 (in Russian).

[14] Meshalkin L.D. and Kagan A.B. (1972) A contribution to the discussion upon the paper "Regression models and life tables" by D.R. Cox *J.R.Statist.Soc.* (B) 2:.

[15] Pollack M.M., Ruttiman U.E., Getson P.R. *et al.* (1987) Accurate Prediction of the outcome of pediatric intensive care: A new Quantitative Method, *N. Engl. J. Med.* 316: 134-139.

[16] Sklar M., Lachova E.F., Lomovskich V.E. and Goldberg S.I. (1992) Computer Technology in work of regional pediatric center. *Public Health of Russian Federation* 2: 25-27 (in Russian).

[17] Spiegelhalter D.J., David A. Ph., Lauritzen S.L. and Cowell R.G. (1992) *Bayesian Analysis in Expert Systems*. Reprint, 46 p.

[18] Tsybulkin E.K., Menshugin I.N. and Jukovskiy K.A. (1984) Clinical aspects of work of children reanimation and consulting centers with atomized consulting systems. *Computer diagnostics in acute and specialized aid in pediatry* Leningrad, 41-48 (in Russian).

[19] Wilkinson J.D., Pollack M.M., Ruttiman E.U. *et al.* (1968) Outcome of pediatric patients with multiple organ system failure. *Crit. Care Med.* 14: 271-4.

12
A Higher Order Bayesian Neural Network for Classification and Diagnosis

A. Holst, A. Lansner

12.1 The Classification Task

In the type of classification task we are interested in here, we are given a set of attributes (*e.g.* symptoms), and we want to decide which of a number of classes that set belongs to (*e.g.* the cause of the symptoms). The goal is to train a neural network to perform the classification, given a set of previous example patterns, each consisting of a set of attributes and their corresponding class. The kind of domain we will focus on is a technical system, which we want to diagnose in case of malfunctioning.

The approach to classification will be to estimate $P(q|\boldsymbol{x})$, the probability of the class q, given a vector $\boldsymbol{x}$ of measurements on the system. If we can do this, and then select the class with the highest probability, we have an optimal Bayesian classifier [2]. The learning task is thus to estimate the above probability distribution from the given database of patterns.

We will in the following concentrate on how to estimate $P(\boldsymbol{x})$, since $P(\boldsymbol{x}|q)$ can be achieved in a similar way, and then Bayes rule can be applied to get the desired result:

$$P(q|\boldsymbol{x}) = P(q)\frac{P(\boldsymbol{x}|q)}{P(\boldsymbol{x})} \qquad (12.1)$$

Unfortunately, $P(\boldsymbol{x})$ cannot be estimated from the database directly, since there are too many possible outcomes of $\boldsymbol{x}$ (exponential in the number of attributes), and there is usually a very limited set of training examples. Given an arbitrary classification task, where we have no additional information about the domain, this means that we have very little or no information about the distribution in most points. Thus the general classification task is impossible to solve in practice, if training has to be done from examples only (see *e.g.* [11]).

However, in practice there is no need to be that pessimistic. Usually there are additional assumptions that can be made about the domain, which helps overcoming this problem. One common situation is that there are invariances of different kinds, for example translation, size or tilt invariances when recognizing written characters. These invariances can decrease the number of degrees of freedom considerably [12]. An extreme case is the invariance found in the parity problem, where only the number of active units matter, and not their position. If we know that we have this type of invariance, the parity problem is trivial and can be solved with a training set that grows linearly with the number of attributes, whereas if we do not know this invariance, it might be impossible to detect that there is a parity relation with less than an exponentially large training set.

In the above invariance cases there are typically an array of similar attributes, for example pixels on a retina. Another situation that is common and which we will focus upon in the following, is when each attribute represents some measurement or property on some specific part of for example, a technical system. Each attribute bears some information about some well defined aspect of the system, and the kind of pure invariance discussed above is more unusual. Instead there is another assumption that is reasonable. This is that the attributes are independent, or mainly governed by low order correlations. If we can express the joint probability distribution in probabilities of variables of a much lower order, then the classification task is tractable again.

If the attributes are all completely independent, the calculation of $P(\boldsymbol{x})$ is straightforward:

$$P(\boldsymbol{x}) = P(x_1)P(x_2)\ldots P(x_n) \qquad (12.2)$$

However, usually this is of course not the case. Many components of the system might depend on each other, and thus measurement on different components can be correlated.

One method to solve this, is to "merge" correlated attributes into groups, or joint variables y_j, such that attributes from different groups are independent. Then the total probability is instead calculated from these (independent) joint variables:

$$P(\pmb{x}) = P(y_1)P(y_2)\ldots P(y_m) \qquad (12.3)$$

The larger a group of attributes gets, the more possible outcomes y_j will have, and the more uncertain will the estimation of probabilities be. To get a good generalization capability, we would like to take care of the strongest correlations while keeping the order of the joint variables as low as possible. Therefore, when many attributes are correlated, another approach has to be taken.

In many domains it is meaningful to talk about *causal structure*, or *dependency graph*, a graph that describes how the attributes are causally connected [1, 10, 13]. Even if all attributes are indirectly affected by most other attributes, the direct dependencies are probably of a lower order. If we know the causal structure, it is possible to utilize it to calculate $P(\pmb{x})$.

Assuming that there are no loops in the causal structure, the attributes can be sorted into the chain rule for joint probabilities, such that they are conditioned only on variables upon which they depend. By writing the conditional probabilities as fractions, and multiplying both numerator and denominator with some factors, it can be rewritten as a non-directed expression. This means that to use it there is no need to know the causal direction, but only the statistical dependence between attributes. This process of rewriting is illustrated in Figure 12.1.

With this method (see [4] for details), the resulting expression will be a product expansion of the form:

$$P(\pmb{x}) = P(x_1)P(x_2)\ldots P(x_n)\Psi(y_1)\Psi(y_2)\ldots \Psi(y_m) \qquad (12.4)$$

where y_j corresponds to "edges" in the graph, *i.e.* groups of attributes with direct dependencies between them. $\Psi(y_j)$ are *compensation factors*, fractions of probabilities to compensate (12.2) for dependencies between attributes in y_j.

If we can find the causal structure, we can thus calculate $P(\pmb{x})$ with the help of this. The main idea here is to first use the database to estimate the causal structure in the domain. Then the database is used again to estimate the conditional probabilities of the classes, given that causal structure. The algorithm is well suited for implementation in a Bayesian confidence propagation neural network.

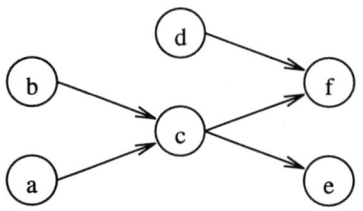 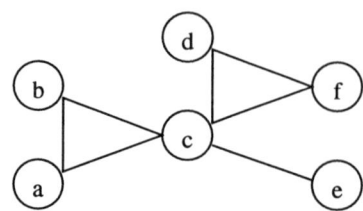

Figure 12.1: Starting from a dependency graph with no loops, it is possible to rewrite the probability distribution over the graph, to a similar expression, using a non-directed generalized graph. First the attributes are sorted into the chain rule for probabilities:
$P(\boldsymbol{x})=P(a)P(b|a)P(c|ab)P(d|abc)P(e|abcd)P(f|abcde)$.
Keep only direct dependencies according to the graph:
$P(\boldsymbol{x}) = P(a)P(b)P(c \mid ab)P(d)P(e \mid c)P(f \mid cd)$.
Rewrite the conditional probabilities as fractions:
$P(\boldsymbol{x}) = P(a)P(b)\frac{P(abc)}{P(ab)}P(d)\frac{P(ce)}{P(c)}\frac{P(cdf)}{P(cd)}$.
Note independencies and rewrite:
$P(\boldsymbol{x}) = P(a)P(b)P(c)P(d)P(e)P(f) \left(\frac{P(abc)}{P(a)P(b)P(c)}\right) \left(\frac{P(ce)}{P(c)P(e)}\right) \left(\frac{P(cdf)}{P(c)P(d)P(f)}\right)$.
The result is of the form (12.4), *i.e.* all first order factors and then additional factors to compensate for the dependencies among these.

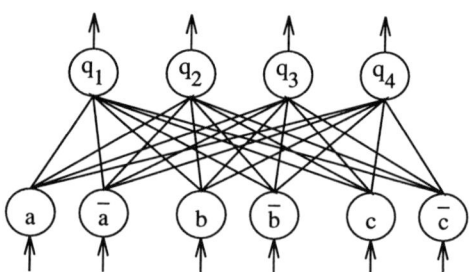

Figure 12.2: A one-layer BCPNN with three binary input attributes and four output classes. See the text for details.

12.2 The Bayesian Confidence Propagation Neural Network

The Bayesian neural network model [6, 5] (or Bayesian Confidence Propagation Neural Network (BCPNN) as it will be called in the following) is designed to calculate the probabilities of some output attributes (classes) given some input attributes. If the attributes are independent it is possible to use a

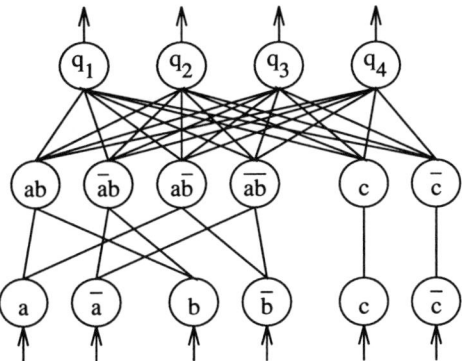

Figure 12.3: A multi-layer BCPNN with a hidden layer consisting of one first-order column c and one second-order complex column ab.

one-layer network [9], then implement a naive Bayesian classifier. In the network there is one unit for each outcome of each input attribute x_i, as well as one unit for each class q (see Figure 12.2). If Bayes' rule (12.1) is combined with the independence assumption (12.2) (and a corresponding assumption conditionally given q) we get:

$$P(q|x) = P(q) \frac{P(x_1|q)}{P(x_1)} \frac{P(x_2|q)}{P(x_2)} \cdots \frac{P(x_n|q)}{P(x_n)} \qquad (12.5)$$

When taking the logarithm of (12.5) the right hand side becomes a sum. Each term $\log(P(x_i|q)/P(x_i))$ in this sum corresponds to the weight from the corresponding input unit x_i to the class unit q. Each class unit sums its inputs and exponentiates the result, yielding the desired probability.

If there are dependencies between attributes, we have to introduce a hidden layer in the network. This hidden layer consists in the BCPNN of *complex columns*, i.e. groups with units for every possible outcome of some joint variables y_j. If we base our calculation on expression (12.3) these variables will represent groups of attributes that are dependent (see Figure 12.3). Let us call the corresponding columns *partitioning* complex columns. Analogous with the above case, the weights are set to the logarithms of the factors $P(y_i|q)/P(y_i)$.

The alternative is to use expression (12.4), in which case the hidden layer will consist of first-order columns for all input attributes (corresponding to the factors $P(x_i)$) and higher order columns for all "edges" in the dependency graph (corresponding to the compensation factors $\Psi(y_j)$). These latter columns are called *overlapping* complex columns (see Figure 12.4). The weight from a unit in an overlapping column is in analogy with the previous cases calculated as $\log(\Psi(y_i|q)/\Psi(y_i))$.

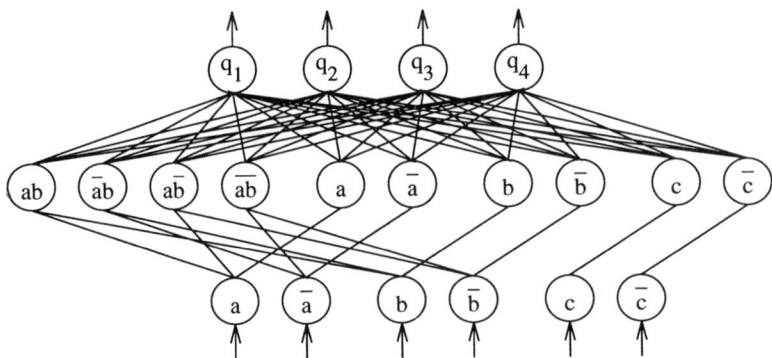

Figure 12.4: A multi-layer BCPNN with a hidden layer consisting of units for the input attributes, plus an overlapping column for a dependency between a and b.

With this neural network architecture the dependency graph is thus embedded in the hidden layer, and the probabilities are propagated through the neural network in a single step, rather than being sequentially propagated through the dependency graph itself.

To construct the hidden layer, we have to find which attributes are correlated. The approach taken here is to first search for low order correlations, and then search for higher order correlations only where we have already found lower order correlations. The reason for this is that an exhaustive search for correlations is infeasible due to its exponential complexity.

The measure used to check if a set of attributes are correlated comes from information theory. The mutual information between two variables X and Y is defined as:

$$I(X,Y) = \sum_{x_i, y_j} P(x_i, y_j) \log \left(\frac{P(x_i, y_j)}{(P(x_i) P(y_j)} \right) \qquad (12.6)$$

If the mutual information between two attributes are above some threshold, they are considered dependent (see, *e.g.*, [8] for why mutual information is suitable for this), and a complex column is created for them (unless this would create a loop in the corresponding graph). All pairs of attributes are considered in this way, in order of decreasing mutual information.

When used only one pass, this is the same method for generating a dependency tree as in [1]. However, here this process is continued a few passes, then considering the mutual information between individual variables and joint variables created in previous passes. This creates successively higher order columns compensating for higher order correlations.

Note that this will find the dependency graph in X, which is not necessarily the same as the dependencies in $X|Q$. An advantage with this method is that it is unsupervised, which means that we can also utilize unlabelled examples. It also means that the whole training set is used for estimation of the same structure, rather than dividing it in separate groups for each class, which would give less confident values of the mutual information. If the causal structure is supposed to depend strongly upon the class, the conditional probabilities given the class can of course be used instead in (12.6).

Although the chain rule works as long as there are no *directed* loops in the dependency graph, and it thus is possible to write the joint probability over the graph in the form (12.4), even when there are some undirected loops; we here only search for pure tree structures. This is because we never bother about the direction of causality, and thus cannot distinguish between directed and undirected loops. If the directions are known it is possible to treat undirected loops too, with only a slight modification in how $\Psi(y_j)$ is calculated. Also, small (directed or undirected) loops can be treated by merging all variables in a loop into one joint variable.

After creating the hidden layer, the weights between the hidden units and the output units are determined in one pass by estimating the required joint probabilities $P(y_i)$ and $P(y_i|q)$ from the database and calculating the weights from them according to the above prescriptions.

12.3 The Diagnosis Application

This BCPNN has been tested on a realistic task: diagnosis of telephone exchange computers. In this case, one malfunctioning circuit card out of 36 is to be identified from an error vector of 122 bits. The training set consists of 442 real examples and the test set of 112 examples.

Classification of these data is done with the one-layer BCPNN as well as both versions of the multi-layer BCPNN. The first multi-layer network has 303 hidden units organized into 73 partitioning complex columns of up to fourth order. The second multi-layer network has 831 hidden units, and uses 123 overlapping columns, also of up to fourth order. The results are compared to those from another investigation of the same database with a back propagation network. This network had a hidden layer of 30 units, which was found to give the best results on this task in that study [3]. Also, in connection with the telephone exchange computer system is a conventional diagnosis program which manages to classify about half of the examples correctly.

Table 12.1 summarizes the results with the neural networks. The fraction of correct answers among the first, second, and third alternatives from the networks are shown. Compared to the one-layer and multi-layer BCPNNs, back propagation seems a little better when tested on the training data, but for generalization the Bayesian neural networks achieve better results. If the three first alternatives are considered together, the BCPNNs come better off on the training set as well. The second version of the multi-layer BCPNN

Table 12.1: Classification results. The different neural networks are tested with both the training and the test set. The percent of correct answers among the first, second and third alternatives from the networks are shown.

	Tested with training set			
	Back-prop Network	One-layer Bayes	Partitioning Columns	Overlapping Columns
First	91.2%	75.1%	83.5%	86.7%
Second	3.4%	15.4%	10.6%	8.1%
Third	0.5%	4.3%	2.7%	2.3%
Other	5.0%	5.2%	3.2%	2.9%
	Tested with test set			
	Back-prop Network	One-layer Bayes	Partitioning Columns	Overlapping Columns
First	67.0%	66.1%	70.5%	75.0%
Second	8.9%	12.5%	10.7%	5.4%
Third	0.9%	8.0%	5.4%	8.0%
Other	23.2%	13.4%	13.4%	11.6%

(which uses overlapping complex columns) gives a result on the training data close to those of back propagation, and the best results for generalization.

The BCPNNs with overlapping complex columns are better than the BCPNNs with partitioning complex columns on both the training set and the test set. This is because expression (12.4) gives a better approximation to the distribution in the training set than expression (12.3), while at the same time using joint probabilities of lower order. The partitioning columns in this example are mainly of order four, whereas most of the overlapping columns are of order two and only a few of higher orders.

Figure 12.5 shows the causal structure found and compensated for by the two versions of multi-layer BCPNNs, when trained on the telephone exchange computer database. Dependencies are represented by arcs between units. It is clear that in this case the overlapping complex columns compensate for far more dependencies than the partitioning complex columns. In the few cases where the former method does not compensate for a correlation that is found by the latter. This is because this would have created a loop in the dependency graph.

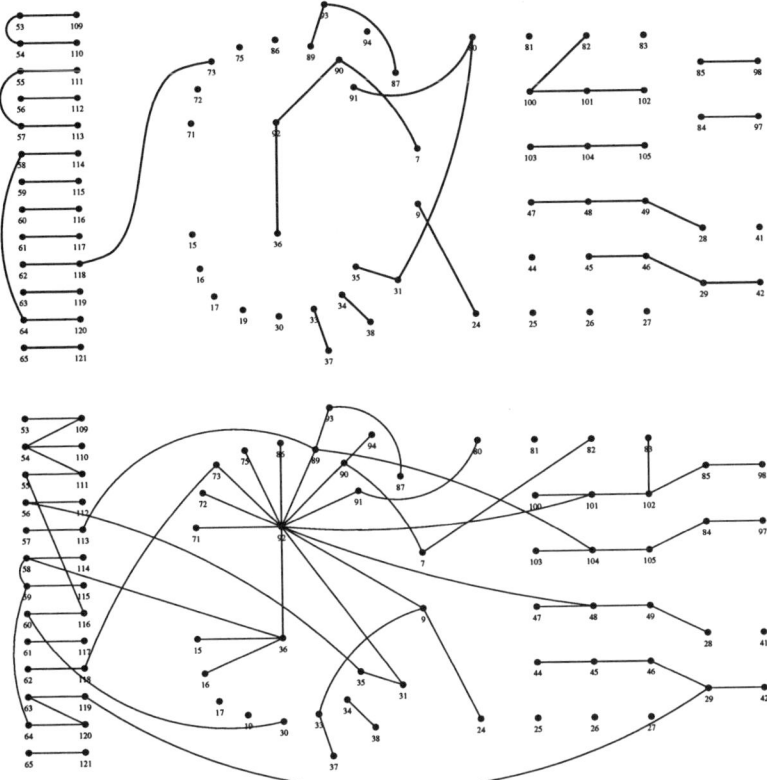

Figure 12.5: The dependencies accounted for by the two versions of multi-layer BCPNNs. The arcs in the first figure connect groups of units that are gathered together in partitioning complex columns in the first multi-layer BCPNN. The arcs in the second figure represent second or higher order correlations between units, and correspond to the overlapping complex columns in the second multi-layer BCPNN.

12.4 Discussion and Conclusions

Complicated technical systems which are hard to analyse is a good example of an area in which neural networks can be useful. The Bayesian confidence propagation neural network is especially suited for use in this kind of domains, since it often involves attributes connected in a causal structure. This study has shown good results when applying BCPNNs to the diagnosis of a telephone exchange computer. The performance is comparable to that of a back-propagation neural network, but the BCPNNs tend to be slightly better at generalization, especially when overlapping complex columns are used. The possibility to depict the found dependency graph may also in itself provide a valuable tool in many domains.

Training of the BCPNN is relatively fast, since this is done mainly in one pass, preceded by a few passes for construction of complex columns.

We have here only treated binary input features. It is, however, also possible to handle continuous attributes by using mixture densities, thus introducing a kind of soft interval coding [7].

The BCPNN model has a large potential for further development. There are, for example, several ways to improve the selection of which primary events to merge into complex columns. But already this network model has proven itself a powerful and general tool for classification and diagnosis.

Acknowledgements

This work was supported by Ellemtel Telecommunication Systems Laboratories (Ellemtel Utvecklings AB) and the Swedish Research Council for Engineering Sciences (TFR) under grant TFR 93-672. We thank M. Boda and H. Brandt for useful comments and discussion.

References

[1] Chow C.K. and Liu C.N (1968) Approximating Discrete Probability Distributions with Dependency Trees. *IEEE Trans. Information Theory* 14: 462–467.

[2] Duda R.O. and Hart P.E. (1973) *Pattern Classification and Scene Analysis*. Wiley, New York.

[3] Gustafsson E. (1991) Investigation of the possibility to use neural networks for fault diagnosis in telecommunications. Master's thesis, Dept. of Telecommunications Theory, Royal Institute of Technology, Stockholm, Sweden.

[4] Holst A. and Lansner A. (1993) A Bayesian Neural Network with Extensions. Tech. Rep. TRITA-NA-P9325, Dept. of Numerical Analysis and Computing Science, Royal Institute of Technology, Stockholm, Sweden.

[5] Kononenko I. (1989) Bayesian Neural Networks. *Biol. Cybernetics* 61: 361–370.

[6] Lansner A. and Ekeberg Ö. (1989) A One-layer Feedback, Artificial Neural Network with a Bayesian Learning Rule. *Int. J. Neural Systems*, 1: 77–88.

[7] Lansner A. and Holst A. (1994) A Higher Order Bayesian Neural Network with Spiking Units. (Submitted).

[8] Lewis II P.M. (1959) Approximating Probability Distributions to Reduce Storage Requirements. *Information and Control* 2: 214–225.

[9] Minsky M.L. and Papert S.A. (1988) *Perceptrons*. MIT Press. Expanded Edition.

[10] Pearl J. (1988) *Probabilistic Reasoning in Intelligent Systems: Networks of Plausible Inference*. Morgan Kaufmann, San Mateo.

[11] Schmidhuber J. (1994) Discovering problem solutions with low Kolmogorov complexity and high generalization capability. Tech. Rep. FKI-194-94, Fakultät für Informatik, Technische Universität München, Germany.

[12] Tråvén H.G.C. (1993) On Pattern Recognition Applications of Artificial Neural Networks. PhD thesis, Dept. of Numerical Analysis and Computing Science, Royal Institute of Technology, Stockholm, Sweden.

[13] Wedelin D. (1993) Efficient Algorithms for Probabilistic Interference, Combinatorial Optimization and the Discovery of Causal Structure from Data. PhD thesis, Dept. of Computer Sciences, Chalmers University of Technology, Göteborg, Sweden.

13

Genetic Algorithms Applied to Bayesian Networks

P. Larrañaga, C.M.H. Kuijpers, R.H. Murga,
Y. Yurramendi, M. Graña, J.A. Lozano, X. Albizuri,
A. D'Anjou, F. J. Torrealdea

13.1 Introduction

Bayesian Networks (BNs) constitute a reasoning method based on probability theory. They model causal relations between events. A BN consists of a set of vertices and a set of arcs which together constitute a Directed Acyclic Graph (DAG). The vertices represent random variables, all of which have, in general, a finite set of states. The arcs indicate the existence of direct causal connections between the linked variables, and the strengths of these connections are expressed in terms of conditional probabilities.

To specify the probability distribution of a BN, $P(x_1, \ldots, x_n)$, one must give prior probabilities for all root nodes (nodes without predecessors) and conditional probabilities for all other nodes, given all possible combinations of their direct predecessors. These numbers, in conjunction with the DAG, specify the BN completely. The joint probability of any particular instantiation of all n variables in a BN can be calculated as follows: $P(x_1, \ldots, x_n) = \prod_{i=1}^{n} P(x_i | \pi_i)$, where x_i represents the instantiation of the variable X_i and π_i represents the instantiation of the parents of X_i. Excellent introductions on BNs can be found in [1, 3].

```
begin GA
    Make initial population at random
    WHILE NOT stop DO
      BEGIN
      Select parents from the population.
      Produce children by the selected parents.
      Mutate the individuals.
      Extend the population by adding the children to it.
      Reduce the extended population.
      END
      Output the best individual found.
end GA
```

Figure 13.1: Pseudo code of a genetic algorithm.

This chapter is structured as follows. In Section 13.2 we introduce the genetic algorithms. We give a review of genetic operators that were used for tackling the Travelling Salesman Problem (TSP) in Section 13.3. Sections 13.4, 13.5 and 13.6 treat, respectively, the problems of the optimal decomposition, of the structure learning and of the fusion. Two of these problems we model like problems that resemble the TSP. Conclusions are given in a final Section 13.7.

13.2 Genetic Algorithms

Evolutionary algorithms [4, 5] are probabilistic search algorithms that simulate natural evolution. Roughly speaking, three different types of evolutionary algorithms exist: genetic algorithms [6, 7], evolutionary programming [8] and evolution strategies [9]. In this chapter we consider the Genetic Algorithms (GAs). GAs are search algorithms based on the mechanics of natural selection and natural genetics. They combine survival of the fittest among string structures with a structured yet randomized information exchange to form a search algorithm that, under certain conditions, evolves to the optimum with probability 1 [10, 11].

In GAs the search space of a problem is represented as a collection of individuals. The individuals are represented by character strings, which are often referred to as chromosomes. The purpose of the use of a GA is to find the individual from the search space with the best "genetic material". The quality of an individual is measured with an objective function. The part of the search space to be examined is called the population.

Roughly, a genetic algorithm works as follows, see Figure 13.1. First, the initial population is chosen, and the quality of each of its individuals is determined. Next, in every iteration parents are selected from the population. These parents produce children, which are added to the population. For all newly created individuals of the resulting population a probability near zero exits that they "mutate", *i.e.* they change their hereditary distinctions. After that, some individuals are removed from the population according to a selection criterion in order to reduce the population to its initial size. One iteration of the algorithm is referred to as a generation. The operators that define the offspring production process and the mutation process are called the *crossover operator* and *mutation operator*, respectively.

13.3 Genetic Operators in Relation to the TSP

The problem of the optimal decomposition as well as the problem of the structure learning can be modelled as problems that resemble the intensively studied Travelling Salesman Problem (TSP). The TSP is, given a collection of cities, to determine the shortest tour that visits each city precisely once and then returns to its starting point. The problems related to BNs, however, are not symmetrical and search for acyclic orderings.

Several representations and operators have been used in tackling the TSP with genetic algorithms, like the *binary* representation [6, 12, 14], the *adjacency* representation [15, 17], the *ordinal* representation [15], the *matricial* representations [18, 20] and the *path* representation. See [21] for a review on representations used in relation to the TSP. We choose to use the *path* representation. Therefore, we represent an ordering of a set of elements $\{x_1, x_2, \ldots, x_n\}$ by a permutation of the natural numbers $\{1, 2, \ldots, n\}$. If in the permutation i appears before j, then in the ordering x_i comes before x_j. Since, in combination with the path representation, the *classical* genetic operators [6] cannot be used, other genetic operators were developed [21].

13.3.1 Crossover Operators

The **partially-mapped crossover operator (PMX)** [22] transmits ordering and value information from the parents strings to the offspring. A portion of one parent string is mapped onto a portion of the other parent string and the remaining information is exchanged. Consider, for example, the following two parents: (1 2 3 4 5 6 7 8) and (3 7 5 1 6 8 2 4). The PMX operator creates an offspring in the following way. It begins by selecting uniformly at random two cut points along the strings, which represent the parents. Suppose, for example, that the first cut point is selected between the third and the fourth string elements, and the second one between the sixth and the seventh string elements. Hence, (1 2 3 | 4 5 6 | 7 8) and (3 7 5 | 1 6 8 | 2 4). The substrings between the cut points are called the mapping sections. In our example they define the mappings $4 \leftrightarrow 1$, $5 \leftrightarrow 6$ and $6 \leftrightarrow 8$. Now the mapping section of the first parent is copied into the second offspring, and the mapping section of the second parent is copied into the first offspring — offspring 1: $(x\,x\,x\,|\,1\,6\,8\,|\,x\,x)$ and offspring 2: $(x\,x\,x\,|\,4\,5\,6\,|\,x\,x)$. Then offspring i ($i = 1, 2$) is filled up by copying the elements of the i-th parent. In case a number is already present in the offspring it is replaced according to the mappings. For example, the first element of offspring 1 would be a 1, like the first element of the first parent. However, there is already a 1 present in offspring 1. Hence, because of the mapping $1 \leftrightarrow 4$ we choose the first element of offspring 1 to be a 4. The second, third and seventh elements of offspring 1 can be taken from the first parent. However, the last element of offspring 1 would be an 8, which is already present. Because of the mappings $8 \leftrightarrow 6$ and $6 \leftrightarrow 5$, it is chosen to be a 5. Hence, offspring 1: $(4\,2\,3\,|\,1\,6\,8\,|\,7\,5)$. Analogously, we find offspring 2: $(3\,7\,8\,|\,4\,5\,6\,|\,2\,1)$. The absolute positions of some elements of both parents are preserved.

The **cycle crossover operator (CX)** [23] attempts to create an offspring from the parents where every position is occupied by a corresponding element from one of the parents. For example, consider again the parents $(1\,2\,3\,4\,5\,6\,7\,8)$ and $(2\,4\,6\,8\,7\,5\,3\,1)$. Now we choose the first element of the offspring equal to either the first element of the first parent string or the first element of the second parent string. Hence, the first element of the offspring has to be a 1 or a 2. Suppose we choose it to be 1, $(1 * * * * * * *)$. Now consider the last element of the offspring. Since this element has to be chosen from one of the parents, it can only be an 8 or a 1. However, if a 1 were selected, the offspring would not represent a legal individual. Therefore, an 8 is chosen, $(1 * * * * * * 8)$. Analogously, we find that the fourth and the second element of the offspring also have to be selected from the first parent, which results in $(1\,2 * 4 * * * 8)$. The positions of the elements chosen up to now are said to be a cycle. Now consider the third element of the offspring. This element we may choose from any of the parents. Suppose that we select it to be from parent 2. This implies that the fifth, sixth and seventh elements of the offspring also have to be chosen from the second parent, as they form another cycle. Thus, we find the following offspring: $(1\,2\,6\,4\,7\,5\,3\,8)$. The absolute positions of on average half the elements of both parents are preserved.

The **order crossover operator (OX1)** [24] constructs an offspring by choosing a substring of one parent and preserving the relative order of the elements of the other parent. For example, consider the following two parent strings: $(1\,2\,3\,4\,5\,6\,7\,8)$ and $(2\,4\,6\,8\,7\,5\,3\,1)$, and suppose that we select a first cut point between the second and the third bit and a second one between the fifth and the sixth bit. Hence, $(1\,2\,|\,3\,4\,5\,|\,6\,7\,8)$ and $(2\,4\,|\,6\,8\,7\,|\,5\,3\,1)$. The offspring are created in the following way. Firstly, the string segments between the cut point are copied into the offspring, which gives $(* * |\,3\,4\,5\,| * * *)$ and $(* * |\,6\,8\,7\,| * * *)$. Next, starting from the second cut point of one parent, the rest of the elements are copied in the order in which they appear in the other parent, also starting from the second cut point and omitting the elements that are already present. When the end of the parent string is reached, we continue from its first position. In our example this gives the following children: (8 7 | 3 4 5 | 1 2 6) and (4 5 | 6 8 7 | 1 2 3).

The **order-based crossover operator (OX2)** [25], which was suggested in connection with schedule problems, is a modification of the OX1 operator. The OX2 operator selects at random several positions in a parent string, and the order of the elements in the selected positions of this parent is imposed on the other parent. For example, consider again the parents (1 2 3 4 5 6 7 8) and (2 4 6 8 7 5 3 1), and suppose that in the second parent the second, third and sixth positions are selected. The elements in these positions are 4, 6 and 5, respectively. In the first parent these elements are present at the fourth, fifth and sixth positions. Now the offspring is equal to parent 1 except in the fourth, fifth and sixth positions: (1 2 3 * * * 7 8). We add the missing elements to the offspring in the same order in which they appear in the second parent. This results in (1 2 3 4 6 5 7 8). Exchanging the role of the first parent and the second parent gives, using the same selected positions, (2 4 3 8 7 5 6 1).

The **position-based crossover operator (POS)** [25], a second modification of the OX1 operator, was also suggested in connection with schedule problems. It also starts with selecting a random set of positions in the parent strings. However, this operator imposes the position of the selected elements on the corresponding elements of the other parent. For example, consider the parents (1 2 3 4 5 6 7 8) and (2 4 6 8 7 5 3 1), and suppose that the second, third and the sixth positions are selected. This leads to the following offspring: (1 4 6 2 3 5 7 8) and (4 2 3 8 7 6 5 1).

The **voting recombination crossover operator (VR)** [26] can be seen as a p-sexual crossover operator, where p is a natural number greater than, or equal to, 2. It starts by defining a threshold, which is a natural number smaller than, or equal to, p. Next, for every $i \in \{1, 2, \ldots, n\}$ the set of i-th elements of all the parents is considered. If in this set an element occurs at least the threshold number of times, it is copied into the offspring. For example, if we consider the parents (p=4) (1 4 3 5 2 6), (1 2 4 3 5 6), (3 2 1 5 4 6), (1 2 3 4 5 6) and we define the threshold to be equal to 3 we find (1 2 x x x 6). The remaining positions of the offspring are filled with mutations. Hence, our example might result in (1 2 4 5 3 6).

The **alternating-position crossover operator (AP)** [27] simply creates an offspring by selecting alternately the next element of the first parent and the next element of the second parent, omitting the elements already present in the offspring. For example, if parent 1 is (1 2 3 4 5 6 7 8) and parent 2 is (3 7 5 1 6 8 2 4), the AP operator gives the following offspring (1 3 2 7 5 4 6 8). Exchanging the parents results in (3 1 7 2 5 4 6 8).

13.3.2 Mutation Operators

The **displacement mutation operator (DM)** (*e.g.* [28]) first selects a substring at random. This substring is removed from the string and inserted in a random place. For example, consider the string (1 2 3 4 5 6 7 8), and suppose that the substring (3 4 5) is selected. Hence, after removal of the substring we have (1 2 6 7 8). Suppose that we randomly select element 7 to be the element after which the substring is inserted. This gives (1 2 6 7 3 4 5 8).

The **exchange mutation operator (EM)** (*e.g.* [29]) randomly selects two elements in the string that represents the individual and exchanges them. For example, consider the string (1 2 3 4 5 6 7 8), and suppose that the third and the fifth element are randomly selected. This results in (1 2 5 4 3 6 7 8).

The **insertion mutation operator (ISM)** (*e.g.* [28]) randomly chooses an element in the string that represents the individual, removes it from this string, and inserts it in a randomly selected place. For example, consider again the string (1 2 3 4 5 6 7 8), and suppose that the insertion mutation operator selects element 4, removes it, and randomly inserts it after element 7. The resulting offspring is (1 2 3 5 6 7 4 8).

The **simple-inversion mutation operator (SIM)** (*e.g.* [6]) selects randomly two cut points in the string that represents the individual, and it reverses the substring between these two cut points. For example, consider the string (1 2 3 4 5 6 7 8), and suppose that the first cut point is chosen between element 2 and element 3, and the second cut point between the fifth and the sixth element. This results in (1 2 5 4 3 6 7 8).

The **inversion mutation operator (IVM)** (*e.g.* [30]) randomly selects a substring, removes it from the string and inserts it, in reversed order, in a randomly selected position. Consider again (1 2 3 4 5 6 7 8), and suppose that the substring (3 4 5) is chosen, and that this substring is inserted immediately after element 7. This gives (1 2 6 7 5 4 3 8).

The **scramble mutation operator (SM)** (*e.g.* [25]) selects a random substring and scrambles the elements in it. For example, consider the string (1 2 3 4 5 6 7 8), and suppose that the substring (4 5 6 7) is chosen. This might result in (1 2 3 5 6 7 4 8).

13.4 Decomposition of Bayesian Networks

13.4.1 Introduction

One of the most famed problems, in the context of the BNs, is related to the propagation of evidence. It consists of, once the value of some variables are known, the assignment of probabilities to the values of the rest of the variables. This problem, and even the more general problem of finding approximate solutions, belong to the class of NP-hard problems [31, 32].

Fundamentally, the problem of the propagation of evidence has been tackled in two ways: with *exact (or deterministic) algorithms* [1, 33, 35] and with *approximate algorithms* [36, 44].

Our interest in obtaining an optimal decomposition originates in the evidence propagation algorithm proposed in [34]. As can be read in [3, p. 104]

> ... the only problematic step in the process from DAG to junction tree is the triangulation. Since any elimination sequence will produce a triangulation it may not seem as a problem, but for the propagation algorithm it is. In probability propagation the cliques in the junction graph shall have joint probability tables attached to them. The size of the table is the product of the number of states of the variables. So, the size increases exponentially with the size of the clique. A good triangulation, therefore, is a triangulation yielding small cliques; or to be more precise, yielding small probability tables.

The basic technique used to triangulate a moral graph G is through successive elimination of the vertices of G, where the elimination of a vertex v consists of adding edges to the graph in such a way that all vertices adjacent to v become pairwise adjacent, and subsequently deleting v and its adjacent edges. In this way the graph triangulation problem is equivalent to the search for an optimal vertex elimination sequence. We define a vertex elimination sequence $\#$ to be optimal if minimizes $w(G_\#^t) = \log_2 \sum_C 2^{w(C)}$, where $w(G_\#^t)$ is the weight of the triangulated graph $G_\#^t$ obtained by eliminating the vertices of G in the order defined by $\#$, and where $w(C)$ is the weight of clique C defined as $w(C) = \sum_{i=1}^{k} w(v_i)$, where $w(v_i) = \log_2 n_i$ is the weight of vertex v_i as a function of the number of its states n_i.

The search for an optimal triangulation is NP-hard [45].

13.4.2 Description of the Experiments

For the experiments we use two artificial graphs which were introduced in [46]: the dense graph *Dense* and the sparse graph *Sparse*. Both these graphs were obtained by simulation and contain 50 vertices, all of which have a number of states chosen at random between 2 and 5.

We use a genetic algorithm that is based on the principles of GENITOR [47]: in every generation two orderings are selected for crossover, where the probability of an ordering to be selected depends upon the rank of its objective function value. The new created offspring substitutes, in case it is better, the worst ordering in the population.

13.4.3 Results and Conclusions

In Table 13.1 the results of [46] and [51] are presented, while in Tables 13.2 and 13.3 the best results obtained with the combinations of the crossover and

Table 13.1: Optimal decomposition. Results presented in [46, 51].

	Sparse	Dense
Algorithm	w(G)	w(G)
Random elimination	32.86	62.84
Max. cardinality search [48]	27.24	59.14
Lexicographic search [49]	26.41	54.39
Ext. random elimination [50]	24.72	53.72
FMINT (random)	25.01	52.88
FMINT (max. card.)	24.62	56.42
Min. size heuristic	23.05	55.01
Min. fill heuristic	24.82	54.40
Min. weight heuristic	24.82	53.48
Simulated annealing	22.61	50.88

Table 13.2: Optimal decomposition. Best results (Graph Sparse).

	PMX	CX	OX1	OX2	POS	VR	AP
DM	22.66	22.61	22.64	22.61	22.61	22.73	22.66
EM	22.61	22.61	22.66	22.61	22.61	22.79	22.65
ISM	22.61	22.61	22.63	22.61	22.61	22.72	22.66
SIM	22.66	22.61	22.78	22.61	22.61	22.72	23.20
IVM	22.61	22.61	22.61	22.61	22.61	22.77	22.63
SM	22.67	22.61	22.72	22.61	22.61	22.97	23.23

mutation operators that we have introduced in Section 13.3 are given. More detailed results can be found in [27].

We see that the results obtained with our genetic algorithm improve, for most combinations of operators, the results of the other known triangulation methods. Only the results obtained with simulated annealing are comparable to our results.

In combination with [27], it can be observed that especially the results found with the CX operator are very good. This operator leads to a very robust algorithm that, almost independently of the values of the other parameters, manages to find the best result obtained with simulated annealing. Other operators (*e.g.* PMX, OX2, POS) give faster algorithms than the CX operator, but in combination with small population sizes these algorithms in general converge too fast. The results obtained with the distinct mutation operators are not very different, although statistically significant differences

Table 13.3: Optimal decomposition. Best results (Graph Dense).

	PMX	CX	OX1	OX2	POS	VR	AP
DM	50.88	50.88	50.92	50.88	50.88	51.50	50.88
EM	50.88	50.88	50.90	50.88	50.88	51.07	51.00
ISM	50.88	50.88	50.89	50.88	50.88	51.19	50.95
SIM	50.88	50.88	51.05	50.88	50.88	51.47	51.31
IVM	50.88	50.88	51.01	50.88	50.88	51.13	50.91
SM	50.91	50.88	52.01	50.88	50.88	51.04	51.63

exist.

13.5 Structure Learning

13.5.1 Introduction and Related Work

The construction of a BN exclusively from the information provided by an expert is time-consuming and subject to mistakes. Therefore, and due to the fact that large databases become more accessible, algorithms for automatic learning, that occasionally use the information provided by an expert, can be of great help.

With respect to the type of induced network, we can divide the research on structure learning in *trees and polytrees* [52, 55], *multiple connected networks, assuming an ordering between the nodes* [56, 62] and *multiple connected networks, without an ordering between the nodes* [63, 68]. Others authors have worked on the induction of other graphical models, for instance [69, 74].

13.5.2 Searching in the Space of Network Structures

13.5.2.1 Notation and Representation

Denoting with D the set of BN structures for a fixed domain with n variables, and the alphabet S being $\{0,1\}$, a Bayesian network structure can be represented by an $n \times n$ connectivity matrix C, where its elements, c_{ij}, verify:

$$c_{ij} = \begin{cases} 1 & \text{if } j \text{ is a parent of } i, \\ 0 & \text{otherwise.} \end{cases}$$

13.5.2.2 Assuming an Ordering between the Nodes

If an ordering between the variables of the BN is assumed, which means that a node x_i can only have node x_j as a parent node if in the ordering

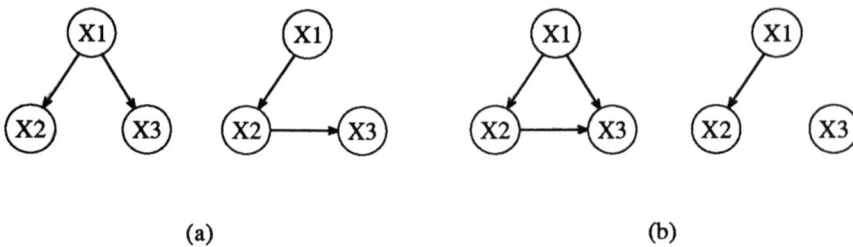

(a) (b)

Figure 13.2: With order assumption: Crossing over two BN structures.

(a) (b)

Figure 13.3: With order assumption: Mutating a BN structure.

node x_j comes before node x_i, then it is easy to verify that the connectivity matrices of the network structures are triangulated and that, therefore, the genetic operators are closed operators with respect to the DAG conditions. We represent an individual of the population by the string:

$$c_{21}c_{31}c_{41}\ldots c_{n1},\ldots c_{32}c_{42}\ldots c_{n2},\ldots c_{n-1,n-2}, c_{n,n-2}, c_{n,n-1}.$$

With this representation in mind, we show how the crossover and mutation operators work, by using simple examples.

Example 1. Consider a domain of three variables on which the two BN structures of Figure 13.2(a) are defined. Using the above described representation, the networks are represented by the strings 110 and 101. Suppose now that the two network structures are crossed over and that the crossover point is chosen between the second and the third bit. This gives the offspring strings 111 and 100. Hence, the offspring structures created are the ones presented in Figure 13.2(b).

Example 2. Consider the DAG of Figure 13.3(a). It is represented by the

GENETIC ALGORITHMS APPLIED TO BAYESIAN NETWORKS

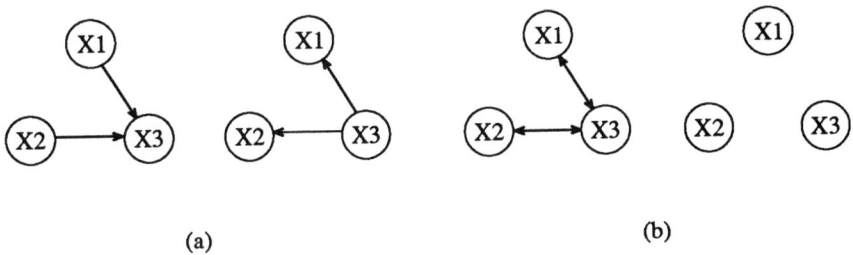

Figure 13.4: Without order assumption: The crossover operator is not a closed operator.

string 100. Suppose that the third bit is altered by mutation. This gives the string 101, which corresponds with the graph of Figure 13.3(b).

13.5.2.3 Without Assuming an Ordering between the Nodes

If no ordering assumption on the variables is made, we represent an individual of the population by the string:

$$c_{11}c_{21}\ldots c_{n1}c_{12}c_{22}\ldots c_{n2}\ldots c_{1n}c_{2n}\ldots c_{nn}.$$

As can be seen in the following examples, the genetic operators are no closed operators with respect to the DAG conditions.

Example 1. Consider a domain of three variables on which the two BN structures of Figure 13.4(a) are defined. Using the above described representation, the networks are represented by the strings 001001000 and 000000110. Suppose now that the two network structures are crossed over and that the crossover point is chosen between the sixth and the seventh bit. This gives the offspring strings 001001110 and 000000000. Hence, the offspring structures created are the ones presented in Figure 13.4(b). We see that the first offspring structure is not a DAG.

Example 2. Consider the DAG of Figure 13.5(a). It is represented by the string 010001000. Suppose that the seventh bit is altered by mutation. This gives the string 010001100, which corresponds with the cyclic graph of Figure 13.5(b).

To ensure the closeness of the genetic operators, we introduce a *repair operator*, which transforms the child structures that do not verify the DAG conditions into DAGs, by randomly eliminating the edges that invalidate the DAG conditions.

Figure 13.5: Without order assumption: The mutation operator is not a closed operator.

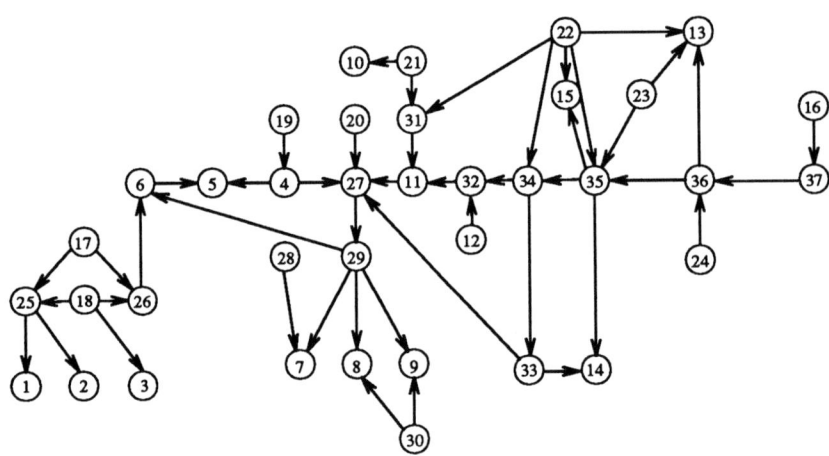

Figure 13.6: The ALARM network structure.

13.5.2.4 Experiments

For the experiments we use a simulation, consisting of the 3000 first cases of the database obtained by Herskovits [75], of the ALARM network (see Figure 13.6), which was designed by Beinlinch et al. [76] as a prototype for modelling potential anesthesia problems in the operating room. The ALARM network and its simulation have become a benchmark for evaluating the performance of newly proposed algorithms. In case we assume an ordering between the variables the cardinality of the search space is given by the formula $2^{\binom{n}{2}}$, where n is the number of variables; otherwise, the cardinality of the

Table 13.4: Structure learning. Best results.

	$-\log P(D|B_S)$
$B_S=$ ALARM	1.4412e04
With order	1.4404e04
Without order	1.4417e04

search space is [77]:

$$f(n) = \sum_{i=1}^{n}(-1)^{i+1}\binom{n}{i}2^{i(n-i)}f(n-i), f(0)=1, f(1)=1.$$

In our experiments $n = 37$, and the cardinality of the search spaces are, respectively, 3.061e200 and 3.008e237.

The characteristics of the genetic algorithm to be used are the following: the initial population of λ individuals is generated at random, subject to the restriction that a node never has more than four parent nodes. The *objective function* to be used to evaluate the quality of a structure is based on the formula of [58].

Each individual is selected to be a parent with a probability proportional to the rank of its objective function. The purpose of this transformation is to avoid the *premature convergence* of the algorithm caused by *superindividuals*.

After applying crossover and mutation, the created structures do not necessarily fulfil the restriction that the nodes all have at most four parents. To maintain this restriction, we hybridize the genetic algorithm with a *local optimizer*. This optimizer selects, for every node in a child structure, the best subset of at most four elements from the set of its parents nodes. The process of generating child structures and the application of the local optimizer is repeated in every iteration of the algorithm. The reduction criteria is elitist of degree λ, which means that the λ best structures among the parent and children structures of the current generation will be in the next generation. The best evaluations obtained with and without assuming an ordering between the variables are presented in Table 13.4. More results can be found in [67].

13.5.3 Searching for the Best Ordering

In this approach we search for the best ordering between the variables using the genetic operators described in Section 13.3. The individuals of the population are orderings whose fitness is computed by applying the formula of [58] to the structure that is induced by applying the K2 algorithm to it. Now, the cardinality of the search space is $n!$.

Table 13.5: Structure learning. Best evaluations obtained with the different combinations of genetic operators. Population size 50.

	AP	CX	OX1	OX2
DM	1.4422e04	1.4422e04	1.4422e04	1.4422e04
EM	1.4422e04	1.4423e04	1.4424e04	1.4423e04
ISM	1.4422e04	1.4423e04	1.4417e04	1.4423e04
IVM	1.4423e04	1.4422e04	1.4417e04	1.4423e04
SIM	1.4426e04	1.4423e04	1.4427e04	1.4423e04
SM	1.4442e04	1.4417e04	1.4430e04	1.4423e04

	PMX	POS	VR
DM	1.4423e04	1.4423e04	1.4424e04
EM	1.4423e04	1.4423e04	1.4424e04
ISM	1.4423e04	1.4423e04	1.4424e04
IVM	1.4417e04	1.4423e04	1.4424e04
SIM	1.4424e04	1.4423e04	1.4433e04
SM	1.4427e04	1.4423e04	1.4432e04

We use again a genetic algorithm based on the principles of GENITOR. Hence, in every generation one new ordering is created that substitutes, in case it is better, the worst ordering in the population.

The other parameters of the algorithm are equal to the ones of the algorithms in the previous section.

In Table 13.5 we present the results of the experiments carried out with the simulation of the ALARM network used before, in terms of the $-\log P(D|B_s)$.

We note that the genetic algorithm of Section 13.5.2.2 could be applied to the best orderings found with the method described above, in order to try to find even better structures than those obtained with K2.

13.5.4 Remarks and Conclusions

We see that the different methods presented constitute powerful tools for the automatic construction of BNs.

With assuming an ordering between the variables we have obtained a network with an even better evaluation than the ALARM network. This means that we have been able to construct a network that even better reflects the relations in a database than the network that was used for generating this database.

Without assuming an ordering and using a local optimizer we have obtained a network with an evaluation that is only slightly worse than the evaluation of the ALARM network.

Searching for the best ordering between the variables we have not found better evaluations than the one of the ALARM network. However, for all combinations of genetic operators the results were quite good. In [78] can be seen that again the use of the CX operator leads to a very robust algorithm. Even with a population size of 10 the CX operator gives results that are similar to those presented in Table 13.5. The other crossover operators give in combination with small population sizes in general algorithms that suffer from premature convergence.

13.6 Fusion

13.6.1 Introduction

Most research in BNs is focused upon *single-author* models. An interesting research topic, however, is the question of how to combine BNs that represent the knowledge of different authors into one unique BN [79, 82]. The reason why such fusions of BNs are desirable can be various. The BNs to be combined can be the result of a distributed design, or they can be designed by different authors who do not know about each other's work. It is also possible that local BNs have to be combined into a global network.

13.6.2 Proposed Approach

We consider the problem of the fusion of BNs as a structure learning problem. For tackling this search problem we propose the procedure that consists of the following steps:

1. Simulation of the BNs coming from different authors. The size of the simulations to be carried out depend upon the confidence that is given to the authors of the networks.

2. The concatenation of the databases that contain the results of the simulations.

3. The search for a BN structure that best fits the resulting database.

4. The estimation of the conditional probabilities that belong to the network structure obtained by calculating their maximum likelihood estimators from the database which was the result of the concatenation.

13.6.3 An Example

We consider two BNs developed by two different authors that model the same domain variables (see Figure 13.7). Both networks are simulated by means

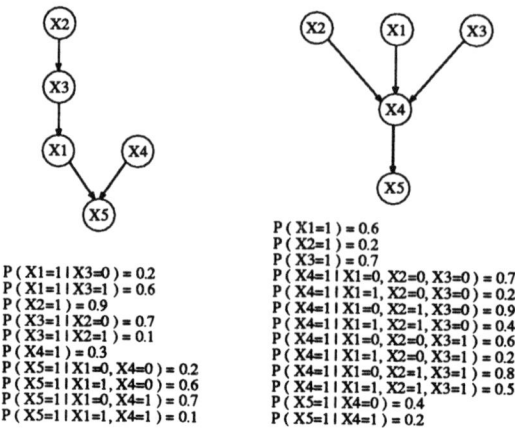

P(X1=1 | X3=0) = 0.2
P(X1=1 | X3=1) = 0.6
P(X2=1) = 0.9
P(X3=1 | X2=0) = 0.7
P(X3=1 | X2=1) = 0.1
P(X4=1) = 0.3
P(X5=1 | X1=0, X4=0) = 0.2
P(X5=1 | X1=1, X4=0) = 0.6
P(X5=1 | X1=0, X4=1) = 0.7
P(X5=1 | X1=1, X4=1) = 0.1

P(X1=1) = 0.6
P(X2=1) = 0.2
P(X3=1) = 0.7
P(X4=1 | X1=0, X2=0, X3=0) = 0.7
P(X4=1 | X1=1, X2=0, X3=0) = 0.2
P(X4=1 | X1=0, X2=1, X3=0) = 0.9
P(X4=1 | X1=1, X2=1, X3=0) = 0.4
P(X4=1 | X1=0, X2=0, X3=1) = 0.6
P(X4=1 | X1=1, X2=0, X3=1) = 0.2
P(X4=1 | X1=0, X2=1, X3=1) = 0.8
P(X4=1 | X1=1, X2=1, X3=1) = 0.5
P(X5=1 | X4=0) = 0.4
P(X5=1 | X4=1) = 0.2

Figure 13.7: BNs coming from two different authors.

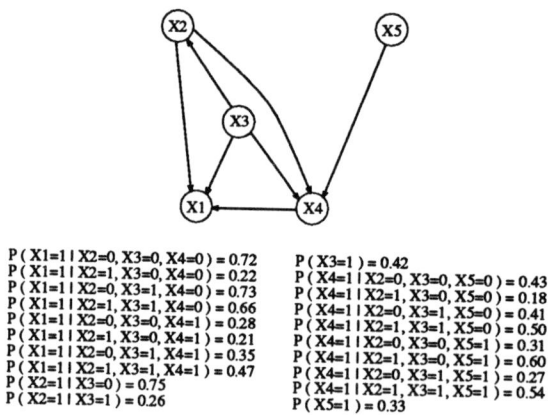

P(X1=1 | X2=0, X3=0, X4=0) = 0.72
P(X1=1 | X2=1, X3=0, X4=0) = 0.22
P(X1=1 | X2=0, X3=1, X4=0) = 0.73
P(X1=1 | X2=1, X3=1, X4=0) = 0.66
P(X1=1 | X2=0, X3=0, X4=1) = 0.28
P(X1=1 | X2=1, X3=0, X4=1) = 0.21
P(X1=1 | X2=0, X3=1, X4=1) = 0.35
P(X1=1 | X2=1, X3=1, X4=1) = 0.47
P(X2=1 | X3=0) = 0.75
P(X2=1 | X3=1) = 0.26

P(X3=1) = 0.42
P(X4=1 | X2=0, X3=0, X5=0) = 0.43
P(X4=1 | X2=1, X3=0, X5=0) = 0.18
P(X4=1 | X2=0, X3=1, X5=0) = 0.41
P(X4=1 | X2=1, X3=1, X5=0) = 0.50
P(X4=1 | X2=0, X3=0, X5=1) = 0.31
P(X4=1 | X2=1, X3=0, X5=1) = 0.60
P(X4=1 | X2=0, X3=1, X5=1) = 0.27
P(X4=1 | X2=1, X3=1, X5=1) = 0.54
P(X5=1) = 0.33

Figure 13.8: BN obtained after fusion.

of *probabilistic logic sampling* [39]. We assume that an equal amount of confidence is given to both authors. Therefore, the simulations of both networks have the same size: we choose the simulation sizes to be 2000 cases each.

Using simulations, we obtain a 4000 cases database which is exposed to structure learning. This structure learning is carried out in the ways described in Sections 13.5.2.2 and 13.5.2.3. In our example, both approaches result in the same network structure (see Figure 13.8).

13.6.4 Remarks and Conclusions

We have converted the problem of the fusion of Bayesian networks into a structure learning problem. We justify this conversion with the good results of Section 13.5. Since the databases resulting from the simulations of the BNs to fuse are supposed to be good representations of the existing relations between the system variables, and since the results of Section 13.5 indicate that our algorithms uncover well the underlying structure in a database of cases, we expect that our fusion method results in a Bayesian network that is a meaningful representation of the knowledge expressed in the initial networks before fusion.

13.7 Conclusions

We have shown how genetic algorithms can be applied to three combinatorial optimization problems that are related to Bayesian networks: the optimal decomposition of the moral graph that belongs to the network, the structure learning from a database of cases and the fusion of Bayesian networks developed by multiple authors. The second problem we have tackled first assuming an ordering between the system variables, after which we have tried to release this assumption.

The results obtained are very satisfactory. Moreover, the genetic algorithms developed turn out to have robust behaviours.

In case a genetic algorithm is used for tackling a problem in which an optimal acyclical permutation has to be searched for, the CX operator seems to be a good crossover operator to choose.

Acknowledgements

We thank Gregory F. Cooper for providing his simulation of the ALARM network, and Uffe Kjærulff for the data of the graphs Sparse and Dense.

This work was supported by the *Diputación Foral de Gipuzkoa*, under grant OF 94/0762, by the *Fondo de Investigación Sanitaria, Ministerio de Sanidad y Consumo*, under grant 94/1370, and by the *Gobierno Vasco*, under grant PI94/78.

References

[1] Pearl J. (1988) *Probabilistic Reasoning in Intelligent Systems*, Morgan Kaufmann, Palo Alto, CA.

[2] Neapolitan R.E. (1990) *Probabilistic Reasoning in Expert Systems. Theory and Algorithms*. John Wiley & Sons.

[3] Jensen F.V. (1994) Introduction to Bayesian networks, Technical Report R 94-2008, Department of Mathematics and Computer Science, University of Aalborg, Denmark.

[4] Bremermann H.J., Rogson M. and Salaff S. (1965) Search by evolution. In *Biophysics and Cybernetic Systems*, Maxfield M., Callahan A. and Fogel L.J. (eds) 157–167, Spartan Books, Washington.

[5] Rechenberg I. (1973) *Optimierung technischer Systeme nach Prinzipien der biologischen Information*, Frommann Verlag, Stuttgart (in German).

[6] Holland J.H. (1975) *Adaptation in Natural and Artificial Systems*. University of Michigan Press, Ann Arbor.

[7] Davis L. (1991) *Handbook of Genetic Algorithms*. Van Nostrand Reinhold, New York.

[8] Fogel L.J. (1962) Atonomous automata. *Ind. Res.* 4: 14–19.

[9] Schwefel H.-P. (1977) *Numerische Optimierung von Computer-Modellen mittels der Evolutionsstrategie*. Birkhäuser, Basel (in German).

[10] Eiben A. E., Aarts E.H.L. and Van Hee K.M. (1990) Global convergence of genetic algorithms: An infinite Markov chain analysis. *Computing Science Notes*. Eindhoven University of Technology, The Netherlands.

[11] Rudolph G. (1994) Convergence analysis of canonical genetic algorithms. *IEEE Trans. on Neural Networks* (5) 1: 96–101.

[12] Lidd M.L. (1991) The traveling salesman problem domain application of a fundamentally new approach to utilizing genetic algorithms. Technical Report, MITRE Corporation.

[13] Whitley D., Starkweather T. and Fuquay D. (1989) Scheduling problems and travelling salesman: The genetic edge recombination operator. *Proc. International Conference on Genetic Algorithms* 3: 133–140.

[14] Whitley D., Starkweather T. and Shaner D. (1991) The traveling salesman and sequence scheduling: Quality solutions using genetic edge recombination. In *Handbook of Genetic Algorithms*, Davis L. (ed) 350–372, Van Nostrand Reinhold, New York.

[15] Grefenstette J., Gopal J.R., Rosmaita B. and Van Gucht D. (1985) Genetic algorithms for the traveling salesman problem. *Proc. International Conference on Genetic Algorithms and their Applications* 160–165.

[16] Jog P., Suh J.Y. and Van Gucht D. (1989) The effects of population size, heuristic crossover and local improvement on a genetic algorithm for the traveling salesman problem. *Proc. International Conference on Genetic Algorithms* 3: 110–115.

[17] Suh J.Y. and Van Gucht D. (1987) Incorporating heuristic information into genetic search, *Proc. International Conference on Genetic Algorithms and Their Applications* 2: 100–107.

[18] Fox B.R. and McMahon M.B. (1991) Genetic operators for sequencing problems. *Workshop on the Foundations of Genetic Algorithms and Classifier Systems* 1: 284–300.

[19] Homaifar A. and Guan S. (1991) A new approach on the traveling salesman problem by genetic algorithms. Technical Report, North Carolina A & T State University.

[20] Seniw D. (1991) A genetic algorithm for the traveling salesman problem. M.Sc. Thesis, University of North Carolina at Charlotte.

[21] Larrañaga P., Kuijpers C.M.H. and Murga R.H. (1994) Evolutionary algorithms for the travelling salesman problem: A review of representations and operators. *Artificial Intelligence Review* (submitted).

[22] Goldberg D.E. and Lingle Jr.R. (1985) Alleles, loci and the traveling salesman problem. *Proc. International Conference on Genetic Algorithms and Their Applications* 154–159.

[23] Oliver I.M., Smith D.J. and Holland J.R.C. (1987) A study of permutation crossover operators on the TSP. *Proc. International Conference on Genetic Algorithms and Their Applications* 2: 224–230.

[24] Davis L. (1985). Applying adaptive algorithms to epistatic domains. *Proc. International Joint Conference on Artificial Intelligence* 162–164.

[25] Syswerda G. (1991) Schedule optimization using genetic algorithms. In *Handbook of Genetic Algorithms*, Davis L. (ed) 332–349, Van Nostrand Reinhold, New York.

[26] Mühlenbein H. (1989) Parallel genetic algorithms, population genetics and combinatorial optimization. *Proc. International Conference on Genetic Algorithms* 3: 416–421.

[27] Larrañaga P., Kuijpers C.M.H., Poza M. and Murga R.H. (1994) Optimal decomposition of Bayesian networks by genetic algorithms. Internal Report EHU-KZAA-IKT-3-94, Dept. of Computer Science and Artificial Intelligence, University of the Basque Country, Spain.

[28] Michalewicz Z. (1992) *Genetic Algorithms + Data Structures = Evolution Programs*. Springer-Verlag, Berlin.

[29] Banzhaf W. (1990) The "molecular" traveling salesman. *Biological Cybernetics* 64: 7–14.

[30] Fogel D.B. (1990) A parallel processing approach to a multiple traveling salesman problem using evolutionary programming. *Annual Parallel Processing Symposium* 4: 318–326.

[31] Cooper G.F. (1990) The computational complexity of probabilistic inference using Bayesian belief networks. *Artificial Intelligence* (42) 2-3: 393–405.

[32] Dagum P. and Luby M. (1993) Approximating probabilistic inference in Bayesian belief networks is NP-hard. *Artificial Intelligence* (60) 1: 141–153.

[33] Shachter R.D. (1988) Probabilistic inference and influence diagrams. *Operations Research* (36) 4: 589–604.

[34] Lauritzen S.L. and Spiegelhalter D.J. (1988) Local computations with probabilities on graphical structures and their application on expert systems. *Journal of the Royal Statistical Society, Series B* (50) 2: 157–224.

[35] Jensen F. (1994) Implementation aspects of various propagation algorithms in Hugin. Technical Report R 94-2014, Department of Mathematics and Computer Science, University of Aalborg, Denmark.

[36] Chavez R.M. and Cooper G.F. (1990) A randomized approximation algorithm for probabilistic inference on Bayesian belief networks. *Networks* 20: 661–85.

[37] Dagum P. and Horvitz E. (1993) A Bayesian analysis of simulation algorithms for inference in Belief networks. *Networks* 23.

[38] Fung R.M. and Chang K.C. (1990) Weighing and integrating evidence for stochastic simulation in Bayesian networks. *Uncertainty in Artificial Intelligence* 5: 209–220.

[39] Henrion M. (1990) Propagating uncertainty in Bayesian networks by probabilistic logic sampling. *Uncertainty in Artificial Intelligence* 2: 149–163.

[40] Hryceij T. (1990) Gibbs Sampling in Bayesian Networks. *Artificial Intelligence* 46: 351–363.

[41] Jensen C.S., Kong A. and Kjærulff U. (1993) Blocking Gibbs sampling in very large probabilistic expert systems. Technical Report R 93-2031, Department of Mathematics and Computer Science, University of Aalborg, Denmark.

[42] Pearl J. (1987) Evidential reasoning using stochastic simulation of causal models. *Artificial Intelligence* (32) 2: 245–257.

[43] Shachter R.D. and Peot M.A. (1990) Simulation approaches to general probabilistic inference on belief networks. *Uncertainty in Artificial Intelligence* 5: 221–234.

[44] Shwe M. and Cooper G. (1991) An empirical analysis of likelihood-weighting simulation on a large, multiply connected medical belief network. *Computer and Biomedical Research* 24: 453–475.

[45] Wen W.X. (1991) Optimal decomposition of belief networks. *Uncertainty in Artificial Intelligence* 6: 209–224.

[46] Kjærulff U. (1990) Triangulation of graphs – Algorithms giving small total state space, Technical Report R 90-09. Department of Mathematics and Computer Science, University of Aalborg, Denmark.

[47] Whitley D. (1989) The GENITOR algorithm and selection pressure: Why rank-based allocation of reproductive trials is best. *Proc. International Conference on Genetic Algorithms* 3: 116–121.

[48] Tarjan R.E. and Yannakakis M. (1984) Simple linear-time algorithms to test chordality of graphs, test acyclicity of hypergraphs, and selectively reduce acyclic hypergraphs. *SIAM Journal on Computing* 13: 566–579.

[49] Rose D.J., Tarjan R.E. and Lueker G.S. (1976) Algorithmic aspects of vertex elimination on graphs. *SIAM Journal on Computing* (5) 2: 266–283.

[50] Fujisawa T. and Orino H. (1974) An efficient algorithm of finding a minimal triangulation of a graph. *IEEE International Symposium on Circuits and Systems* 172-5.

[51] Kjærulff U. (1992) Optimal decomposition of probabilistic networks by simulated annealing. *Statistics and Computing* 2: 7–17.

[52] Chow C.K. and Liu C.N. (1968) Approximating discrete probability distributions with dependence trees. *IEEE Trans. on Information Theory* 14: 62–67.

[53] Suzuki J. (1993). A construction of Bayesian networks from databases based on an MDL principle. *Proc. Uncertainty in Artificial Intelligence* 9: 266–73.

[54] Rebane G. and Pearl J. (1989) The recovery of causal polytrees from statistical data. *Uncertainty in Artificial Intelligence* 3: 175–182.

[55] Acid S., De Campos L.M., Gonzalez A., Molina R. and Perez de la Blanca N. (1991) Learning with CASTLE. In *Symbolic and Quantitative Approaches to Uncertainty, Lectures Notes in Computer Science 548,* Kruse R. and Siegel P. (eds), Springer-Verlag.

[56] Srinivas S., Russell S. and Agogino A. (1990) Automated construction of sparse Bayesian networks from unstructured probabilistic models and domain information. *Uncertainty in Artificial Intelligence* 5: 295–308.

[57] Herskovits E. and Cooper G. (1990) KUTATÓ: An entropy-driven system for construction of probabilistic expert systems from databases. Report KSL-90-22, Knowledge Systems Laboratory. Medical Computer Science, Stanford University.

[58] Cooper G.F. and Herskovits E.A. (1992) A Bayesian method for the induction of probabilistic networks from data. *Machine Learning* 9: 309–347.

[59] Aliferis C.F. and Cooper G.F. (1994) An evaluation of an algorithm for inductive learning of Bayesian belief networks using simulated data sets. *Uncertainty in Artificial Intelligence* 10: 8–14.

[60] Chickering D.M., Geiger D. and Heckerman D. (1995) Learning Bayesian networks: Search methods and experimental results. *International Workshop on Artificial Intelligence and Statistics* 5: 112–128.

[61] Bouckaert R.R. (1994) Properties of Bayesian belief networks learning algorithms. *Uncertainty in Artificial Intelligence* 10: 102–109.

[62] Larrañaga P., Murga R.H., Poza M. and Kuijpers C.H.M. (1995) Structure learning of Bayesian networks by hybrid Genetic Algorithms. *Proc. International Workshop on Artificial Intelligence and Statistics* 5: 310–316.

[63] Bouckaert R.R. (1992) Optimizing causal orderings for generating DAGs from data. *Proc. Uncertainty in Artificial Intelligence* 8: 9–16.

[64] Singh M. and Valtorta M. (1993) An algorithm for the construction of Bayesian network structures from data. *Proc. Uncertainty in Artificial Intelligence* 9: 259–265.

[65] Lam W. and Bacchus F. (1994) Learning Bayesian belief networks. An approach based on the MDL principle. *Computational Intelligence* (10) 4:.

[66] Lam W. and Bacchus F. (1993) Using causal information and local measures to learn Bayesian Networks. In *Proc. Uncertainty in Artificial Intelligence* 9: 243–250.

[67] Larrañaga P., Poza M., Yurramendi Y., Murga R.H. and Kuijpers C.M.H. (1995) Structure learning of Bayesian networks by genetic algorithms: A performance analysis of control parameters. *IEEE Trans. on Pattern Analysis and Machine Intelligence* (submitted).

[68] Provan G.M. and Singh M. (1995) Learning Bayesian networks using feature selection. *Proc. International Workshop on Artificial Intelligence and Statistics* 5: 450–456.

[69] Andersen L.R., Krebs J.H. and Andersen J.D. (1991) STENO: an expert system for medical diagnosis based on graphical models and model search. *Journal of Applied Statistics* (18) 1: 139–153.

[70] Fung R.M. and Crawford S.L. (1990) CONSTRUCTOR: A system for the induction of probabilistic models. *Proc. of AAAI* 762–769.

[71] Lauritzen S.L., Thiesson B. and Spiegelhalter D.J. (1993) Diagnostic systems created by model selection methods-A case study. *International Workshop on Artificial Intelligence and Statistics* 4: 93–105.

[72] Madigan D., Raftery A.E., York J.C., Bradshaw J.M. and Almond R.G. (1993) Strategies for graphical model selection. *International Workshop on Artificial Intelligence and Statistics* 4: 331–336.

[73] Mechling R. and Valtorta M. (1993) PaCCIN: A parallel constructor of Markov networks. *International Workshop on Artificial Intelligence and Statistics* 4: 405–410.

[74] Provan G.M. (1995) Model selection for diagnosis and treatment using temporal influence diagrams. *International Workshop on Artificial Intelligence and Statistics* 4: 469–480.

[75] Herskovits E.H. (1991) Computer based probabilistic-network construction. Doctoral dissertation, Medical Information Sciences, Stanford University.

[76] Beinlinch I.A., Suermondt H.J., Chavez R.M. and Cooper G.F. (1989) The ALARM monitoring system: A case study with two probabilistic inference techniques for belief networks. *Proc. European Conference on Artificial Intelligence in Medicine* 2: 247–256.

[77] Robinson R.W. (1977) Counting unlabeled acyclic digraphs in *Lectures notes in mathematics 622: Combinatorial mathematics V*, Little C.H.C. (ed), Springer-Verlag, New York.

[78] Larrañaga P., Kuijpers C.M.H., Murga R.H. and Yurramendi Y. Learning Bayesian network structures by searching for the best ordering by genetic algorithms. *IEEE Trans. on Systems, Man and Cybernetics* (to appear).

[79] Matzkevich I. and Abramson B. (1992) The topological fusion of Bayes Nets. *Proc. Conference on Uncertainty in Artificial Intelligence* 8: 191–198.

[80] Ng K.-C. and Abramson B. (1992) Consensus diagnosis: A simulation study. *IEEE Trans. on Systems* Man and Cybernetics, (22) 5: 916–928.

[81] Ng K.-C. and Abramson B. (1994) Probabilistic multi-knowledge-base systems. *Journal of Applied Intelligence* 4: 219–236.

[82] Ng K.-C. (1990) Probabilistic multi-knowledge-base systems: Automated group decision making in expert systems. Ph.D. Thesis, Computer Science Department, University of Southern California, Los Angeles, CA.

Part IV

Decision-Making, Optimization and Classification

14

Rationality, Conditional Independence and Statistical Models of Competition

J.Q. Smith, C.T.J. Allard

14.1 Introduction

A good way for a statistician to begin the model building process is to try faithfully and formally to encapsulate an expert's verbal description of his problem. Now, at least initially, it is unusual for an expert's coding of his problem to be probabilistic. More likely he will make qualitative statements, perhaps about certain causal relationships, about what information is independent of other information or how various variables might be related to one another. He may also be able to provide strong information about when various decisions need to be made and on the basis of what information. In contrast to elicited probabilistic statements the expert will usually have a degree of confidence in the accuracy of his statements and their more obvious implications. Therefore when it is possible to produce a framework within which this qualitative description can be recorded and used to feed back the less direct implications of a set of statements, the statistician can often begin a fruitful dialogue about the nature of the fundamental structure of a model. Once this structure has been agreed the statistician can then proceed to model the more quantitative aspects of a model, always ensuring that the elicited quantitative model is consistent with the derived qualitative structure.

One currently popular way of representing some of the qualitative features of a problem is to use a graph: for example an influence diagram [7, 20, 9, 23] or a graphical association model ([10, 27]). Here nodes of a graph represent uncertain quantities, decision variables or utilities, whilst the existence or otherwise of edges between these nodes represent various forms of relevance statements between information sets, the timing of decisions and causal relationships between variables.

Now, under the hypothesis that the qualitative statements made by the expert are just a verbal description of certain features of a fully specified probability model which is there in his head but is not yet drawn out, it can be shown that all relationships expressed in influence diagrams and graphical association models correspond to various types of conditional independence statements within the probability model. Conditional independence between a set of random vectors $(\mathbf{X}_1, \mathbf{X}_2, \ldots, \mathbf{X}_n) = \mathcal{X}$ has certain properties that can be deduced from the axioms of probability [5]. Explicitly if $A, B, C \subseteq \mathcal{X}$, read $A \perp\!\!\!\perp B | C$ as "given the value of variables in C, we expect the values of variables in B to be irrelevant to statements about the predicted values of variables in A". Then for arbitary subsets A, B, C, D the following statements must hold:

P1 $A \perp\!\!\!\perp B | A \cup C$

P2 $A \perp\!\!\!\perp B | C \Leftrightarrow B \perp\!\!\!\perp A | C$

P3 $A \perp\!\!\!\perp B \cup C | D \Leftrightarrow \begin{cases} A \perp\!\!\!\perp B | C \cup D \\ A \perp\!\!\!\perp C | D \end{cases}$

On the basis of these properties above it is possible to use the influence diagram or association graph to deduce others — see the d-Separation theorem below.

Of course, the hypothesis that an expert without relevant training nevertheless codes his thoughts probabilistically is a heroic one. However, since the proven results require only that P1, P2 and P3 are valid we need no such strong hypothesis.

If properties P1, P2, P3 are known to hold on subsets of uncertain measurements $\mathcal{X}$, which may or may not be sets of random vectors, then $\mathcal{X}$ is said to admit a graphoid [15] or generalized conditional independence structure [23]. It has been argued (see for example Smith [25]) that *whether or not an expert is thinking probabilistically* these properties, associated with the relevance of one variable to another given a third, should hold for any internally rational or coherent expert. By using the rules above it is possible to deduce new sets of conditional independence statements from conditional independence statements elicited initially.

These deductions are fed back to the expert who then, if he believes it to be necessary, modifies his initial statements. This process continues until such time as a model is derived for which the expert perceives no necessity for further modifications. Such an equilibrium model, called *requisite* by Phillips [17], provides the framework of a *research hypothesis* [27].

It is recommended that only after such a framework has been developed is a full statistical model built. Any quantitative features that are needed for a full analysis, such as probabilities, means and variances, are added only at this later stage.

Such a process of statistical model building is now becoming quite common. Through experience what has become increasingly apparent is that many central statistical ideas such as sufficiency [24], observability [25] and orthogonality [25] can be expressed in terms of the rather general idea of a conditional independence structure. The aim of this chapter is to add to this list. We intend to show how ideas associated with the concept of (external) rationality, borrowed from game theory, can be used by an observer to formulate a research hypothesis in terms of his conditional independence structure about the nature of the moves of players in a game or the purchases of customers in a market.

Traditionally, it has been difficult to use conventional game theory directly to deduce plausible probabilistic models of a competitive market. This has been due to the necessity of providing highly specified and mutually known preference structures and belief structures between players. By defining a qualitative rather than quantitative form of rationality we sidestep many of these specification problems. Assumptions of qualitative rationality allow us to derive a set of conditional independence statements which should be implicit for the informed observer. These then form the basis of a posited research hypothesis which can be tested on data. Thus, game theory allows us to construct a statistical model which is consistent with a normative theory about the way in which people might plausibly handle information as it arises. This is our first small step towards developing game theory ideas to help to structure statistical models of competitive markets. The ideas of rationality developed here complement rather than conflict with the more quantitative ideas of rationalisability [14, 4].

In Section 14.2 we define the construction of the influence diagram on the basis of elicited conditional independence statements and two theorems about the deductions that can be made from influence diagrams are stated. In Section 14.3 we address some ideas of common-knowledge and rationality in terms of conditional independence and link these ideas with the statistical concept of parsimony. Two simple examples are given which illustrate these principles in action. In Section 14.4 the idea of equilibral rationality is introduced which has stronger implications. It is argued that any posited model without this feature is at least normatively inadequate. We conclude the chapter with some simple examples of how the form of an observer's research hypothesis can be framed using rationality in two types of market.

14.2 Making Deductions in Statistical and Decision Problems

Graphs of various kinds are very useful to help make deductions about the relevance structure between arbitary sets of variables given some initial generalised conditional independence statements. Suppose we order a given set of uncertain quantities $(X_1, \ldots, X_n)$ on which we assert that the following $(n-1)$ (possibly degenerate) conditional independence statements are valid

$$X_i \perp\!\!\!\perp R(X_i) | Pa(X_i) \quad 2 \leq i \leq n \quad (14.1)$$

where $R(X_i) = H(X_i) \backslash Pa\{X_i\}$, $Pa\{X_i\} \subseteq H(X_i)$, $H(X_i) = \{X_1, \ldots, X_{i-1}\}$.

A *valid (chance) influence diagram* I is a graph whose nodes are labelled $\{X_1, \ldots, X_n\}$ where each node X_i, $2 \leq i \leq n$ has a set $Pa(X_i)$ of nodes (called *parents*) with edges into X_i such that the $(n-1)$ statements (14.1) are all true.

For any sets $D, E, F \subseteq \mathcal{X}$, it can now be determined whether the conditional independence statement $D \perp\!\!\!\perp E | F$ is deducible from the set of statements (14.1), given only the properties P1, P2 and P3. Define $A(X_j)$, the *ancestral set* of X_j, by

$$A(X_j) = \{X_i \in I : \text{ there exists a directed path from } X_i \text{ to } X_j\}$$

The *ancestral* set of a collection $C = \{X_k : k \in K\}$ of nodes is simply $A(C) = \bigcup_{k \in K} A(X_k)$.

Similarly, the *descendant set* of a collection $C = \{X_k : k \in K\}$ of nodes is

$$F(C) = \bigcup_{k \in K} F(X_k)$$

where $F(X_k) = \{X_j \in I: \text{ there exists a directed path from } X_k \text{ to } X_j\}$. The *non-descendants* of C are just $\mathcal{X} \backslash F(C)$.

The following key theorem was proved by Pearl & Verma [16] and by Lauritzen *et al.* [12].

d-Separation Theorem

If I is a valid influence diagram with nodes $\mathcal{X}$ on a set of conditional independence statements (14.1) and $D, E, F \subseteq \mathcal{X}$ are subsets of variables construct an undirected graph J from I in the following way:

(i) Form the directed subgraph I_1 of I whose nodes consist of the ancestral set $A(D \cup E \cup F)$ and whose directed edges are those in I which lie between these nodes.

(ii) For all $Z \in I_1$, join all unconnected pairs of nodes $(X, Y) \in Pa(Z)$ by an undirected edge. Call this mixed graph I_2.

(iii) Form an undirected graph J by replacing all directed edges in I_2 by undirected edges. Then it can be deduced from the conditional independence statements embedded in I that

$$D \perp\!\!\!\perp E | F$$

if all undirected paths in J between a node $X \in D$ and $Y \in E$ must pass through a node $Z \in F$.

Note that in various senses (see Lauritzen et al. [12]) these are *all* the statements that can be validated by just using the conditional independence structure so this is a powerful result which will be used later in this chapter.

The following discussion is a brief resumé of some results given more formally in Allard & Smith [1] for the case of a game with a single decision-maker Π. To describe how to draw an influence diagram which represents decisions and preferences as well as uncertain quantities it is most straightforward to describe the construction for a Bayesian analysis for discrete variables and decision spaces. Here variables consist of one utility variable, a set of decision variables consisting of decisions that need to be taken at various times and the set of random variables in the problem. First list the variables in any order that is consistent with no variable being listed before another when it occurs after it in time. List the utility variable last.

Square nodes represent decision variables, circular (chance) nodes represent random variables and a diamond node represents the decision maker's utility. Now taking nodes in their listed order add directed edges into these variables as follows:

(a) for a *decision node D* connect a previously listed variable node X to D if and only if X is known to Π at the time D is taken;

(b) for a random quantity W with conditional mass function $g(w)$, conditioned on the values of the set of variables $H(W)$ proceeding it in the list, connect the W node only to those nodes whose variables explicitly appear in $g(w)$;

(c) connect the utility node U to those variables of which U is an explicit function.

The influence diagram so drawn is clearly directed and acyclic with the utility node having no edges out of it. This definition is equivalent to that given by Shachter [20]. For an account of the cases which involve a mixture of discrete and continuous variables see Smith [24].

Now the first point that should be clear is that such an influence diagram generalises away from the Bayesian or probabilistic paradigm. Definition (a) is clearly independent of how inferences are being made. If in (b) we replace $g(w)$ by the decision-maker's *beliefs* about W saying that given $H(W)$ these beliefs depend only on the parent set $Pa(W)$ of W then we need not assume that these beliefs can be represented by a probability distribution. Finally, the utility can be replaced by a preference order where this order across decisions is only an explicit function of the parents of U. So although historically the decision influence diagram was introduced as a succinct representation of a Bayesian decision analysis, it actually represents a much more fundamental picture of a problem, which can be acceptable to a decision-maker, Bayesian or not.

Our second point is that there is an intriguing link between a decision influence diagram as described above and the chance influence diagram based on generalized conditional independence that was defined earlier. Suppose an impartial observer listens to the decision-maker's description of his problem. All the variables the decision-maker includes in this decision influence diagram are uncertain quantities, sometimes degenerate, to that observer. Furthermore, we argue that the decision-maker's decision influence diagram, if thought of as the observer's chance influence diagram should be valid for the observer provided she believes the decision-maker's statements about the dependence of variables as explicitly expressed in the diagram. She can argue as follows. Since the client's utility (preference ordering) is known to be a function of its parents, given the values of the parents none of the other variables in the influence diagram are relevant to her statements about this utility function. If the decision-maker's beliefs about an uncertain quantity W given $H(W)$ depends only on $Pa(W)$, then $Pa(W)$, which to the observer consists of a set of uncertain quantities, should contain all that is relevant to her about W given $H(W)$. And finally if the client can base a decision D only on $Pa(D)$ then within $H(D)$ only $Pa(D)$ can be relevant to the observer's beliefs about which decision the decision-maker will take. This argument is expanded for the Bayesian decision-maker and Bayesian observer in Smith [23], and more generally for games in Allard & Smith [1].

The final issue we discuss in this section is whether any further conditional independences can be deduced by the observer on learning the decision-maker's influence diagram. This happens to be the case provided the observer assumes that the decision-maker will choose his decisions parsimoniously. Consider the following trivial example illustrated in the influence diagram I_1 of Figure 14.1.

Note I_1 states that the decision-maker can make his decision on the basis of X but that his utility does not depend on X. Although the decision-maker could choose to let D depend upon X, for example to randomize between two optimal decisions were X to represent the outcome of a coin toss, it will not help him. Therefore it is not unreasonable for the observer to assume that the decision-maker will act parsimoniously and ignore the value of X when making his decisions. So the observer can *deduce* the influence diagram I'_0 and the client be assumed to be acting with a reduced information base with influence diagram I'_1. This argument is central to later sections when we go on to discuss games.

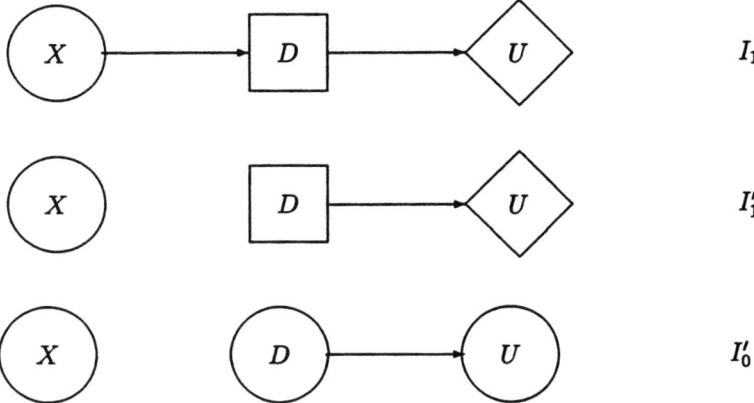

Figure 14.1: A simple sufficiency theorem reduction.

Say an influence diagram I' derives from an influence diagram I if:

(i) the nodes $(Y_1, \ldots, Y_n)$ on I and I' agree;

(ii) the edges into the utility node and chance nodes agree on I and I';

(iii) the sets of parents $P'_a(D)$, $Pa(D)$, in I' and I respectively of any decision node D in $\{Y_1 \ldots Y_n\}$ satisfy $P'_a(D) \subseteq Pa(D)$.

In any problem represented by an influence diagram I, Π will have beliefs about his chance nodes and have preferences, represented in a Bayesian model through a utility function. Call I' a derivative *associated with* I if I' represents the same problem as I with the same beliefs and preferences but which implicitly denies Π information in the set $Pa(D)\backslash Pa'(D)$ when he makes decision D, for each decision node D in I. Call a derivative I' associated with I *valid* if there exists a decision rule consistent with I' which is *optimal* for the problem as posed in I. Put simply, I' is valid if regardless of Π's beliefs and preferences, he can disregard without loss any information available in $Pa(D)\backslash Pa'(D)$ but not in $Pa'(D)$, this being so for each decision node D in I.

The following result can then be proved for Bayesian clients (Smith [23]) and more generally (Allard & Smith [1]). For any decision node D write $Q(D) = Pa(D)\backslash Pa'(D)$.

Sufficiency Theorem
Suppose that it is possible to deduce from an observer's chance influence diagram I_0 of a decision maker's influence diagram I_1 that

$$U \perp\!\!\!\perp Q(D)|\{D\} \cup Pa'(D)$$

where $Q(D)$ is defined above. Then I'_1, the influence diagram derived from I by deleting the set of edges $(Q(D), D)$, is a valid derivative of I_1. □

14.3 Common-Knowledge and Rationality

To move from a single decision making frame to one involving many supposedly rational players, it is first necessary to define a common-knowledge base for the problems. Informally an assertion X is said to be common-knowledge if all players know X and *furthermore* they all know that each other knows it, know that all know that each other knows it, and so on. Formally, if K denotes the set of assertions X that are *common knowledge* to players $\Pi_1, \ldots, \Pi_n$ then $X \in K \Leftrightarrow$ "Π_i knows X" $\in K$, $1 \leq i \leq n$.

Now we argued in the last section that in a decision problem the decision-maker's decisions would be treated as uncertain quantities by an observer and that the corresponding chance influence diagram, agreeing in its node and edge set was "valid" for the observer. In exactly the same way, in a competitive market with several players, for each player the other's decisions are uncertain quantities. Thus each player can be treated like an observer of the other's decisions in the sense of the last section.

In this instance the most natural starting point from which to define a qualitative form of common-knowledge is to assume that Π_i's influence diagram I_i of the problem is common-knowledge. That is, any two players Π_i and Π_j have influence diagrams I_i and I_j that have identical node sets and edge sets and only differ in terms of the identification of nodes as either decision nodes, chance nodes or utility nodes. Thus we say that I_0 is a *c-k-influence diagram* of a game between players $\Pi_1, \ldots, \Pi_n$ if

(i) I_0 contains only chance nodes,

(ii) the influence diagrams $I_1, \ldots, I_n$ of players $\Pi_1, \ldots, \Pi_n$ respectively agree with I_0 in their node and edge sets, and

(iii) this fact is common-knowledge.

The assumption that an influence diagram is common-knowledge amongst a set of players and an observer is quite strong. One implication is that for $j = 1, 2, \ldots, n$, all players other than Π_j state that

$$U_j \perp\!\!\!\perp \mathcal{X} \backslash \{U_j \cup Pa(U_j)\} | Pa(U_j) \qquad (14.2)$$

Thus they believe that once $Pa(U_j)$ is known they can learn about Π_j's preferences only through observing U_j itself — all other variables are irrelevant. For Π_j to have the same set of edges into U_j he would need to be able to state that U_j was an explicit function of variables in $Pa(U_j)$ only.

A simple case when (14.2) could be asserted immediately is when $Pa(U_j)$ is exactly those variables appearing in a Bayesian's utility function U_j where U_j is *known* to all other players, and Π_j takes only one decision. On the other hand if U_j were only partially known to another player Π_i then statement (14.2) for Π_i would mean that the only variables that help Π_i obtain information about U_j are $Pa(U_j)$. This may or may not be an appropriate condition depending on the context of the game being played and the meaning of the nodes in the influence diagram. A fuller discussion of how to decide whether utilities in a game can be legitimately expressed in terms of a common knowledge influence diagram, by introducing a new type of node called a predisposition node, is given in Allard & Smith [1].

In the aforementioned paper, we also argue that all conditional independence statements implied by an influence diagram can be deduced from the equivalent *c-k*-influence diagram. For example, if a decision D is taken by Π_j, we would expect all players to agree that

$$D \perp\!\!\!\perp R(D) | Pa(D) \tag{14.3}$$

Furthermore, all deductions made via the *d*-separation theorem are valid for all players.

Also, if the observer knows the two decisions D_1 and D_2, then learning the value of one payoff does not provide any additional information about the other payoff. Furthermore, these relationships between the variables are common knowledge to Π_1 and Π_2, so I_0 is a *c-k*-influence diagram.

The second ingredient that is necessary before any normative deductions can be made is that all players will reason predictably to the other players. The most natural way to do this in this context is just to assume that it is common-knowledge that each player treated as a single decision-maker reasons as if their influence diagram is valid and makes valid deductions according to that influence diagram (call this $c - k$-*rationality*).

Thus it is assumed that whereas players may not be reasoning probabilistically and acting to maximize their expected utility nonetheless they will make deductions about relevance as a Bayesian would with the corresponding conditional independence statements. Note that we do not require any quantitative statements by the players either about their own beliefs and utilities, *or* about their thoughts on how the others will play. Instead we make the much weaker and much more plausible assumption that they are agreed about the structure of the game and about how certain deductions can be made. For instance in the example of Figure 14.2 no assumption is made about the equality of each player's beliefs about the distribution of X or of

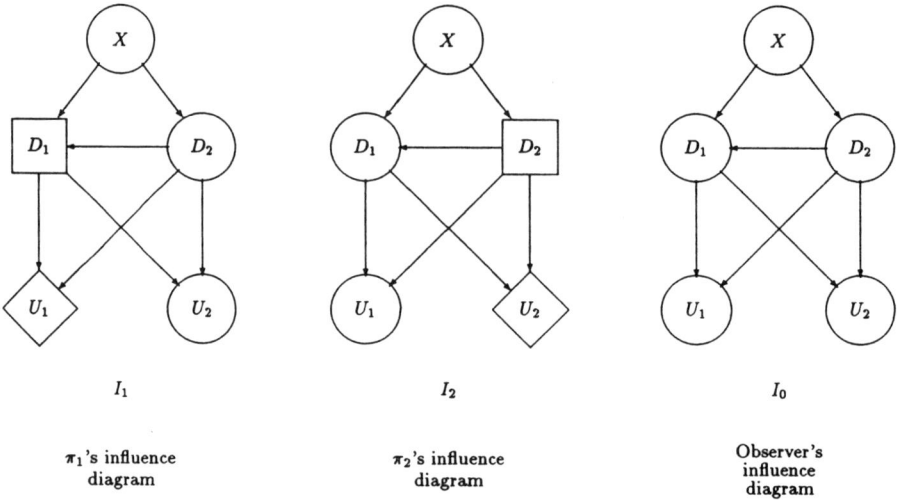

Figure 14.2: Common-knowledge influence diagrams in a simple game.

the functional forms of the utilities concerned. Indeed, Π_1 could be playing a minimax strategy or be using belief functions whilst Π_2 could be a Bayesian, and the common-knowledge base above would not necessarily be violated.

Within this general structure we would like to make deductions using the sufficiency theorem. For this we need to make an additional tie-breaking assumption between two strategies that are equally good. We assume that the following assertion, called *(graphical) parsimony*, is common-knowledge. If any player needs to choose between two options d_1 and d_2 which by evoking the d-separation and sufficiency theorems can be shown to be otherwise equally preferable where d_1 is a decision rule based on the set of nodes Pa' in the diagram and d_2 is a decision rule based on the set of nodes Pa, and $Pa' \subset Pa$, then action will be taken on the basis of d_1.

Assumptions of parsimony are often implicit in game models — indeed without them such models would become hopelessly cumbersome (see Allard & Smith [1]). In a probabilistic setting parsimony essentially states that it is common-knowledge that all players will employ decision rules which are minimal within the coding of the commonly-known influence diagram of the problem. One objection to parsimony might be that it disallows the explicit use of mixed strategies, since such a strategy makes use of some random mechanism, which is known only to the player concerned, and is otherwise

RATIONALITY, CONDITIONAL INDEPENDENCE 247

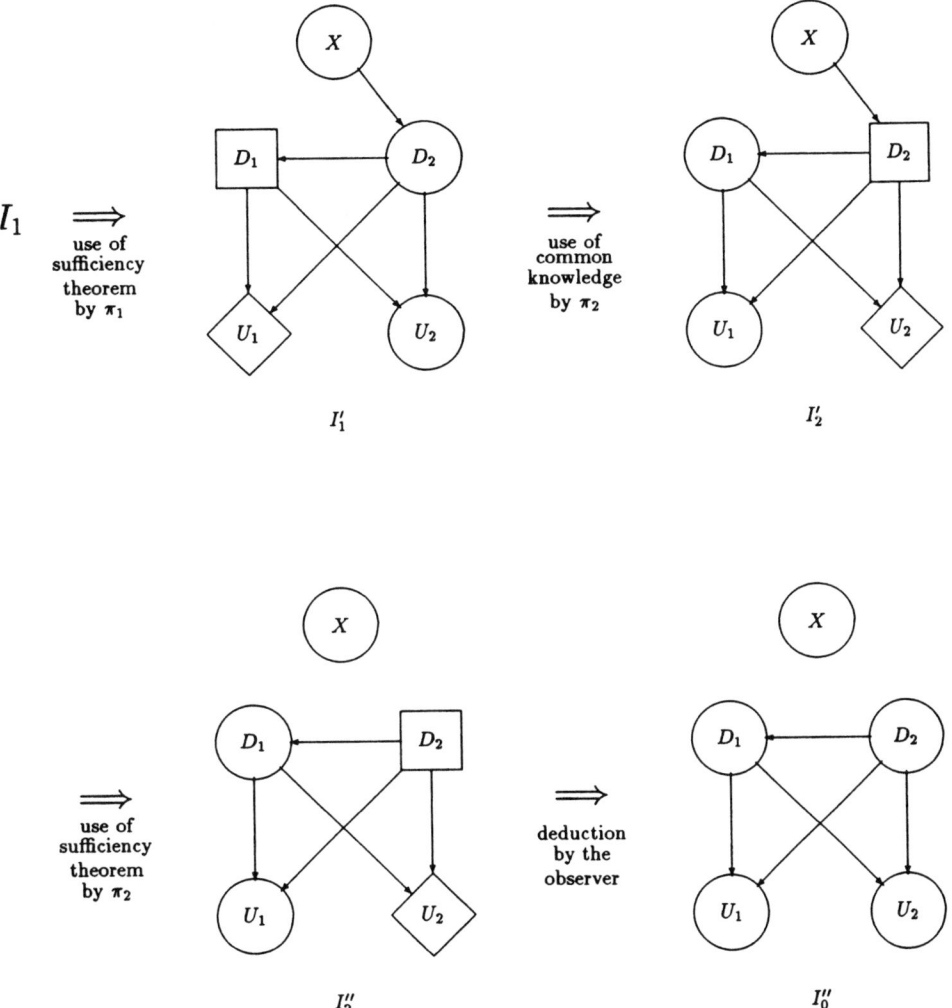

Figure 14.3: Using common-knowledge to deduce implied conditional independence.

inconsequential as regards the outcome of the game. We would argue that the importance of the mixed strategy in game theory relates not to the explicit use of such a device, but to the uncertainty of other players regarding which pure strategy is being employed. Such an argument has been used by Rubinstein [18] to describe the mixed strategy and, defined in this way, it does not contradict the parsimony principle.

Example 1.
Consider the game described above and defined by Figure 14.2.
From I_1, Π_1 can use the sufficiency theorem to deduce under parsimony she will act as if I'_1 is a valid derivative of I_1, i.e. take no regard for the value of X. By c-k-rationality Π_2 will know that Π_1 will reason in this way. Because of this Π_2 can himself use the sufficiency theorem to delete the edge (X, D_2). So in particular by common-knowledge the observer can conclude that neither player will allow their strategy to depend upon X and that consequently (D_1, D_2, U_1, U_2) is independent of X as illustrated by Figure 14.3. Note that without parsimony as common-knowledge this argument would break down. A more substantial example of this line of argument is given in Section 14.5.

14.4 Equilibral Rationality

Another concept predominant in game theory is that of an equilibrium. Here players are said to be in equilibrium if no advantage can be had by any one player deviating from their chosen decision rule if all other players hold to theirs'. In this section a type of rationality, based on influence diagrams, is defined which is stronger than the rationality developed in the last section and which is closely related to the idea of an equilibrium.

Suppose at a given point in a game that n players $\Pi_1, \ldots, \Pi_n$ make simultaneous decisions $D_1, \ldots, D_n$ based on the respective sets of variables $W_1, W_2, \ldots, W_n$ known to them at that time. So D_i is not in the ancestor set of D_j, $1 \leq i, j \leq n$. The sets $W_1, \ldots, W_n$ are arbitary in this definition and need not be disjoint. Suppose each set W_i contains v variables from the graph and write

$$W_i = (W_i(1), \ldots, W_i(v)), \qquad 1 \leq i \leq n.$$

We now assume a further piece of common-knowledge; that all pairs of players Π_i and Π_j acknowledge $W_i(k)$ and $W_j(k)$ as "analogous variables" $1 \leq i, j \leq n$ for all indices k, $1 \leq k \leq v$. More precisely we assume that it is common-knowledge that if the rôles of any two players Π_i and Π_j were permuted then Π_i would permute $W_i(k)$ by $W_j(k)$ in her new description of the problem (and vice versa). So, for example, $W_i(k)$ may be the same variable for each player Π_i, or perhaps their own last move or the result of their latest experiment. Under this extension of the common-knowledge base it is possible to ask the

RATIONALITY, CONDITIONAL INDEPENDENCE

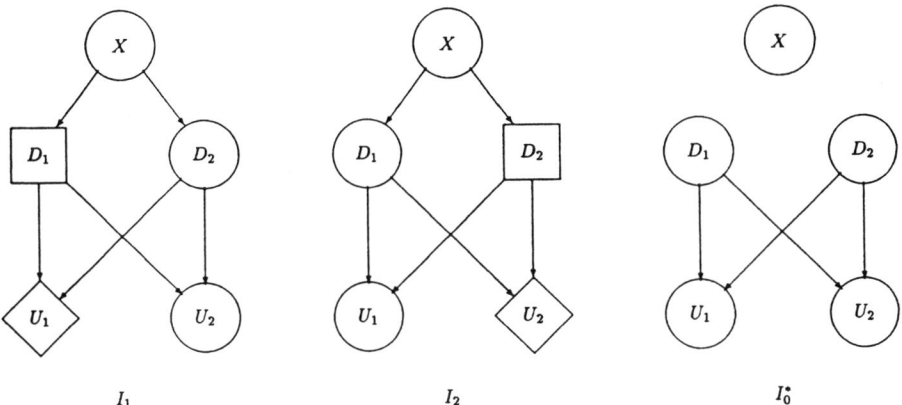

Figure 14.4: Reduction using equilibral rationality.

following question. Suppose Π_j believes that all players Π_i other than Π_j decide not to let their decision depend upon

$$W_i(C) = \{W_i(k) : k \in C\} \quad C \subseteq \{1, 2, \ldots, n\} \quad 1 \leq i \neq j \leq n.$$

If Π_j were being parsimonious would he then decide not to let his decision depend upon $W_j(C)$ where $W_j(C) = \{W_j(k) : k \in C\}$?

If under the sufficiency theorem the answer to the above question is "yes" for each player Π_j, $1 \leq j \leq n$, then the model represented by the observer's influence diagram of this corresponding conditional independence is said to be a *rational equilibrium reduction* of the game. In graphical terms, Π_j deletes the edges $(D_i, W_i(k))$, $1 \leq i \neq j \leq n$, $1 \leq k \leq v$ in Π_j to form a new influence diagram $\hat{I}_j$. If the sufficiency theorem then allows him to assert that I' is a valid derivative of $\hat{I}_j$ where I' is identical to I_j except for the omission of the edges $(D_j, W_j(k))$, $1 \leq k \leq v$ and this is true for all $1 \leq j \leq n$ then I' is a rational equilibrium reduction.

Example 2.

In the game depicted in Figure 14.4, Π_1 and Π_2 make their respective moves D_1 and D_2 simultaneously based on both knowing the value of their joint payoff will not depend upon X.

Setting $W_1(C) = W_2(C) = \{X\}$, then clearly I_0^* is a rational equilibrium reduction of this game since if Π_i, $(i = 1, 2)$ assumes his opponent will ignore

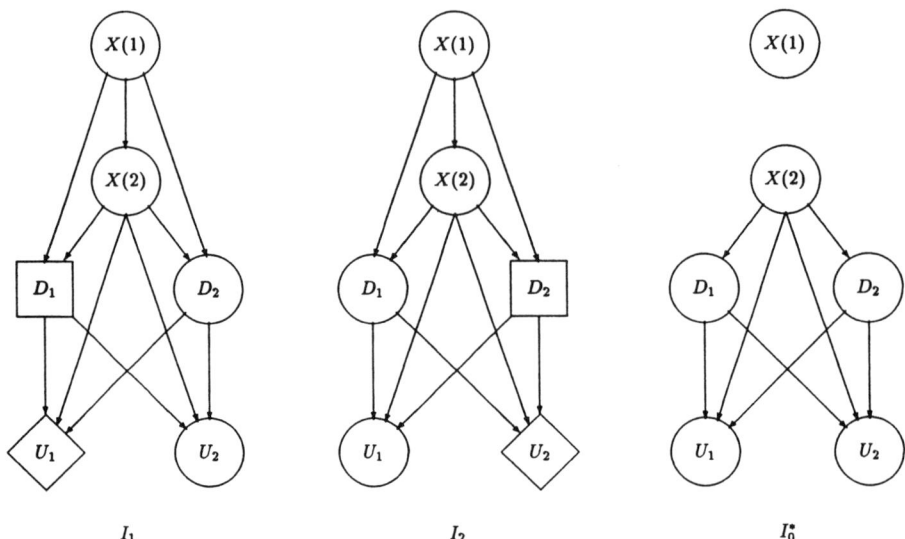

Figure 14.5: Another example of reduction.

X it is safe to ignore X himself. Notice, however, that this conclusion cannot be deduced from the type of rationality defined in Section 14.3 above.

Example 3.

In this slightly more complex example depicted in Figure 14.5 by setting $C = \{1\}$ and $W_1(1) = W_2(1) = \{X(1)\}$ the sufficiency theorem implies that a reduction of the game is given by I_0^*. Note that this theorem does not prescribe the deletion of either the edge $(X(2), D_1)$ or $(X(2), D_2)$ since because $X(2)$ may influence the expected utility of either player it may be necessary for either or both of D_1 and D_2 to depend upon it.

Rational equilibrium reductions are clearly a function of posited common-knowledge about analogous variables and are typically non-unique. Furthermore, when co-operation or co-ordination is mutually advantageous then such a reduction may not be expedient even when possible. For example, a well-known concept in game theory is that of a correlated equilibrium (see Aumann [3]). Rational equilibrium reduction may remove from the game some information, commonly known by two or more players which could act as the correlating mechanism for a correlated equilibrium — see, for example, the description of "the battle of the sexes" in Allard & Smith [1].

On the other hand, it is a commonly held position in game theory that models which describe behaviour in terms of two chosen strategies *not* in equilibrium, where the equilibrium is appropriately defined, are unstable and therefore suspect. Thus if the answer to the question above is — no — that is, in a game with the types of symmetries discussed above, player Π_j would not choose to ignore certain information even when $i \neq j$ $1 \leq i, j \leq n$ chose not to then it could be asserted that such a structure would be untenable as a normative model. For example Young & Smith [29] show that if player Π_1 assumes Π_2 is using a strategy based only on the latest n pairs of moves in a potentially infinitely repeating symmetric game he can improve on Π_2's performance by using all past data to estimate Π_2's probability of moves given the n-back history. Thus in the Bayesian setting it is proved that strategies based on the lastest n pairs of moves cannot form a rational equilibrium reduction and the models of Wilson [28] are suspect as normative models of rational behaviour. Player Π_1 is assuming behaviour on the part of Π_2 which in his shoes would be assessed as suboptimal and therefore irrational. On the otherhand it is shown in Smith & Young [21] and Smith [22] that in a Prisoner's Dilemma game, if two players believe the other's strategy depends only upon:

(i) a function of a set of statistics, which in their respective models are sufficient for each player and current total pay-off pair;

(ii) a form of utility for each player which is increasing in total pay-off;

then such a model is in fact a rational equilibrium reduction. A graphical derivation of this result is given in Smith [22].

14.5 Statistical Models Structured from Evoking Rationality

Usually the statistician is cast in the rôle of an observer of a game or competitive market. It is now illustrated how, from the assumptions of qualitative rationality developed above, that statistician can deduce research hypotheses about the conditional independences she might expect to be evident in the data. A trivial setting where such deductions can be made is given below.

Example 4.

The uncertain quantities $X(1)$ and $X(2)$ would be considered independent by the observer were she to know the players' (Π_1, Π_2's) decisions (D_1, D_2) respectively. Her problem is that she does not observe these values. However exactly the same arguments as those given in Example 1 allow her to conclude that $X(1)$ and $X(2)$ will be independent marginally, see influence

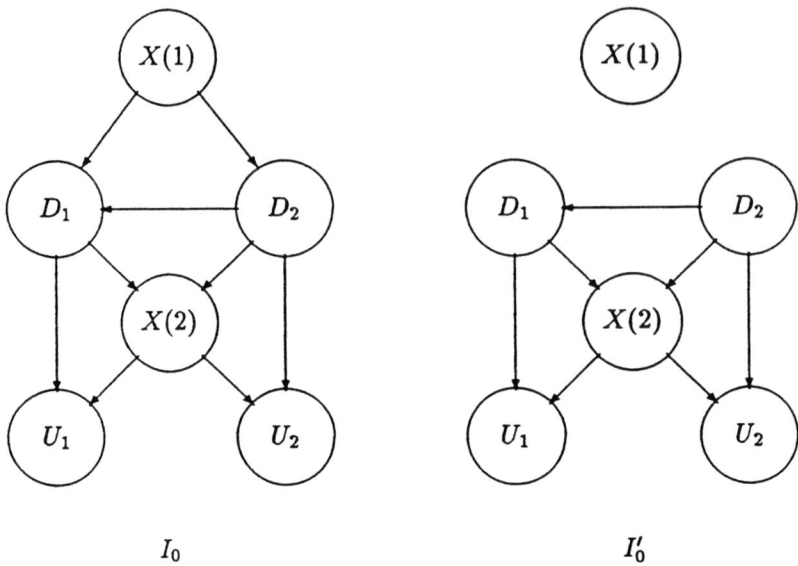

Figure 14.6: An observer's deductions assuming rationality.

diagram I'_0, *provided* both players are acting rationally in the sense described in Section 14.3.

Even when D_1 and D_2 are mutually ignorant of each other's moves, so that the edge (D_1, D_2) is missing in I_0, by the arguments of Example 2. the observer can deduce that the model which assumes $X(1)$ and $X(2)$ are marginally independent is at least a plausible one.

The next example, whilst idealized, is slightly less simple.

Example 5.

Consider a market supplied by three companies, Π_1, Π_2, and Π_3, whose sales in month t are denoted by $X_1(t)$, $X_2(t)$, and $X_3(t)$ respectively. Company Π_2 is the market leader, and attempts to sell to every customer. Π_1 supplies an "upmarket" brand, while Π_3 caters for the "downmarket" end. There is a population (assumed to be fixed) consisting of two types of consumer, Type-a and Type-b. Type-a consumers choose between the products of companies Π_1 and Π_2 only, while Type-b choose between those of Π_2 and Π_3. This is an example of a partially-segmented market.

The sales of Π_2's product in month t to Type-a and Type-b consumers are respectively $X_2^{(a)}(t)$ and $X_2^{(b)}(t)$. However, these values are never observed directly but only their sum $X_2(t) = X_2^{(a)}(t) + X_2^{(b)}(t)$. A simple influence diagram for these variables for the month t is given by Figure 14.7.

After the first month each company can choose whether or not to pay for

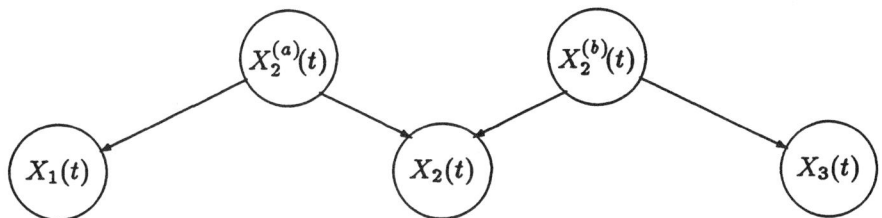

Figure 14.7: The market at time t.

additional advertising to increase its sales at the expense of its competitor(s). Company Π_2 as market leader has to act early in order to preserve its competitive advantage, and so has only its own sales figures $X_2(1)$ on which to base its decision D_2. In contrast, Π_1 and Π_3 wait to see what Π_2 does before taking their own decisions D_1 and D_3. In the meantime, the monthly sales figures for all three companies $X(1) = (X_1(1), X_1(2), X_1(3))$ are published. So both know $X(1)$ as well as D_2 when they come to make their decisions.

Discussions with Π_1's market research director suggest that the distribution of $X_2^{(a)}(2)$ depends on $(X_2(1), D_1, D_2)$ only, and a similar model is proposed for $X_2^{(b)}(2)$. Since we are working within a partial-segmentation model, we can assume that D_2 only affects the competitors' sales $X_1(2)$ and $X_3(2)$ via its effects on $X_2^{(a)}(2)$ and $X_2^{(b)}(2)$, and that the decisions by Π_1 and Π_3 have no effect on the other's market segment. Thus we postulate that the distributions of $X_1(2)$ and $X_3(2)$ depend only on $\left(X_1(1), D_1, X_2^{(a)}(2)\right)$ and $\left(X_3(1), D_3, X_2^{(b)}(2)\right)$. (Note that we are acting as 'the observer' for the purposes of this analysis.) Now let U_i be the second month profits for company Π_i ($i = 1, 2, 3$). So U_i depends only on D_i and $X_i(2)$.

The complete problem (from the point of view of company Π_1) is represented by the influence diagram I_1 given in Figure 14.8.

Now suppose that the $c - k$-influence diagram I_0 with the same sets of nodes and edges is valid for this problem (in other words, all three companies agree that the conditional independence structure as given by I_1 is accurate.) We can now make some deductions about the companies' behaviour, and thus simplify our model. Firstly, the d-separation theorem tells us that

$$U_1 \perp\!\!\!\perp X_3(1) | D_1, D_2, X_1(1), X_2(1).$$

So assuming parsimony on the part of Π_1, we can delete the edge $(X_3(1), D_1)$ to form the valid derivative, I_0'.

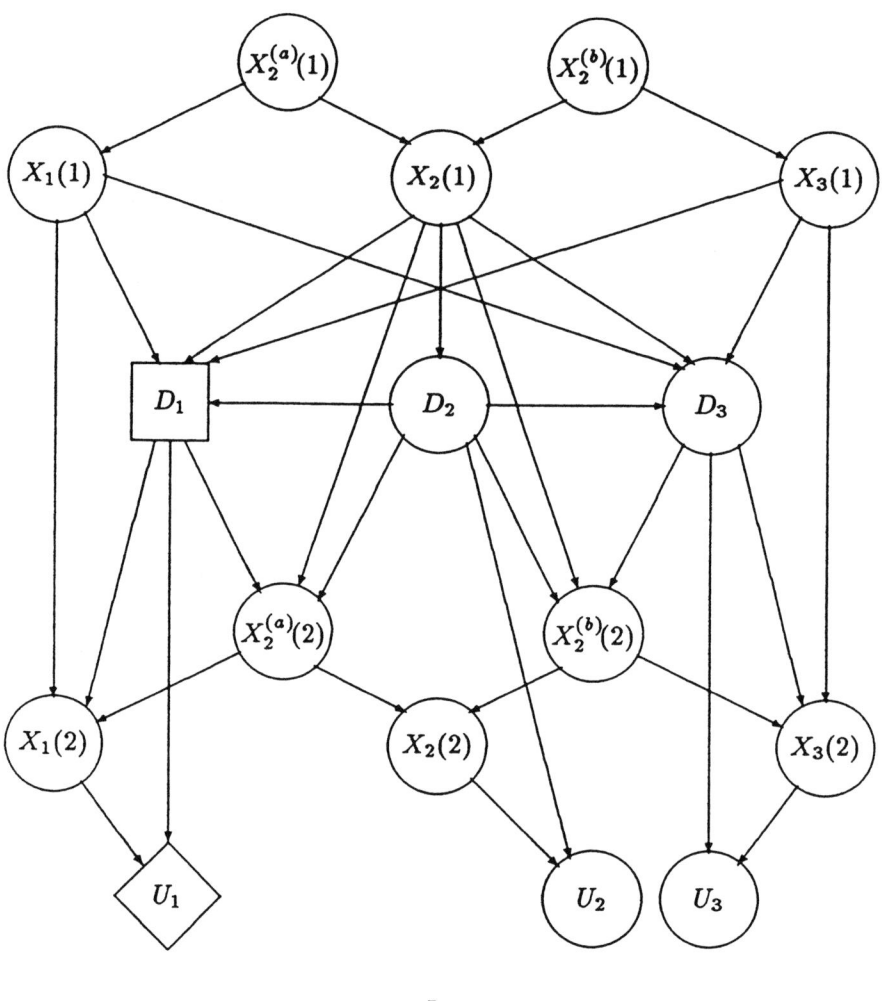

Figure 14.8: Π_1's influence diagram of a game played on a partially segmented market.

Consider an individual who is outside the market, but who has some interest in it, for instance a market regulator, or a consumer. (Although consumers as a whole can influence the market, we presume that an individual cannot, so is not properly a player in our model.) He will only know those variables which are in the public domain, namely the sales figures, $(X(1), X(2))$. Nevertheless, he can immediately deduce from our qualitative model the following relation on those variables,

$$X_1(2) \perp\!\!\!\perp X_3(1) | X_1(1), X_2(1),$$

which was not previously apparent. An analogous argument based on parsimony by Π_3 gives,

$$X_3(2) \perp\!\!\!\perp X_1(1) | X_3(1), X_2(1).$$

To take things a little further, a company could measure or estimate most of the variables in the model (using market research), and compare the historic data with the above model. Typically, additional conditional independence statements might be inferred from the data. (For a practical example of this process, see Wermuth and Lauritzen [27]). These could then be included in the model by deleting the appropriate edge(s) from *that company's* influence diagram.

Now if it could be assumed that the same analysis could be done by all three companies (for example if the relevant historic data was in the public domain), then such edge(s) could also be deleted from the $c - k$-influence diagram. We might then find that the model could be simplified again, just by re-applying the sufficiency theorem. It is worth noting that, in practice, this will often not be the case, since a company's internal decisions are generally kept secret from its competitors. However, the inferences to be drawn based on that assumption can be tested against the data, and can throw some light on whether such assumptions may be valid.

For example, suppose that an analysis of the data and/or logical deduction were to suggest that the edge $(X_1(1), X_1(2))$ could in practice be ignored. We then have:

$$U_1 \perp\!\!\!\perp X_1(1) | D_1, D_2, X_2(1),$$

so the edge $(X_1(1), D_1)$ can be deleted. This can be interpreted as saying that if, given all the other information, $X_1(1)$ is of no use in predicting the value of $X_1(2)$, then this will be true also for company Π_1, given only the subset of variables $Pa(D_1)$, so it can afford to ignore it's previous sales when deciding whether or not to advertise.

Finally, suppose that the same relationship is observed for Π_3's sales, so the edges $(X_3(1), X_3(2))$ and $(X_3(1), D_3)$ disappear, and that further calculations lead to the deletion of the edges $(X_2(1), X_2^{(a)}(2))$ and $(X_2(1), X_2^{(b)}(2))$. Then repeated application of the sufficiency theorem allows the deletion of all edges from the sales nodes $X(1)$ into the decision nodes. So an outsider who was privy to this information could deduce that even when controlling decisions are taken which are unknown to him, he might nevertheless conclude that

$$X(1) \perp\!\!\!\perp X(2).$$

The point we make here is that an appeal to rationality can give a plausible collection of research hypotheses which are nested and can be tested one against another. In particular, the type of rationality we define here together with a posited qualitative model of relationships between variables gives rise to statistical hypotheses about the conditional independence of various combinations of variables. These can then be investigated using standard statistical techniques and, on the evidence of observed data sets, be deemed either as plausible or unlikely descriptions of the mechanism of a given market.

14.6 Conclusions

By using some very weak definitions of rationality we have illustrated by some simple examples how statistical models can be structured to have features of conditional independence which are consistent with such a normative model. The great advantage of this use of game theory to suggest plausible families of statistical models is that it does not require the heroic assumption that mutual beliefs exist at least implicitly, as quantifiable and quantified objects on the part of the players of the game.

Acknowledgements

One of the authors was supported by an S.E.R.C. grant whilst engaged in this research.

References

[1] Allard C. and Smith J.Q. (1992). A rational approach to game theory using graphical models. Warwick Univ. Statistics Res. Rep. 245.

[2] Aumann R.J. (1974) Subjectivity and Correlation in Randomised Strategies. *Journal of Mathematical Economics* 1: 67–96.

[3] Aumann R.J. (1987) Correlated Equilibrium as an expression of Bayesian Rationality. *Econometrica* 55: 1–18.

[4] Bernheim D. (1984) Rationalizable Strategic Behaviour. *Econometrica* 52: 1007–1028.

[5] Dawid A.P. (1979) Conditional independence in statistical theory. *J.R. Statist. Soc. B* 41: 1–31.

[6] Fudenberg D. and Tirole J. (1991) *Game Theory*. MIT Press.

[7] Howard R.A. and Matheson J.E. (1981) Influence diagrams. In Howard R.A. and Matheson J.E. (eds) *Readings on the principles and applications of decision analysis*, Vol. II, Strategic Decision Group, Menlo Park, Calif., 719–762.

[8] Kuhn H.W. (1953) Extensive games and the problem of information. In Kuhn H.W. and Tucker A.W. (eds) *Contributions to the Theory of Games 2*, Princeton University Press, 193–216.

[9] Lauritzen S.L. and Spiegelhalter D.J. (1988) Local computations with probabilities on graphical structures and their application to expert systems (with discussion). *J.R. Statist. Soc. B* 50: 157–224.

[10] Lauritzen S.L. (1989) Mixed graphical associated models. *Scand. J. Statist.* 16: 273–306.

[11] Lauritzen S.L. and Wermuth N. (1989) Graphical models for associations between variables, some of which are qualitative and some quantitative. *Annals of Statistics* 17: 31–54.

[12] Lauritzen S.L., Dawid A.P., Larson B.N. and Leimer H.G. (1990) Independence properties of directed Markov fields. *Networks* 20: 491–505.

[13] Oliver R.M. and Smith J.Q. (eds) (1990) *Influence Diagrams, Belief Nets and Decision Analysis*. Wiley, Chichester.

[14] Pearce D. (1984) Rationalisable Strategic Behaviour and the Problem of Perfection. *Econometrica* 52: 1029–1050.

[15] Pearl J. (1988) *Probabilistic Reasoning in Intelligent Systems: Networks of Plausible Inference*. Morgan Kaufmann, San Mateo.

[16] Pearl J. and Verma T.S. (1987) The logic of representing dependencies by directed graphs. In *Proc., 6th Natl. Conf. on AI(AAAI-87)*, Seattle, 374–379.

[17] Phillips L.D. (1984) A theory of requisite decision models. *Acta Psychologica* 56: 29–48.

[18] Rubinstein A. (1991) Comments on the Interpretation of Game Theory. *Econometrica* 59: 909–924.

[19] Samli A.C. (1989) Heterogeneity of markets and segmentation in retailing. In Samli A.C. (ed) *Retail Marketing Strategy* 135–153. New York, Quorum.

[20] Shachter R.D. (1986) Evaluating Influence Diagrams. In Basu A.P. (ed) *Reliability and Quality Control*, 321–344. Elsevier (North-Holland).

[21] Smith J.Q. and Young S.C. (1988) Stochastic Experimental Games—A Bayesian Approach. Warwick Univ. Statistics Res. Rep. 128.

[22] Smith J.Q. (1988) Models, Optimal Decisions & Influence Diagrams. Bayesian Statistics 3. In Bernardo J.M., DeGroot M.H., Lindley D.V. and Smith A.F.M. (eds). Oxford University Press, 765–776.

[23] Smith J.Q. (1989) Influence Diagrams for Bayesian Decision Analysis. *European Journal of Operations Research* (40)3: 363–376.

[24] Smith J.Q. (1989) Influence diagrams for statistical modelling. *Annals of Statistics* 17: 654–672.

[25] Smith J.Q. (1990) Statistical Principles on Graphs (with discussion). In Smith J.Q. and Oliver R.M. (eds) *Influence Diagrams, Belief Nets and Decision Analysis*, 89–120. Wiley, Chichester.

[26] Smith J.Q. (1994) Decision Influence Diagrams and their uses. In *Proceedings of Decision Making: Towards the 21st Century*. Theory and Decision Library, Klumer (to appear).

[27] Wermuth N. and Lauritzen S.L. (1990) On substantive research hypotheses, conditional independence graphs and graphical chain models. *J.R. Statist. Soc. B* (52)1: 21–50.

[28] Wilson J.G. (1986) Subjective probability and the Prisoner's Dilemma. *Management Science* (32)1: 45–55.

[29] Young S.C. and Smith J.Q. (1992) Suboptimality of M-step back strategies in Bayesian games. *Int. J. of Game Theory* 21: 57–74.

15

Axioms for Dynamic Programming

P.P. Shenoy

15.1 Introduction

The main objective of this chapter is to describe axioms that permit the use of dynamic programming methodology. We describe an abstract framework for representing and solving discrete optimization problems. The abstract framework is called a valuation network (VN). In VNs, information is represented by a collection of functions called valuations. The system includes two operators called combination and marginalization that operate on valuations. Combination tells us how to combine the valuations. Marginalization tells us how to reduce valuations by deleting variables. Solving a VN can be described simply as finding the marginal of the joint valuation for the empty set. The joint valuation is the valuation obtained by combining all valuations. We describe a fusion algorithm for solving a VN using local computation. We describe three axioms for combination and marginalization that make local computation possible. We compare these axioms with the axioms proposed by Mitten [9]. There are several reasons why this is useful.

First, I initially proposed VNs for managing uncertainty in expert systems (Shenoy [14], Shenoy and Shafer [23], Shenoy [16]). Here I show that these systems also have the expressive power to represent and solve discrete optimization problems. Two of the three axioms described here are exactly the same as the axioms described in Shenoy [16]. One axiom is slightly stronger than the corresponding axiom in Shenoy [16].

Second, problems in Bayesian decision analysis involve managing probability and optimization. That these problems can be solved in a common framework suggests that Bayesian decision problems also can be represented and solved in the VN framework. Indeed, Shenoy [17, 18] show that this is true. In fact, the fusion algorithm is always computationally more efficient than the arc-reversal method of influence diagrams. And for symmetric decision problems, the fusion algorithm is more efficient than the backward recursion method of decision trees Shenoy [19].

Third, the fusion algorithm when applied to optimization problems results in a method called non-serial dynamic programming (Nemhauser [10], Bertele and Brioschi [4]). Thus, in an abstract sense, the local computational algorithms that have been described by Pearl [12], Shenoy and Shafer [21], Dempster and Kong [5], Lauritzen and Spiegelhalter [7], and Shafer and Shenoy [13] are just instances of dynamic programming.

Fourth, we provide an answer to the question: What is dynamic programming? Dynamic programming is commonly viewed as an optimization technique. This is how Bellman [3] described it. However, it is also recognized that dynamic programming is more than an optimization technique. For example, Aho, Hopcroft and Ullman [1] refer to dynamic programming as a "divide-and-conquer" methodology. In this chapter, we give an abstract definition of a problem and an abstract method for solving the problem. The abstract method for solving the problem, the fusion algorithm, can be thought of as a general definition of the dynamic programming method.

Fifth, we provide an answer to the question: When does dynamic programming work? We describe some simple axioms for combination and marginalization that enable the use of dynamic programming for solving optimization problems. These axioms are new. They are weaker than those proposed by Mitten [9].

Sixth, the VN described here can be easily adapted to represent propositional logic (Shenoy [15, 20]) and constraint satisfaction problems (Shenoy and Shafer [22]).

An outline of this chapter is as follows. In Section 15.2, we describe the VN framework and show how a discrete optimization problem fits in this framework. In Section 15.3, we state some simple axioms that justify the use of local computation in solving VNs. In Section 15.4, we describe a fusion algorithm for solving a VN using local computation. Throughout the chapter, we use one example to illustrate all definitions and to illustrate the fusion algorithm. In Section 15.5, we compare our axioms to those proposed by Mitten [9] for serial dynamic programming. In Section 15.6, we make some concluding remarks. Finally, in Section 15.7, we provide proofs for all theorems in the chapter.

Table 15.1: Factors of the Objective Function, ϕ_1, ϕ_2, and ϕ_3

$\Omega_{\{E, A, C\}}$			ϕ_1
e	a	c	1
e	a	~c	5
e	~a	c	2
e	~a	~c	2
~e	a	c	3
~e	a	~c	8
~e	~a	c	6
~e	~a	~c	4

$\Omega_{\{B, A\}}$		ϕ_2
b	a	4
b	~a	0
~b	a	8
~b	~a	5

$\Omega_{\{E, B, D\}}$			ϕ_3
e	b	d	0
e	b	~d	6
e	~b	d	5
e	~b	~d	4
~e	b	d	5
~e	b	~d	3
~e	~b	d	1
~e	~b	~d	3

15.2 Valuation Networks and Optimization

A valuation network consists of valuations, combination, and marginalization. We will discuss each of these in detail. We will illustrate all definitions using an optimization problem from Bertele and Brioschi [4].

An Optimization Problem.
There are five variables labelled A, B, C, D, and E. Each variable has two possible values. Let a and $\neg a$ denote the possible values of A, etc.. The global objective function ϕ for variables A, B, C, D, and E factors additively as follows: $\phi = \phi_1 + \phi_2 + \phi_3$, where ϕ_1 is a function for E, A, and C, ϕ_2 is a function for B and A, and ϕ_3 is a function for E, B, and D. Table 15.1 shows the details of these three functions. The problem is to find the minimum value of ϕ and to find a configuration of all variables where the minimum value is achieved.

Variables, State Spaces, and Configurations.
We use the symbol Ω_X for the set of possible values of a variable X, and we call Ω_X the *state space* for X. We are concerned with a finite set Θ of variables, and we assume that all the variables in Θ have finite state spaces.

Given a finite non-empty set h of variables, let Ω_h denote the Cartesian product of Ω_X for X in h, i.e. $\Omega_h = \times \{\Omega_X \mid X \in h\}$. We call Ω_h the *state space* for h. We call elements of Ω_h *configurations* of h. Lower-case bold-faced letters, **x**, **y**, *etc.* will denote configurations. If **x** is a configuration of g, **y** is a configuration of h, and $g \cap h = \emptyset$, then $(\mathbf{x}, \mathbf{y})$ will denote the configuration of $g \cup h$ obtained by concatenating **x** and **y**.

It is convenient to allow the set of variables h to be empty. We adopt

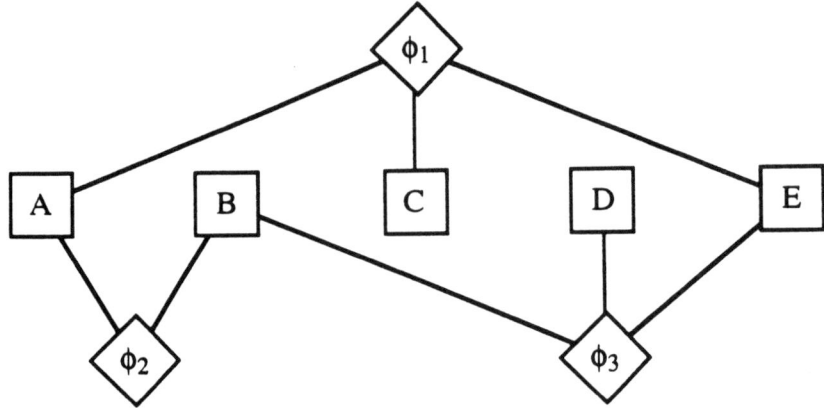

Figure 15.1: A valuation network for the optimization problem

the convention that the state space for the empty set $\emptyset$ consists of a single element, and we use the symbol $\Diamond$ to name that element; $\Omega_\emptyset = \{\Diamond\}$. If $\mathbf{x}$ is a configuration of g, then $(\mathbf{x}, \Diamond)$ is simply $\mathbf{x}$.

Values and Valuations.
We are concerned with a set Δ whose elements are called *values*. Δ may be finite or infinite. Given a set h of variables, we call any function $\sigma : \Omega_h \to \Delta$, a *valuation for* h, and we call h the *domain* of σ. Note that to specify a valuation σ for $\emptyset$, we need to specify only a single value, $\sigma(\Diamond)$. We will use lower-case Greek letters to denote valuations.

In our optimization problem, the set Δ is the set of real numbers, and we have three valuations ϕ_1, ϕ_2 and ϕ_3. ϕ_1 is a valuation for $\{E, A, C\}$, ϕ_2 is a valuation for $\{B, A\}$, and ϕ_3 is a valuation for $\{E, B, D\}$. Figure 15.1 shows a graphical depiction of the qualitative features of the optimization problem. We call such a graph a *valuation network*. In a valuation network, rectangular nodes represent variables, and diamond-shaped nodes represent valuations. Each valuation is connected to the variables in its domain by undirected edges.

Let ϑ_h denote the set of valuations for h, and let ϑ denote the set of valuations, *i.e.* $\vartheta = \cup \{\vartheta_h \mid h \subseteq \Theta\}$.

Projection of Configurations.
Projection of configurations simply means dropping extra coordinates; if $(\neg a, b, \neg c, d, e)$ is a configuration of $\{A, B, C, D, E\}$, for example, then the projection of $(\neg a, b, \neg c, d, e)$ to $\{A, C, E\}$ is simply $(\neg a, \neg c, e)$, which is a configuration of $\{A, C, E\}$.

If g and h are sets of variables, $h \subseteq g$, and $\mathbf{x}$ is a configuration of g, then let $\mathbf{x}^{\downarrow h}$ denote the projection of $\mathbf{x}$ to h. The projection $\mathbf{x}^{\downarrow h}$ is always a configuration of h. If $h = g$ and $\mathbf{x}$ is a configuration of g, then $\mathbf{x}^{\downarrow h} = \mathbf{x}$. If

$h = \emptyset$, then, of course, $\mathbf{x}^{\downarrow h} = \Diamond$.

Combination.

We assume there is a mapping $\copyright : \Delta \times \Delta \to \Delta$ called *combination* so that if u, v $\in \Delta$, then u$\copyright$v is the value representing the combination of u and v. We define a mapping $\oplus : \vartheta \times \vartheta \to \vartheta$ in terms of $\copyright$, also called *combination*, such that if γ and η are valuations for g and h, respectively, then $\gamma \oplus \eta$ is the valuation for $g \cup h$ given by

$$(\gamma \oplus \eta)(\mathbf{x}) = \gamma(\mathbf{x}^{\downarrow g})\copyright\eta(\mathbf{x}^{\downarrow h}) \tag{15.1}$$

for all $\mathbf{x} \in \Omega_{g \cup h}$. We call $\gamma \oplus \eta$ the *combination* of γ and η.

In our optimization problem, $\copyright$ is simply addition, *i.e.*

$$(\gamma \oplus \eta)(\mathbf{x}) = \gamma(\mathbf{x}^{\downarrow g}) + \eta(\mathbf{x}^{\downarrow h}). \tag{15.2}$$

Using (15.2), we can express the global objective function ϕ as follows: $\phi = \phi_1 \oplus \phi_2 \oplus \phi_3$.

Marginalization.

We assume that for each $h \subseteq \Theta$, and for each $X \in h$, there is a mapping $\downarrow (h - \{X\}) : \vartheta_h \to \vartheta_{h-\{X\}}$, called *marginalization to* $h - \{X\}$, such that if η is a valuation for h and $X \in h$, then $\eta^{\downarrow(h-\{X\})}$ is a valuation for $h - \{X\}$. We call $\eta^{\downarrow(h-\{X\})}$ the *marginal* of η for $h - \{X\}$.

For our optimization problem, we define marginalization as follows:

$$\eta^{\downarrow(h-\{X\})}(\mathbf{y}) = MIN\{\eta(\mathbf{y}, \mathbf{x}) \mid \mathbf{x} \in \Omega_X\} \tag{15.3}$$

for all $\mathbf{y} \in \Omega_{h-\{X\}}$.

If γ is a valuation for g, and $h \subseteq g$, then $\gamma^{\downarrow h}$ will denote the marginal of γ for h obtained by sequentially marginalizing all variables in $g - h$ out of γ in some sequence. In the next section, we will state an axiom that says that the sequence in which variables are marginalized out of a valuation does not affect the final answer. This axiom allows us to use this notation. Note that if ϕ is a global objective function for Θ, then, using this notation, $\phi^{\downarrow \emptyset}(\Diamond)$ represents the minimum value of ϕ.

In an optimization problem, besides computing the minimum value of the joint objective function, we are usually also interested in finding a configuration where the minimum is achieved. This motivates the following definition.

Solution for a Valuation.

Suppose η is a valuation for h. We call $\mathbf{x} \in \Omega_h$ a *solution* for η if $\eta(\mathbf{x}) = \eta^{\downarrow \emptyset}(\Diamond)$.

Solution for a Variable.

As we will see, once we have computed the minimum value of a valuation, computing a solution for the valuation is a matter of bookkeeping. Each time we eliminate a variable from a valuation using minimization, we store a table of configurations of the eliminated variable where the minimums are achieved. We can think of this table as a function. We call this function "a solution for the variable". Formally, we define a solution for a variable as follows. Suppose

X is a variable, suppose h is a subset of variables containing X, and suppose η is a valuation for h. We call a function $\Psi_X : \Omega_{h-\{X\}} \to \Omega_X$ a *solution* for X (with respect to η) if

$$\eta^{\downarrow(h-\{X\})}(\mathbf{c}) = \eta(\mathbf{c}, \Psi_X(\mathbf{c})) \tag{15.4}$$

for all $\mathbf{c} \in \Omega_{h-\{X\}}$.

In summary, a VN consists of a set of variables Θ, a state space for each variable $\{\Omega_X\}_{X \in \Theta}$, a set of values Δ, a collection of valuations $\{\sigma_1, \ldots, \sigma_k\}$, a definition of combination $\copyright$, and a definition of marginalization $\downarrow$. Given a VN, there are two problems of interest. First, we would like to compute $(\sigma_1 \oplus \ldots \oplus \sigma_k)^{\downarrow \emptyset}(\diamondsuit)$. (In an optimization problem, this represents the minimum value of the joint objective function.) Second, we would like to compute a solution for $\sigma_1 \oplus \ldots \oplus \sigma_k$. (In an optimization problem, this represents an optimal solution.)

If Θ is a large set of variables, and $\sigma = \sigma_1 \oplus \ldots \oplus \sigma_k$ is a valuation for Θ, then a brute force computation of σ and an exhaustive search of the set of all configurations of Θ to determine a solution for σ are not computationally tractable. In the next section we will state three axioms for combination and marginalization that allow the use of local computation to compute the minimum value of σ and to compute a solution for σ.

15.3 The Axioms

We will state three axioms. Axiom A1 is for combination. Axiom A2 is for marginalization. And Axiom A3 is for combination and marginalization.

Axiom A1 (*Commutativity and associativity of combination for values*): Suppose u, v, w $\in \Delta$. Then

$$u \copyright v = v \copyright u, \text{ and } u \copyright (v \copyright w) = (u \copyright v) \copyright w.$$

Axiom A2 (*Order of deletion does not matter*): Suppose γ is a valuation for g, and suppose $X_1, X_2 \in g$. Then

$$(\gamma^{\downarrow(g-\{X_1\})})^{\downarrow(g-\{X_1,X_2\})} = (\gamma^{\downarrow(g-\{X_2\})})^{\downarrow(g-\{X_1,X_2\})}.$$

Axiom A3 (*Distributivity of marginalization over combination*): Suppose γ and η are valuations for g and h, respectively, suppose $X \in h$, and suppose $X \notin g$. Then

$$(\gamma \oplus \eta)^{\downarrow((g \cup h) - \{X\})} = \gamma \oplus (\eta^{\downarrow(h-\{X\})}).$$

It follows from Axiom A1 that $\oplus$ is commutative and associative. Therefore, the combination of several valuations can be written without using parentheses. For example, $(\ldots((\sigma_1 \oplus \sigma_2) \oplus \sigma_3) \oplus \ldots \oplus \sigma_k)$ can be simply written as

$\oplus\{\sigma_1,\ldots,\sigma_k\}$ or as $\sigma_1 \oplus \ldots \oplus \sigma_k$ without specifying the order in which the combination is done.

If we regard marginalization as a coarsening of a valuation by deleting variables, then Axiom A2 says that the order in which the variables are deleted does not matter. One implication of this axiom is that $(\gamma^{\downarrow(g-\{X_1\})})^{\downarrow(g-\{X_1,X_2\})}$ can be written simply as $\gamma^{\downarrow(g-\{X_1,X_2\})}$, i.e. we need not indicate the order in which the variables are deleted.

Axiom A3 is the crucial axiom that makes local computation possible. Axiom A3 states that computation of $(\gamma \oplus \eta)^{\downarrow((g \cup h)-\{X\})}$ can be accomplished without having to compute $\gamma \oplus \eta$. The combination operation in $\gamma \oplus \eta$ is on the state space for $g \cup h$, whereas the combination operation in $\gamma \oplus (\eta^{\downarrow(h-\{X\})})$ is on the state space for $(g \cup h) - \{X\}$.

For our optimization problem, it is easy to see that the definition of combination in (15.2) and the definition of marginalization in (15.3) satisfy the three axioms.

15.4 A Fusion Algorithm

In this section, we describe a fusion algorithm for solving a VN using local computation, i.e. for computing exactly the marginal of the joint valuation for the empty set without explicitly computing the joint valuation.

Suppose $\rho = \{\Theta, \{\Omega_X\}_{X \in \Theta}, \Delta, \{\sigma_1,\ldots,\sigma_k\}, \copyright, \downarrow\}$ is a VN, and suppose $\copyright$ and $\downarrow$ satisfy Axioms A1-A3. We will describe a fusion algorithm for computing the marginal $(\sigma_1 \oplus \ldots \oplus \sigma_k)^{\downarrow \emptyset}$ and for computing a solution for $\sigma_1 \oplus \ldots \oplus \sigma_k$ using local computation.

The basic idea of the method is to successively delete all variables from the VN. Any sequence may be used. Axiom A2 tells us that all deletion sequences will lead to the same answers. But different deletion sequences may involve different computational costs. We will comment on good deletion sequences at the end of this section.

When we delete a variable, we have to do a "fusion" operation on the valuations. Consider a set of m valuations $\{\alpha_1,\ldots,\alpha_m\}$. Suppose α_i is a valuation for a_i for $i = 1,\ldots,m$. Let $Fus_{X_j}\{\alpha_1,\ldots,\alpha_m\}$ denote the collection of valuations after fusing the valuations in the set $\{\alpha_1,\ldots,\alpha_k\}$ with respect to variable X_j. Then

$$Fus_{X_j}\{\alpha_1,\ldots,\alpha_m\} = \{\alpha^{\downarrow(g_j-\{X_j\})}\} \cup \{\alpha_i \mid X_j \notin a_i\}, \qquad (15.5)$$

where $\alpha = \oplus\{\alpha_i \mid X_j \in a_i\}$, and $g_j = \cup\{a_i \mid X_j \in a_i\}$. After fusion, the set of valuations is changed as follows. All valuations that have X_j in their domain are combined, and the resulting valuation is marginalized such that X_j is eliminated from its domain. The valuations that do not have X_j in their domains remain unchanged.

When we compute the marginal $\alpha^{\downarrow(g_j - \{X_j\})}$ in (15.5), assume that we store a solution for X_j with respect to α, $\Psi_{X_j} : \Omega_{g_j - \{X_j\}} \to \Omega_{X_j}$. We will describe a method for constructing a solution for the joint valuation using these solutions.

We are ready to state the main theorem which describes the fusion algorithm.

Theorem 15.1 *Fusion Algorithm.*

Suppose $\rho = \{\Theta, \{\Omega_X\}_{X \in \Theta}, \Delta, \{\sigma_1, \ldots, \sigma_k\}, \copyright, \downarrow\}$ is a VN satisfying Axioms A1-A3. Suppose $X_1 X_2 \ldots X_n$ is a sequence of variables in Θ. Then

$$(\sigma_1 \oplus \ldots \oplus \sigma_k)^{\downarrow \emptyset} = \oplus Fus_{X_n}\{\ldots Fus_{X_2}\{Fus_{X_1}\{\sigma_1, \ldots, \sigma_k\}\}\}.$$

We give a simple proof of Theorem 15.1 in Section 15.7. The essence of the fusion algorithm is to combine valuations on smaller state spaces instead of combining all valuations on the global state space associated with Θ.

Figure 15.2 shows a graphical depiction of the fusion algorithm for the optimization problem using deletion sequence CDABE. In this figure, the valuation network labelled 0 is the original network. Valuation network 1 is the result after fusion with respect to C. Since there is only one valuation with C in its domain, there is no combination here, only marginalization of ϕ_1 to $\{A, E\}$. Valuation network 2 is the result after fusion with respect to D. Again, since there is only one valuation with D in its domain, there is no combination here, only marginalization of ϕ_3 to $\{B, E\}$. Valuation network 3 is the result after fusion with respect to A. Since A is in the domain of $\phi_1^{\downarrow\{A,E\}}$ and ϕ_2, we first combine these two valuations and then marginalize A out of the resulting valuation. Valuation network 4 is the result after fusion with respect to B. And finally, valuation network 5 is the result after fusion with respect to E. Theorem 15.1 tells us that $[(\phi_1^{\downarrow\{A,E\}} \oplus \phi_2)^{\downarrow\{B,E\}} \oplus \phi_3^{\downarrow\{B,E\}}]^{\downarrow \emptyset} = (\phi_1 \oplus \phi_2 \oplus \phi_3)^{\downarrow \emptyset}$. Table 15.2 shows the details of the computations in the fusion algorithm. The minimum value of the objective function is 2.

When we implement the fusion algorithm, each time we marginalize a variable, assume that we store a solution for that variable. If we use deletion sequence $X_1 X_2 \ldots X_n$, then at the end of the fusion algorithm, we have for each variable X_j, a solution $\Psi_{X_j} : \Omega_{g_j - \{X_j\}} \to \Omega_{X_j}$, where g_j is as defined in (15.5). Note that $g_1 = \cup \{h_i \mid X_1 \in h_i\}$. The precise definition of g_2 will depend upon the valuations in the set $Fus_{X_1}\{\sigma_1, \ldots, \sigma_k\}$. However, since X_1 has been deleted, $g_2 \subseteq \{X_2, \ldots, X_n\}$ and $X_2 \in g_2$. In general, $g_i \subseteq \{X_i, \ldots, X_n\}$, and $X_i \in g_i$ for $i = 1, \ldots, n$. Note that $g_n = \{X_n\}$.

Table 15.2: Numerical computations in the fusion algorithm for the optimization problem

$\Omega_{\{E, A, C\}}$			ϕ_1	$\phi_1^{\downarrow\{E, A\}}$	Ψ_C
e	a	c	1	1	c
e	a	~c	5		
e	~a	c	2	2	c or ~c
e	~a	~c	2		
~e	a	c	3	3	c
~e	a	~c	8		
~e	~a	c	6	4	~c
~e	~a	~c	4		

$\Omega_{\{E, B, D\}}$			ϕ_3	$\phi_3^{\downarrow\{E, B\}}$	Ψ_D
e	b	d	0	0	d
e	b	~d	6		
e	~b	d	5	4	~d
e	~b	~d	4		
~e	b	d	5	3	~d
~e	b	~d	3		
~e	~b	d	1	1	d
~e	~b	~d	3		

$\Omega_{\{E, B, A\}}$			$\phi_1^{\downarrow\{E, A\}}$	ϕ_2	$\phi_1^{\downarrow\{E, A\}} \oplus \phi_2$	$(\phi_1^{\downarrow\{E, A\}} \oplus \phi_2)^{\downarrow\{E, B\}}$	Ψ_A
e	b	a	1	4	5	2	~a
e	b	~a	2	0	2		
e	~b	a	1	8	9	7	~a
e	~b	~a	2	5	7		
~e	b	a	3	4	7	4	~a
~e	~d	~a	4	0	4		
~e	~b	a	3	8	11	9	~a
~e	~b	~a	4	5	9		

$\Omega_{\{E, B\}}$		$(\phi_1^{\downarrow\{E, A\}} \oplus \phi_2)^{\downarrow\{E, B\}}$	$\phi_3^{\downarrow\{E, B\}}$	$(\phi_1^{\downarrow\{E, A\}} \oplus \phi_2)^{\downarrow\{E, B\}} \oplus \phi_3^{\downarrow\{E, B\}}$	$[(\phi_1^{\downarrow\{E, A\}} \oplus \phi_2)^{\downarrow\{E, B\}} \oplus \phi_3^{\downarrow\{E, B\}}]^{\downarrow\{E\}}$	Ψ_B
e	b	2	0	2	2	b
e	~b	7	4	11		
~e	b	4	3	7	7	b
~e	~b	9	1	10		

Ω_E	$[(\phi_1^{\downarrow\{E, A\}} \oplus \phi_2)^{\downarrow\{E, B\}} \oplus \phi_3^{\downarrow\{E, B\}}]^{\downarrow\{E\}}$	$[(\phi_1^{\downarrow\{E, A\}} \oplus \phi_2)^{\downarrow\{E, B\}} \oplus \phi_3^{\downarrow\{E, B\}}]^{\downarrow\varnothing}(\blacklozenge)$	$\Psi_E(\blacklozenge)$
e	2	2	e
~e	7		

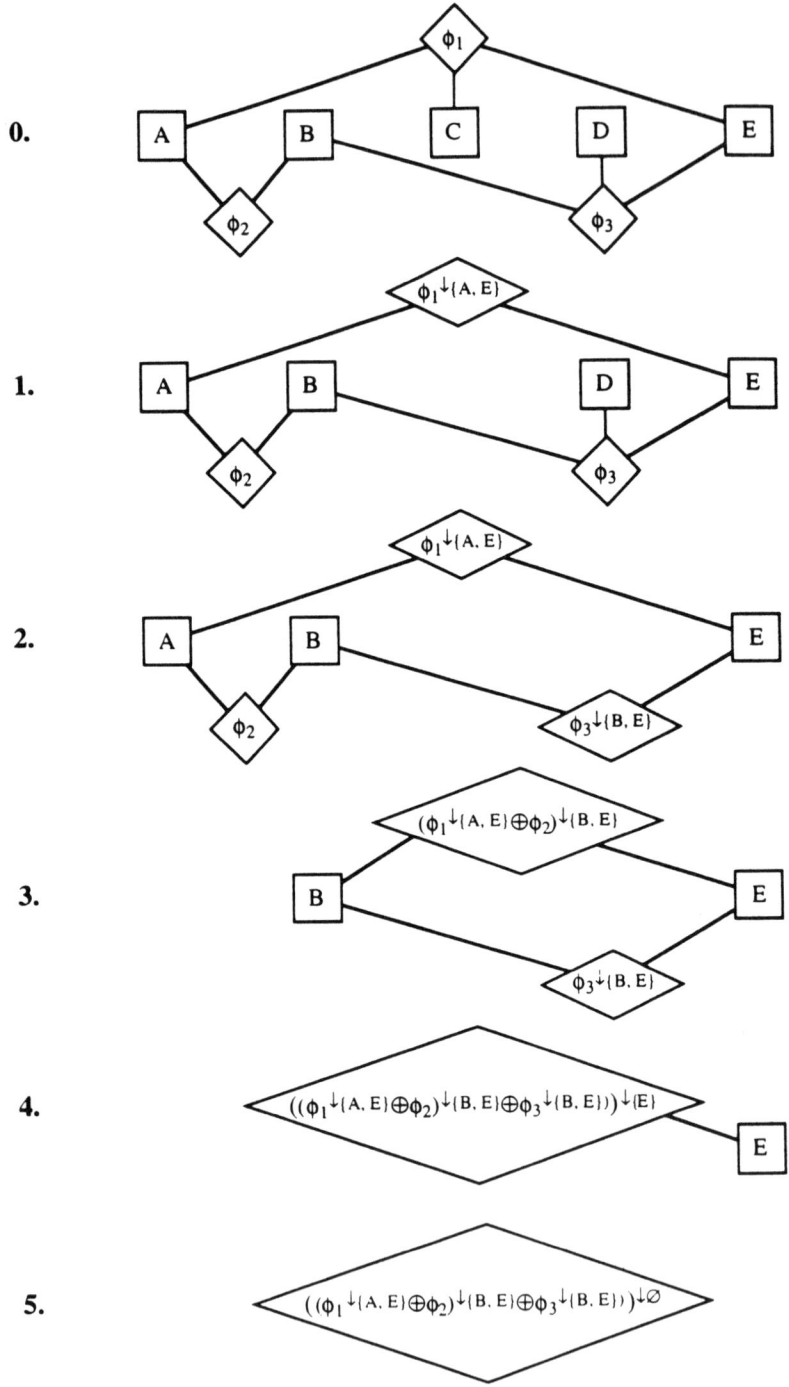

Figure 15.2: The fusion algorithm for the optimization problem using deletion sequence CDABE

Theorem 15.2 describes a recursive method for constructing a solution for the joint valuation. The solution is constructed piecemeal starting with the component in Ω_{X_n} and working sequentially opposite to the deletion sequence.

Theorem 15.2 *Constructing a Solution.*
Suppose $\rho = \{\Theta, \{\Omega_X\}_{X \in \Theta}, \Delta, \{\sigma_1, \ldots, \sigma_k\}, \copyright, \downarrow\}$ *a VN satisfying Axioms A1-A3. Suppose* $X_1 X_2 \ldots X_n$ *is a sequence of variables in* Θ. *Suppose* $\Psi_{X_j} : \Omega_{g_j - \{X_j\}} \to \Omega_{X_j}$ *is a solution for* X_j *computed during fusion of* $Fus_{X_{j-1}}\{\ldots Fus_{X_2}\{Fus_{X_1}\{\sigma_1, \ldots, \sigma_k\}\}\}$ *with respect to* X_j, *for* $j = 1, \ldots, n$. *Then* $\mathbf{z} \in \Omega_\Theta$ *given by*

$$\mathbf{z}^{\downarrow \{X_j\}} = \Psi_{X_j}(z^{\downarrow(g_j - \{X_j\})}) \text{ for } j = n, n-1, \ldots, 1$$

is a solution for $\sigma_1 \oplus \ldots \oplus \sigma_k$.

To illustrate Theorem 15.2, consider the optimization problem. We computed the minimum value of $\phi_1 \oplus \phi_2 \oplus \phi_3$ using deletion sequence CDABE. In the process, we have a solution for $C, \Psi_C : \Omega_{\{A,E\}} \to \Omega_C$, a solution for $D, \Psi_D : \Omega_{\{E,B\}} \to \Omega_D$, a solution for $A, \Psi_A : \Omega_{\{E,B\}} \to \Omega_A$, a solution for $B, \Psi_B : \Omega_E \to \Omega_B$, and a solution for $E, \Psi_E : \Omega_\emptyset \to \Omega_E$. Theorem 15.2 tells us we can construct a solution as follows. Working opposite to the deletion sequence, first, $\Psi_E(\Diamond) = e$. Next, $\Psi_B(e) = b$. Next, $\Psi_A(e, b) = \neg a$. Next, $\Psi_D(e, b) = d$. And finally, $\Psi_C(e, \neg a) = c$ or $\neg c$. Thus, configurations $(\neg a, b, \neg c, d, e)$ and $(\neg a, b, \neg c, d, e)$ are both solutions for ϕ.

Deletion Sequences. The sequence in which we delete variables in the fusion algorithm is called the *deletion sequence*. Which deletion sequence should one use? First, note that all deletion sequences lead to the same final result. This is implied in the statement of Theorem 15.1. Second, different deletion sequences may involve different computational efforts. For example, consider the VN representation of the optimization problem shown in Figure 15.1. In this example, all deletion sequences starting with variable E involve more computational effort than sequences that do not start with E, as the former involves combination on the state space of all five variables. Finding an optimal deletion sequence is a secondary optimization problem that has shown to be NP-complete [2]. But, there are several heuristics for finding good deletion sequences [11, 6, 8].

One such heuristic is called one-step-look-ahead [11, 6]. This heuristic tells us which variable to delete next. As per this heuristic, the variable that should be deleted next is one that leads to combination over the smallest state space with ties resolved arbitrarily. For example, in the VN of Figure 15.2, for the first deletion, this heuristic would pick either C or D since no combination is involved with these deletions. After deletion of C and D, any remaining variables can be used for successive deletions as they all lead to combinations over state spaces of equal sizes.

15.5 Mitten's Axioms for Dynamic Programming

For optimization problems, the fusion algorithm described in Section 15.4 reduces to the method of non-serial dynamic programming [10, 4]. Bellman's dynamic programming methodology appealed to a principle of optimality that translates into Axiom A3 with combination interpreted as addition and marginalization interpreted as maximization over the deleted variables [3]. Mitten [9] has described a more general framework for discrete dynamic programming. In this section, we will describe Mitten's framework using our notation, and compare his axiom with ours.

Values and Valuations. The *value space* is $\mathbf{R}$, the set of real numbers. A *valuation for h* is a real-valued function on Ω_h.

Combination. There is a mapping $\copyright : \mathbf{R} \times \mathbf{R} \to \mathbf{R}$ that is commutative and associative. Define a mapping $\oplus : \vartheta \times \vartheta \to \vartheta$ such that whenever γ and η are valuations for g and h, respectively, $\gamma \oplus \eta$ is a valuation for $g \cup h$ given by

$$(\gamma \oplus \eta)(\mathbf{x}) = \gamma(\mathbf{x}^{\downarrow g}) \copyright \eta(\mathbf{x}^{\downarrow h})$$

for all $\mathbf{x} \in \Omega_{g \cup h}$.

Monotonicity of Combination. We shall say that $\copyright$ is *monotonic* if for any $u, v_1, v_2 \in \mathbf{R}$, $u \copyright v_1 \geq u \copyright v_2$ whenever $v_1 \geq v_2$. Suppose η_1 and η_2 are valuations for h. We shall say that $\eta_1 \geq \eta_2$ if $\eta_1(\mathbf{x}) \geq \eta_2(\mathbf{x})$ for all $\mathbf{x} \in \Omega_h$. Note that if $\copyright$ is monotonic, then for all valuations γ, $\gamma \oplus \eta_1 \geq \gamma \oplus \eta_2$ whenever $\eta_1 \geq \eta_2$.

Marginalization. Define a mapping $\downarrow (h - \{X\}) : \vartheta_h \to \vartheta_{h-\{X\}}$ such that whenever η is a valuation for h, $\eta^{(h-\{X\})}$ is a valuation for $h - \{X\}$ given by

$$\eta^{\downarrow(h-\{X\})}(\mathbf{y}) = \max\{\eta(\mathbf{y}, \mathbf{x}) \mid \mathbf{x} \in \Omega_X\} \qquad (15.6)$$

for all $\mathbf{y} \in \Omega_{h-\{X\}}$.

Theorem 15.3 *Suppose the value space is $\mathbf{R}$, suppose marginalization is defined as in (15.6), and suppose $\copyright$ is monotonic. If γ is a valuations for g, η is a valuation for $h, X \in h$, and $X \notin g$, then $(\gamma \oplus \eta)^{\downarrow((g \cup h) - \{X\})} = \gamma \oplus (\eta^{\downarrow(h-\{X\})})$.*

Thus monotonicity of $\copyright$ implies Axiom A3. The other condition that Mitten requires in his framework is called *separability* and it amounts to a serial factorization of the joint objective function. In our VN framework, we do not require any particular structure for the factorization of the joint valuation.

15.6 Conclusion

In the introduction, we raised two questions: What is dynamic programming? And, when does dynamic programming work? The main contribution of this chapter is the abstract VN framework and three axioms that permit the use of local computation in solving a VN. We can think of the framework and the fusion algorithm as the answer to the first question. The three axioms constitute one answer to the second question.

15.7 Proofs

In this section, we will provide proofs for Theorems 15.1 and 15.2 stated in Section 15.4 and Theorem 15.3 stated in Section 15.5. We start with a lemma we need to prove Theorem 15.1.

Lemma 1. Suppose $\{\Theta, \{\Omega_X\}_{X\in\Theta}, \Delta, \{\sigma_1, \ldots, \sigma_k\}, \copyright, \downarrow\}$ is a VN where σ_i is a valuation for h_i for $i = 1, \ldots, k$, $\Theta = h_1 \cup \ldots \cup h_k$, and $\copyright$ and $\downarrow$ satisfy Axioms A1-A3. Suppose $X \in \Theta$. Then

$$(\sigma_1 \oplus \ldots \oplus \sigma_k)^{\downarrow(\Theta-\{X\})} = \oplus Fus_X\{\sigma_1, \ldots, \sigma_k\}.$$

Proof of Lemma 1: Let $g = \cup\{h_i \mid X \notin h_i\}$, and let $h = \cup\{h_i \mid X \in h_i\}$. Let $\gamma = \oplus\{\sigma_i \mid X \notin h_i\}$, and $\eta = \oplus\{\sigma_i \mid X \in h_i\}$. Note that $X \in h$, and $X \notin g$. Then

$$\begin{aligned}
(\sigma_1 \oplus \ldots \oplus \sigma_k)^{\downarrow(\Theta-\{X\})} &= (\gamma \oplus \eta)^{\downarrow((g\cup h)-\{X\})} \\
&= \gamma \oplus (\eta^{\downarrow(h-\{X\})}) \qquad \text{(using Axiom A3)} \\
&= (\oplus\{\sigma_i \mid X \notin h_i\}) \oplus (\oplus\{\sigma_i \mid X \in h_i\})^{\downarrow(h-\{X\})} \\
&= \oplus Fus_X\{\sigma_1, \ldots, \sigma_k\}
\end{aligned}$$

(by definition of $Fus_X\{\sigma_1, \ldots, \sigma_k\}$).
□

Proof of Theorem 15.1: By Axiom A2, $(\sigma_1 \oplus \ldots \oplus \sigma_k)^{\downarrow\emptyset}$ is obtained by sequentially marginalizing all variables out of the joint valuation. A proof of this theorem is obtained by repeatedly applying the result of Lemma 1. At each step, we delete a variable and fuse the set of all valuations with respect to this variable. Using Lemma 1, after fusion with respect to X_1, the combination of all valuations in the resulting VN is equal to $(\sigma_1 \oplus \ldots \oplus \sigma_k)^{\downarrow(\Theta-\{X_1\})}$. Again, using Lemma 1, after fusion with respect to X_2, the combination of all valuations in the resulting VN is equal to $(\sigma_1 \oplus \ldots \oplus \sigma_k)^{\downarrow(\Theta-\{X_1,X_2\})}$. And so on. When all variables have been deleted, we have the result.
□

Next, we state and prove a lemma we need to prove Theorem 15.2.

Lemma 2. Suppose $\{\Theta, \{\Omega_X\}_{X \in \Theta}, \Delta, \{\sigma_1, \ldots, \sigma_k\}, \copyright, \downarrow\}$ is a VN where σ_i is a valuation for h_i for $i = 1, \ldots, k$, $\Theta = h_1 \cup \ldots \cup h_k$, and $\copyright$ and $\downarrow$ satisfy Axioms A1–A3. Suppose $\Psi_X : \Omega_{g-\{X\}} \to \Omega_X$ is a solution for X computed during the marginalization operation involved in the fusion of $\{\sigma_1, \ldots, \sigma_k\}$ with respect to X, and suppose $\mathbf{c} \in \Omega_{\Theta-\{X\}}$ is a solution for $(\sigma_1 \oplus \ldots \oplus \sigma_k)^{\downarrow(\Theta-\{X\})}$. Then $(\mathbf{c}, \Psi_X(\mathbf{c}^{\downarrow(g-\{X\})}))$ is a solution for $\sigma_1 \oplus \ldots \oplus \sigma_k$.

Proof of Lemma 2: Without loss of generality, suppose that $\sigma_1, \ldots, \sigma_m$ are the only valuations that have X in their domain. Let $\sigma = \sigma_1 \oplus \ldots \sigma_m$, let $g = g_1 \cup \ldots \cup g_m$, and let $\mathbf{c} \in \Omega_{\Theta-\{X\}}$. We need to prove that $(\sigma_1 \oplus \ldots \oplus \sigma_k)(\mathbf{c}, \Psi_X(\mathbf{c}^{\downarrow g(-\{X\})})) = (\sigma_1 \oplus \ldots \oplus \sigma_k)^{\downarrow \emptyset}(\diamondsuit)$. We have

$(\sigma_1 \oplus \ldots \oplus \sigma_k)(\mathbf{c}, \Psi_X(\mathbf{c}^{\downarrow(g-\{X\})}))$

$= \sigma(\mathbf{c}^{\downarrow(g-\{X\})}, \Psi_X(\mathbf{c}^{\downarrow(g-\{X\})}))\copyright \sigma_{m+1}(\mathbf{c}^{\downarrow g_{m+1}})\copyright \ldots \copyright \sigma_k(\mathbf{c}^{\downarrow g_k})$

(by definition of combination)

$= \sigma^{\downarrow(g-\{X\})}(\mathbf{c}^{\downarrow(g-\{X\})})\copyright \sigma_{m+1}(\mathbf{c}^{\downarrow g_{m+1}})\copyright \ldots \copyright \sigma_k(\mathbf{c}^{\downarrow g_k})$

(since Ψ_X is a solution for X)

$= \oplus Fus_X\{\sigma_1, \ldots, \sigma_k\}(\mathbf{c})$

(by definition of $Fus_X\{\sigma_1, \ldots, \sigma_k\}$)

$= (\sigma_1 \oplus \ldots \oplus \sigma_k)^{\downarrow(\Theta-\{X\})})(\mathbf{c})$

(by Lemma 1)

$= (\sigma_1 \oplus \ldots \oplus \sigma_k)^{\downarrow \emptyset}(\diamondsuit)$

(since $\mathbf{c}$ is a solution for $(\sigma_1 \oplus \ldots \oplus \sigma_k)^{\downarrow(\Theta-\{X\})}$).

$\square$

Proof of Theorem 15.2: A proof of this theorem is obtained by repeatedly applying Lemma 2. Consider the VN $\oplus Fus_{X_n}\{\ldots Fus_{X_2}\{Fus_{X_1}\{\sigma_1, \ldots, \sigma_k\}\}\}$. There is only one valuation in this VN and it is for the empty set. From Theorem 15.1, $(\oplus Fus_{X_n}\{\ldots Fus_{X_2}\{Fus_{X_1}\{\sigma_1, \ldots, \sigma_k\}\}\})(\diamondsuit) = (\sigma_1 \oplus \ldots \oplus \sigma_k)^{\downarrow \emptyset}(\diamondsuit)$. Since $\diamondsuit$ is a solution for $(\sigma_1 \oplus \ldots \oplus \sigma_k)^{\downarrow \emptyset}$, by Lemma 2, $(\diamondsuit, \Psi_{X_n}(\diamondsuit)) = \Psi_{X_n}(\diamondsuit) = \mathbf{z}^{\downarrow\{X_n\}}$ is a solution for $(\sigma_1 \oplus \ldots \oplus \sigma_k)^{\downarrow\{X_n\}}$.

Since $\mathbf{z}^{\downarrow\{X_n\}}$ is a solution for $(\sigma_1 \oplus \ldots \oplus \sigma_k)^{\downarrow\{X_n\}}$, and $\Psi_{X_{n-1}} : \Omega_{g_{n-1}} - \{X_{n-1}\} \to \Omega_{X_{n-1}}$ is a solution for X_{n-1}, by Lemma 2, $(\mathbf{z}^{\downarrow\{X_n\}}, \Psi_{X_{n-1}}(\mathbf{z}^{\downarrow(g_{n-1}-\{X_{n-1}\})})) = (\mathbf{z}^{\downarrow\{X_n\}}, \mathbf{z}^{\downarrow\{X_{n-1}\}}) = \mathbf{z}^{\downarrow\{X_n, X_{n-1}\}}$ is a solution for $(\sigma_1 \oplus \ldots \oplus \sigma_k)^{\downarrow\{X_n, X_{n-1}\}}$.

Continuing in this fashion, we get the result that $\mathbf{z}$ is a solution for $\sigma_1 \oplus \ldots \oplus \sigma_k$.

$\square$

Proof of Theorem 15.3: Suppose $\mathbf{w} \in \Omega_{g-h}, \mathbf{y} \in \Omega_{g \cap h}$, and $\mathbf{z} \in \Omega_{h-g-\{X\}}$.

$$\begin{aligned}
(\gamma \oplus \eta)^{\downarrow((g \cup h)-\{X\})}(\mathbf{w},\mathbf{y},\mathbf{z}) &= \max\{(\gamma \oplus \eta)(\mathbf{w},\mathbf{y},\mathbf{z},\mathbf{x}) \mid \mathbf{x} \in \Omega_X\} \\
&= \max\{\gamma(\mathbf{w},\mathbf{y}) \copyright \eta(\mathbf{y},\mathbf{z},\mathbf{x}) \mid \mathbf{x} \in \Omega_X\} \\
&\geq \gamma(\mathbf{w},\mathbf{y}) \copyright (\max\{\eta(\mathbf{y},\mathbf{z},\mathbf{x}) \mid \mathbf{x} \in \Omega_X\}) \\
&= \gamma(\mathbf{w},\mathbf{y}) \copyright (\eta^{\downarrow(h-\{X\})}(\mathbf{y},\mathbf{z})) \\
&= (\gamma \oplus (\eta^{\downarrow(h-\{X\})}))(\mathbf{w},\mathbf{y},\mathbf{z}).
\end{aligned}$$

In other words, $(\gamma \oplus \eta)^{\downarrow((g \cup h)-\{X\})} \geq \gamma \oplus (\eta^{\downarrow(h-\{X\})})$.

Since $\copyright$ is monotonic and $\max\{\eta(\mathbf{y},\mathbf{z},\mathbf{x}) \mid \mathbf{x} \in \Omega_X\} \geq \eta(\mathbf{y},\mathbf{z},\mathbf{x})$ for all $\mathbf{x} \in \Omega_X$, we have $\gamma(\mathbf{w},\mathbf{y}) \copyright (\max\{\eta(\mathbf{y},\mathbf{z},\mathbf{x}) \mid \mathbf{x} \in \Omega_X\}) \geq \gamma(\mathbf{w},\mathbf{y}) \copyright \eta(\mathbf{y},\mathbf{z},\mathbf{x})$ for all $\mathbf{x} \in \Omega_X$. In particular, this inequality must hold for the maximum of the RHS with respect to $\mathbf{x}$, i.e. $\gamma(\mathbf{w},\mathbf{y}) \copyright (\max\{\eta(\mathbf{y},\mathbf{z},\mathbf{x}) \mid \mathbf{x} \in \Omega_X\}) \geq \max\{\gamma(\mathbf{w},\mathbf{y}) \copyright \eta(\mathbf{y},\mathbf{z},\mathbf{x}) \mid \mathbf{x} \in \Omega_X\}$, i.e. $\gamma \oplus (\eta^{\downarrow(h-\{X\})}) \geq (\gamma \oplus \eta)^{\downarrow((g \cup h)-\{X\})}$. Earlier we showed that $(\gamma \oplus \eta)^{\downarrow((g \cup h)-\{X\})} \geq \gamma \oplus (\eta^{\downarrow(h-\{X\})})$. Therefore, $(\gamma \oplus \eta)^{\downarrow((g \cup h)-\{X\})} = \gamma \oplus (\eta^{\downarrow(h-\{X\})})$.

$\square$

Acknowledgements

This work was supported in part by a grant from Hughes Research Laboratories.

References

[1] Aho A.V., Hopcroft J.E. and Ullman J.D. (1974) *The Design and Analysis of Computer Algorithms*. Addison-Wesley, Reading, MA.

[2] Arnborg S., Corneil D.G. and Proskurowski A. (1987), Complexity of finding embeddings in a k-tree. *SIAM Journal of Algebraic and Discrete Methods* 8: 277–284.

[3] Bellman R.E. (1957) *Dynamic Programming*. Princeton University Press, Princeton, NJ.

[4] Bertele U. and Brioschi F. (1972) *Nonserial Dynamic Programming*. Academic Press, New York.

[5] Dempster A.P. and Kong A. (1988) Uncertain evidence and artificial analysis. *Journal of Statistical Planning and Inference* 20: 355–368.

[6] Kong A. (1986) Multivariate belief functions and graphical models. Ph.D. thesis, Department of Statistics, Harvard University, Cambridge, MA.

[7] Lauritzen S.L. and Spiegelhalter D.J. (1988) Local computations with probabilities on graphical structures and their application to expert systems (with discussion). *Journal of the Royal Statistical Society, series B*, 50(2): 157–224.

[8] Mellouli K. (1987) On the propagation of beliefs in networks using the Dempster-Shafer theory of evidence. Ph.D. thesis, School of Business, University of Kansas, Lawrence, KS.

[9] Mitten L.G. (1964) Composition principles for synthesis of optimal multistage processes. *Operations Research* 12: 610–619.

[10] Nemhauser G.L. (1966) *Introduction to Dynamic Programming*. John Wiley & Sons, New York, NY.

[11] Olmsted S.M. (1983) On representing and solving decision problems. Ph.D. thesis, Department of Engineering-Economic Systems, Stanford University, Stanford, CA.

[12] Pearl J. (1986), Fusion, propagation and structuring in belief networks. *Artificial Intelligence* 29: 241–288.

[13] Shafer G. and Shenoy P.P. (1990) Probability propagation. *Annals of Mathematics and Artificial Intelligence* 2(1–4): 327-352.

[14] Shenoy P.P. (1989) A valuation-based language for expert systems. *International Journal for Approximate Reasoning* 3(5): 383–411.

[15] Shenoy P.P. (1990) Valuation-based systems for propositional logic. In Ras Z.W., Zemankova M. and Emrich M.L. (eds) *Methodologies for Intelligent Systems* 5: 305–312, North-Holland, Amsterdam.

[16] Shenoy P.P. (1992) Valuation-based systems: A framework for managing uncertainty in expert systems. In Zadeh L.A. and Kacprzyk J. (eds) *Fuzzy Logic for the Management of Uncertainty* 83–104, John Wiley & Sons, New York, NY.

[17] Shenoy P.P. (1992) Valuation-based systems for Bayesian decision analysis. *Operations Research*, 40(3): 463–484.

[18] Shenoy P.P. (1993) A new method for representing and solving Bayesian decision problems. In Hand D.J. (ed) *Artificial Intelligence Frontiers in Statistics: AI and Statistics III* 119–138, Chapman & Hall, London.

[19] Shenoy P.P. (1994), A comparison of graphical techniques for decision analysis. *European Journal of Operational Research*, 78(1): 1–21.

[20] Shenoy P.P. (1994), Consistency in valuation-based systems. *ORSA Journal on Computing* 6(3): 281–291.

[21] Shenoy P.P. and Shafer G. (1986) Propagating belief functions using local computation. *IEEE Expert* 1(3): 43–52.

[22] Shenoy P.P. and Shafer G. (1988) Constraint propagation. Working Paper No. 208, School of Business, University of Kansas, Lawrence, KS.

[23] Shenoy P.P. and Shafer G. (1990) Axioms for probability and belief-function propagation. In Shachter R.D., Levitt T.S., Lemmer J.F. and Kanal L.N. (eds) *Uncertainty in Artificial Intelligence* 4: 169–198, North-Holland, Amsterdam. Reprinted in Shafer G. and Pearl J. (eds) (1990) *Readings in Uncertain Reasoning*, 575–610, Morgan Kaufmann, San Mateo, CA.

16

Mixture-Model Cluster Analysis Using the Projection Pursuit Method

S. Aïvazian[1]

16.1 Mixture-Model and Cluster Analysis Problem

In this chapter the problem of clustering n individuals on the basis of p-dimensional observations[2]

$$\mathbf{X}^{(n)} = \{X_1, X_2, \ldots, X_n\}, \quad \text{where} \quad X_i = (x_i^{(1)}, \ldots, x_i^{(p)})^\top, \quad (16.1)$$

is studied using a mixture of normal probability density functions (p.d.f.). In this method, the number k of clusters, the mixing proportions $\boldsymbol{\pi}$, the mean vectors $\boldsymbol{\mu}_1, \ldots, \boldsymbol{\mu}_k$, and the covariance matrices $\boldsymbol{\Sigma}_1, \boldsymbol{\Sigma}_2, \ldots, \boldsymbol{\Sigma}_k$ of the class distributions are all not known a *priori*. If we assume that each observation X_i has probability π_m of coming from the m-th population (class), $m =$

---

[1] This work is supported by the International Association for the Promotion of Cooperation with Scientists from the Independent States of the FSU (INTAS), reference number INTAS-93-0725

[2] This problem is known under many names such as: *unsupervised classification, learning without a teacher, mixture-model cluster analysis* (see [1]).

$1, 2, \ldots, k$, then (16.1) is a sample from the population with p.d.f.

$$f(X; \pi, \mu, \Sigma) = \sum_{m=1}^{k} \pi_m \varphi(X; \mu_m, \Sigma_m), \qquad (16.2)$$

where $\pi = (\pi_1, \pi_2, \ldots, \pi_{k-1})$ are $k-1$ independent mixing proportions such that $0 \leq \pi_m \leq 1$ for $m = 1, 2, \ldots, k$ and $\pi_k = 1 - (\pi_1 + \cdots + \pi_{k-1})$, and $\varphi(X; \mu_m, \Sigma_m)$ is the m-th component multivariate normal density function given by

$$\varphi(X; \mu_m, \Sigma_m) = (2\pi)^{-\frac{p}{2}} |\Sigma_m|^{-\frac{1}{2}} \exp\left\{-\frac{1}{2}(X - \mu_m)^\top \Sigma_m^{-1}(X - \mu_m)\right\}. \qquad (16.3)$$

Then the classification procedure of observations (16.1) breaks down into the following stages:

Stage 1: Statistical estimation (based on the sample (16.1)) of the unknown parameters $\Theta = (k; \pi_1, \ldots, \pi_{k-1}; \mu_1, \ldots, \mu_k; \Sigma_1, \ldots, \Sigma_k)$ in the expression (16.2); this problem is known as a very complicated one, and the main difficulty is that k is unknown.

Stage 2: Proper classification of observations (16.1) using the estimated values $\widehat{k}; \widehat{\pi}_1, \ldots, \widehat{\pi}_{k-1}; \widehat{\mu}_1, \ldots, \widehat{\mu}_k; \widehat{\Sigma}_1, \ldots, \widehat{\Sigma}_k$ of parameters and expressions (16.3) for the p.d.f. of each class; this problem is the standard problem of the discriminant analysis: the observation X_i belongs to "class m_0" if

$$\pi_{m_0} \varphi(X_i; \widehat{\mu}_{m_0}, \widehat{\Sigma}_{m_0}) = \max_{1 \leq m \leq k} \widehat{\pi}_m \varphi(X_i; \widehat{\mu}_m, \widehat{\Sigma}_m).$$

In this chapter, we consider the problem of "Stage 1" in the framework of the model (16.1)–(16.3) *under the additional assumption that*

$$\Sigma_1 = \Sigma_2 = \cdots = \Sigma_k = \Sigma. \qquad (16.4)$$

We note that the number N of the unknown parameters in this case will be

$$N = kp + (k-1) + \frac{1}{2} p(p+1) + 1. \qquad (16.5)$$

So we have the N-dimensional vector parameter

$$\Theta = \left(k; \pi_1, \ldots, \pi_{k-1}; \mu_1, \ldots, \mu_k; \Sigma\right). \qquad (16.6)$$

16.2 Estimating the Unknown Parameters of the Mixture-Model

The estimating procedures for the standard multivariate normal mixture-model are described in many papers and books (see, for example, [2, 4, 6] for various versions of the procedure giving the *maximum likelihood estimators* (MLE): the algorithms of type EM, SEM).

We recall here the general iterative scheme of these procedures.

The logarithmic likelihood function $l_k(\Theta)$ of the data (16.1) is

$$l_k(\Theta) = \ln L_k(\Theta|X_1, X_2, \ldots, X_n) = \sum_{i=1}^{n} \ln \left(\sum_{m=1}^{k} \pi_m (2\pi)^{-\frac{p}{2}} |\Sigma|^{-\frac{1}{2}} \right.$$

$$\left. \times \exp \left\{ -\frac{1}{2}(X_i - \boldsymbol{\mu}_m)^\top \Sigma^{-1}(X_i - \boldsymbol{\mu}_m) \right\} \right). \quad (16.7)$$

To obtain the Maximum Likelihood Estimators (MLE) of the unknown parameters Θ (under the condition that the *value of "k" is known a priori*), we use matrix differential calculus and compute the partial derivatives of $l(\theta)$ with respect to $\pi_1, \ldots, \pi_{k-1}, \boldsymbol{\mu}_1, \ldots \boldsymbol{\mu}_k$ and Σ, respectively, and set these equal to zero. After some algebra, for model (16.1)–(16.4) we obtain the following ML equations (for $m = 1, 2, \ldots, k$):

$$\widehat{P}(m \mid X_i) = \frac{\widehat{\pi}_m \varphi(X_i; \widehat{\boldsymbol{\mu}}_m, \widehat{\Sigma})}{\sum_{m=1}^{k} \widehat{\pi}_m \varphi(X_i; \widehat{\boldsymbol{\mu}}_m, \widehat{\Sigma})}, \quad (16.8)$$

$$\widehat{\pi}_m = \frac{1}{n} \sum_{i=1}^{n} \widehat{P}(m \mid X_i), \quad (16.9)$$

$$\widehat{\boldsymbol{\mu}}_m = \frac{1}{n\widehat{\pi}_m} \sum_{i=1}^{n} X_i \widehat{P}(m \mid X_i), \quad (16.10)$$

$$\widehat{\Sigma} = \frac{1}{n} \sum_{i=1}^{n} X_i X_i^\top - \sum_{m=1}^{k} \widehat{\pi}_m \widehat{\boldsymbol{\mu}}_m \widehat{\boldsymbol{\mu}}_m^\top, \quad (16.11)$$

where $\widehat{\pi}_m$ is the estimated mixing proportion π_m, $\widehat{\boldsymbol{\mu}}_m$ is the estimated mean vector $\boldsymbol{\mu}_m$, $\widehat{\Sigma}$ is the estimated covariance matrix Σ, and $\widehat{P}(m \mid X_i)$ is the estimated posterior probability of group membership of the observation X_i in the cluster "m".

To start up this iterative procedure we must choose initial values $\pi_m^0, \boldsymbol{\mu}_m^0$, and Σ^0 of the parameters $\pi_m, \boldsymbol{\mu}_m, \Sigma$. Our experience shows that the best approach to the initialization of this procedure is to apply the "k-means" algorithm (see [1, pp.114–120]).

Convergence properties and the reliability of the iterative MLE within the context of the mixture-model cluster analysis have been established in [9].

16.3 Problem of Estimating the Unknown Number of Classes

The problems of statistical inference on the number of clusters "actually" present in a set of data, and testing for model fit have not yet received much attention, but are becoming increasingly recognized as important (see, for instance, [13]).

The complexity of this problem is explained at least by two circumstances.

The 1-st circumstance. If we adopt natural for this problem a stepwise procedure of the classical test theory approach, then we want to test sequentially:

in step 1   - the null hypothesis $H_0: k = 1$
                versus the alternative hypothesis $H_1: k = 2$
in step 2   - the null hypothesis $H_0: k = 2$
                versus the alternative hypothesis $H_1: k = 3$
....
in step j - the null hypothesis $H_0: k = j$
                versus the alternative hypothesis $H_1: k = j + 1$
.... so on.

If we reject the null hypothesis $H_0: k = j$ at some prescribed level of significance α_j in "step j", then we need to test the hypothesis "$k = j + 1$", which is now the *null* hypothesis in "step $j + 1$" against the alternative "$k = j + 2$" using, say α_{j+1}. In this manner if we continue to test the hypotheses sequentially and stop at some point where we cannot reject any more, then these tests are dependent in the probability sense. This raises the question of how to control the overall error rate which is *always a problem in multiple-decision problems*.

The 2-nd circumstance. Let $\widehat{L}_m$ denote the maximized likelihood function of the data (16.1) for a given number of classes m (see above, the point 2). Then

$$\lambda = \frac{\widehat{L}_k}{\widehat{L}_{k'}}$$

is the Likelihood Ratio (LR) statistic for testing k classes against k' classes ($k < k'$). In the classical test theory approach the statistic -2λ (or $-2c(n, p, k)\lambda$) is usually utilized as the test-statistic which has (asymptotically) the "chi-square" null distribution. *Unfortunately, in our case this fact is not true.* The asymptotic null distribution of this statistic essentially depends upon: (a) the closeness of the parameter values to the boundary of the parameter space (*on the boundary of the parameter space the identifiability condition of the mixture-model fails*); (b) the number of classes k (more exactly, on the interrelation between "k" and "n").

These facts were noticed by many investigators (see, for example, [3, 8, 12, 14]).

The following question arises: what rules and approaches should we use for determination of the number of classes in the standard multivariate normal mixture-model (16.1)–(16.4) on the basis of sample (16.1)?

Basically, there are two approaches to the investigation of this problem. *The first approach* is based on the restrictions for the value of "k" caused by the natural interrelation between "k", "n" and "p".

The semi-heuristic guidelines ("the rules of thumb") of this type [4] are:

$$k < \frac{2n}{(p+1)(p+2)},$$

$$k \sim n^{0,3},$$

$$\left[\frac{n}{\ln n}\right]^{\frac{1}{3}} < k < \left[\frac{n}{2}\right]^{\frac{1}{2}}.$$

The investigator uses these results *as a heuristic guide* to decide k, the hypothesized number of clusters when he fits the mixture-model.

The second approach is based on the introduction of a special criterion ("loss function") of the estimation quality that consists of the two summands: the first summand describes the "**lack of fit**" (it is $-2\ln L(\Theta)$) and second summand describes the "**lack of model's profusion of complexity**". The criteria of this type are considered in [4] (AIC-Akaiki's Information Criterion, CAIC — Consistent Akaiki's Information Criterion, ICOMP — Informational Measure of Complexity).

We give here the expression for $W_k(\Theta) = $ ICOMP in our case (*i.e.* for the model (16.1)–(16.4)):

$$\begin{aligned} W_k(\widehat{\Theta}) &= -2l_k(\widehat{\Theta}) + \left(kp + \frac{p(p+1)}{2}\right) \\ &\times \ln\left[\frac{k\left\{\frac{1}{\pi_k}\operatorname{tr}(\widehat{\Sigma}) + \frac{1}{2}\operatorname{tr}(\widehat{\Sigma}^2) + \frac{1}{2}(\operatorname{tr}\widehat{\Sigma})^2 + \sum_{j=1}^p \widehat{\sigma}_{jj}^2\right\}}{kp + \frac{1}{2}p(p+1)}\right] \\ &\times \left\{(p+2)k\ln|\widehat{\Sigma}| - p\sum_{m=1}^k \ln(n_m)\right\} - (kp)\ln(2n). \end{aligned} \quad (16.12)$$

Here $l_k(\Theta)$ is the logarithmic likelihood function (16.7), $\operatorname{tr} A$ is the trace of the matrix A, $\widehat{\sigma}_{jj}$ is the (j,j)-th element of the matrix $\widehat{\Sigma}$ and

$$\widehat{\Theta} = \widehat{\Theta}(k) = (\widehat{\pi}(k); \widehat{\mu}_1(k), \ldots, \widehat{\mu}_k(k); \widehat{\Sigma}(k)) = \arg\max_{\Theta} \ln l_k(\Theta).$$

Then we determine an integer $\widehat{k}$ such that

$$W_{\widehat{k}}(\widehat{\Theta}) = \min_k W_k(\widehat{\Theta}). \quad (16.13)$$

In the next section we propose one more approach to the problem of finding the unknown number of classes.

16.4 Detecting the Number of Clusters by Means of Projection Pursuit

16.4.1 Projection Pursuit Technique

The Projection Pursuit (PP) technique [2, 5, 7] is used to reveal any particularities of a multivariate data set and to find out a geometric structure of it, for instance, the presence, and the number, of clusters. The PP approach gives us an opportunity of *the visual analysis of "interesting" linear projections of the initial data* (1) for this goal.

We are considering here a step-wise version of the PP-method. Namely, let $C_1 = (c_{11}, c_{12}, \ldots, c_{1p})$ be a linear projection operator from R^p into R^1 and $Q(C_1, \mathbf{X}^{(n)})$ be the sampling value of some *Projection Index* (PI). The PI is constructed so that its value appreciates the quality of presentation of the explored properties of $\mathbf{X}^{(n)}$ in $R^{(1)}$ (*i.e.* after the projection of $X^{(n)}$ into $R^{(1)}$ by means of C_1). So the vector C_1 is determined as the solution of the problem

$$C_1 = \arg\max_C Q(C, \mathbf{X}^{(n)}). \tag{16.14}$$

The following vector $C_2 = (c_{21}, c_{22}, \ldots, c_{2p})$ is defined from (16.14), but under the additional condition that the influence of C_1 is excluded in a suitable way. More formally:

$$\begin{cases} C_2 = \arg\max_C Q(C, \mathbf{X}^{(n)}) \\ \text{under the condition: } C_1 \widehat{S} C_2^\mathsf{T} = 0 \end{cases}, \tag{16.15}$$

where $\widehat{S}$ is an estimate of the covariance matrix S of X on the basis $\mathbf{X}^{(n)}$.

Consider now the linear orthogonal projection operator (the $(2 \times p)$-matrix)

$$A = \begin{pmatrix} C_1 \\ C_2 \end{pmatrix} \tag{16.16}$$

from R^p into the plane R^2.

By applying the operator A to $\mathbf{X}^{(n)}$, we have the points (in the plane)

$$\mathbf{Z}^{(n)} = (Z_1, Z_2, \ldots, Z_n),$$

where

$$Z_i = AX_i = \begin{pmatrix} C_1 X_i \\ C_2 X_i \end{pmatrix} = \begin{pmatrix} z_i^{(1)} \\ z_i^{(2)} \end{pmatrix}.$$

If the criterion $Q(C, X^{(n)})$ is chosen *"in a suitable way"* then the visual analysis of the geometric structure of $\mathbf{Z}^{(n)}$ makes it possible to formulate the reliable hypothesis about the number of classes "k".

16.4.2 Discriminant Subspace

Following Rao [10], we define now the discriminant subspace of the space R^p for our model (16.1)–(16.4). It is well known that the covariance matrix $\mathbf{S}$ of the X may be presented as

$$\mathbf{S} = \mathbf{B} + \mathbf{\Sigma}, \tag{16.17}$$

where $\mathbf{B} = \sum_{j=1}^k \pi_j (\boldsymbol{\mu}_j - \overline{\boldsymbol{\mu}})(\boldsymbol{\mu}_j - \overline{\boldsymbol{\mu}})^\mathsf{T}$ is the *between-components scatter matrix* (here $\overline{\boldsymbol{\mu}} = \sum_{j=1}^k \pi_j \boldsymbol{\mu}_j$) and $\mathbf{\Sigma}$ is the *within-components scatter matrix* (see above (16.3), (16.4)). According to this the variance $s^2(C)$ of the one-dimensional projection along the direction C is

$$s^2(C) = b^2(C) + \sigma^2(C), \tag{16.18}$$

where $b^2(C) = C\mathbf{B}C^\mathsf{T}$ and $\sigma^2(C) = C\mathbf{\Sigma}C^\mathsf{T}$.

The ratio

$$t^2(C) = \frac{b^2(C)}{\sigma^2(C)} \tag{16.19}$$

is a measure of dissimilarity of the mixture (16.2)–(16.4) for the projection with vector-operator $C = (c_1, c_2, \ldots, c_p)$. Maximizing the $t^2(C)$ leads to the solution of the generalized eigenvector problem

$$(\mathbf{B} - \lambda \mathbf{\Sigma})V = \mathbf{O}.$$

There are q eigenvectors $V_1, \ldots, V_q$ with corresponding positive eigenvalues $\lambda_1, \ldots, \lambda_q$ ($q < \min(p, k-1)$, $\lambda_j = t^2(V_j)$).

The discriminant subspace is defined as the subspace

$$\Omega = \operatorname{span}(V_1, V_2, \ldots, V_q)$$

or

$$\Omega = \operatorname{span}(\mathbf{\Sigma}^{-1}\boldsymbol{\mu}_1, \mathbf{\Sigma}^{-1}\boldsymbol{\mu}_2, \ldots, \mathbf{\Sigma}^{-1}\boldsymbol{\mu}_k).$$

The number q depends upon the configuration of the vectors $\boldsymbol{\mu}_1, \boldsymbol{\mu}_2, \ldots, \boldsymbol{\mu}_k$ and Ω contains the complete information about the differences among the mixture (16.2)–(16.4) components. The PP gives a way to get Ω without any information about the matrix $\mathbf{\Sigma}$ unlike discriminant analysis problem where we can estimate $\mathbf{\Sigma}$ and $\mathbf{B}$. Further without of loss generality we consider the case $\overline{\boldsymbol{\mu}} = \mathbf{O}$ (centred data).

16.4.3 Projection Index Suitable for Detecting the Number of Clusters

We consider the one-parametric family of the PI based on Renyi entropy [11, 15]

$$Q_\beta(C, X) = s^\beta(C) \cdot E f^\beta(z(C)) \quad (\beta > 0). \tag{16.20}$$

Here the theoretical quantity of PI is designated by $Q_\beta(C, X)$ in contact to the sampling quantity $Q_\beta(C, \mathbf{X}^{(n)})$, and $s^2(C)$ is the variance of $z(C) = CX$.

Assertion 1 (based on Lemma 19.1 from [2].

Assume that the vectors $C_1, C_2, \ldots, C_q$ are found by the step-wise (sequential) maximization of the $Q_\beta(C, X)$ (see (16.20) and Section 16.4.1; for $j > 2$ the vector C_j is S-orthogonal to the subspace $\Omega_{j-1} = \operatorname{span}(C_1, \ldots, C_{j-1})$). Then every vector C_j belongs to the discriminant subspace Ω and

$$\Omega = \operatorname{span}(C_1, C_2, \ldots, C_q).$$

Assertion 2 (based on Section 19.9 from [2]).
The expression

$$\widehat{Q}_\beta(C, \mathbf{X}^{(n)}) = \frac{1}{n}\widehat{s}^\beta(C) \cdot \left(\frac{2r}{n}\right)^\beta \sum_{i=1}^{n}\left[z(C)_{(i+r)^-} - z(C)_{(i-r)^+}\right]^{-\beta} \quad (16.21)$$

is the consistent and asymptotically normal estimate for the PI $Q_\beta(C, X)$.

In (16.21), $z(C) = CX$, $\widehat{s}^2(C) = \frac{1}{n}\sum_{i=1}^{n}(z_i(C) - \overline{z}(C))^2$, $z(C)_{(j)}$ is the j-th order statistic of the sample $z_1, \ldots, z_n$, r is integer value $(r < \frac{n}{2})$, $(i+r)^- = \min(n, i+r)$ and $(i-r)^+ = \max(1, i-r)$.

The estimate (16.21) depends upon r. But the influence of r is not so great (the experimental fact). In practice, we use the values $r = cn^{\frac{1}{2}}, 0.5 < c < 1.5$.

Remark 1. The proof of Assertion 1 is essentially based on the inequalities

$$\frac{1}{\sqrt{2\pi(1+\beta)}}\left(\sum_{j=1}^{k}\pi_j^{\beta+1}\right)\left(1+t^2(C)\right)^{\frac{\beta}{2}} \leq Q_\beta(C, X)$$

$$\leq \frac{1}{\sqrt{2\pi(1+\beta)}}\left(1+t^2(C)\right)^{\frac{\beta}{2}}.$$

Remark 2. For the case $\beta = 1$ and $\beta = 0$ we have:

$$Q_1(C, X) = \frac{1}{2\sqrt{\pi}}(1+t^2(C))^{\frac{1}{2}}\left(1 + 2\sum_{i>j}^{k}\pi_i\pi_j e^{-t_{ij}^2}\right),$$

where $t_{ij}^2 = \frac{(\boldsymbol{\mu}_i(C)-\boldsymbol{\mu}_j(C))^2}{s^2(C)}$ and

$$Q_0(C, X) = -\int \ln(s(C) \cdot f(z(C))) \, dz(C).$$

16.5 Mixture-Model Cluster Analysis and Bayesian Belief Networks

Stochastic networks, and particularly the *Bayesian Belief Networks* (BBN), provide a useful tool for solving the well-known difficult mathematical and computational problems of decomposition of distribution mixtures. For example, BBNs help one to detect a specific structure of relationships between the components of the vector X under study, and therefore make it possible to use a lower-dimensional parameter description of the unknown covariance Σ. This simplifes greatly the computational algorithms of decomposition of distribution mixtures.

Conversely, the preliminary statistical analysis of the mixture-model is useful for the implementation of BBNs. Consider the following scheme. Assume case i is characterized simultaneously by a "behaviour variable" $X_i =$

$(x_i^{(1)}, \ldots, x_i^{(p)})$ and a "state variable" $Y_i = (y_i^{(1)}, \ldots, y_i^{(m)})$, where the components of Y are determined by experts and constitute an initial (as a rule, redundant) set of variables.

We are interested in finding those components of Y_i that enable us to determine the type of case i in the "behaviour space" (the so-called "type generating variables.") A way to cope with this *BBN task* is as follows. First, the statistical analysis of the mixture-model in the space X is applied. Second, the resulting clusterization of cases is employed, namely, the clusters are used as learning samples in the BBN task.

References

[1] Aïvazian S. A., Bejaeva Z. I. and Staroverov O. V. (1974) *Classification of the multidimensional observations.* Statistika, Moscow (In Russian).

[2] Aïvazian S. A., Buchstaber V. M., Yenukov I. S., and Meshalkin L. D. (1989) Applied statistics: classification and reduction of dimensionality. In Aïvazian S. A. (ed) *Finansy i Statistika*, Moscow. (In Russian, English version – 1996, to appear.)

[3] Binder D. A. (1978) Bayesian cluster analysis. *Biometrika* (65) 31–38.

[4] Bozdogan H. (1993) Mixture-model cluster analysis using model selection criteria and a new informational measure of complexity. In *Multivariate Statistical Modelling, (2) Proceedings of the First US/Japan Conference on the Frontiers of Statistical Modelling.* Kluwer Academic, Dordrecht, Netherlands.

[5] Buchstaber V. M. and Maslov V. K. (1975) Factor analysis on manifolds and the problem of selection of variables in pattern recognition. *Izvestiya Akad. Nauk SSSR*, ser. *Tekhnicheskaya Kibernetika* 6: 194–200.

[6] Celeux G. and Diebolt J. (1985) The SEM algorithm: a probabilistic teacher algorithm derived from the EM algorithm for the mixture problem. *Computational Statistics Quarterly* (2) 73–82.

[7] Friedman J. H. and Tukey J. W. (1974) A projection pursuit algorithm for exploratory data analysis. *IEEE Trans. Comp.* (C-23) 881–889.

[8] Hartigan J. A. (1977) Distribution problems in clustering. In *Classification and Clustering*, J. Van Ryzin (ed), Academic Press, New York, 45–71.

[9] Peters B. C. and Walker H. E. (1978) An iterative procedure for obtaining maximum-likelihood estimates of the parameters for a mixture of normal distributions. *SIAM Journal on Applied Mathematics* (35) 362–378.

[10] Rao C. R. (1965) *Linear Statistical Inferences and its Applications*. Wiley, New York.

[11] Renyi A. (1970) *Probability Theory*. North-Holland, Amsterdam.

[12] Sclove S. L. (1977 Population mixture models and clustering algorithms. *Communications in Statistics (Theory and Methods)* A6: 417–434.

[13] Sokal R. R. (1977) Clustering and classification: background and current directions. In *Classification and Clustering*, J. Van Ryzin (ed), Academic Press, New York.

[14] Wolfe J. H. (1971) A Monte-Carlo study of the sampling distribution of the likelihood ratio for mixture of multinormal distribution. *Research Memorandum (72-2), U.S. Naval Research Activity*. San Diego, California.

[15] Yenyukov I. S. (1987) Detecting outliers and clusters in multivariate data based on projection pursuit. In *Proc. World Congress of Bernoulli Society*. VHU Science Press, Utrecht, (2) 137–141.

17

A Parallel k_n-Nearest Neighbour Classifier for Estimation of Non-linear Decision Regions

A. Kovalenko

A nonparametric classifier is considered; classification depends upon whether or not an observation belongs to an empirical informative set (decision region) constructed from a learning sample by the k_n-nearest neighbour technique. It has linear computational and spatial complexity and is efficiently implemented on a parallel computing system. Its consistency is proved, and the results deduced from the solution of a model problem are discussed.

17.1 Introduction

The use of computational learning and probabilistic reasoning methods for solving applied statistical problems is a rapidly expanding area in informatic science. Its attractiveness largely lies in the possibility of using graphical models in those situations where the search for only classical probabilistic solutions is presently unfruitful. But probabilistic methods can be useful in a problem of computational learning such as model selection. For example, in the case of classification important information about relationships between nodes of graphical model can be received by testing overlap of decision regions in the feature space. The results of the testing can elucidate cause-effect relationships among input variables and between ones and output decisions.

If the number of input variables and the size of learning sample is large, we need to use for testing, the fast algorithms with linear computational and spatial complexity.

We shall study a k_n-nearest neighbour classifier, assuming that a learning sample of size n is given. The elements in the sample set are pairs, which consist of vectors of the co-ordinates of the points in a d-dimensional Euclidean space and their corresponding pattern numbers belonging to a finite set $\Omega = \{\omega_1, \omega_2, \ldots, \omega_m\}$. Given that an observation (d-dimensional vector) arrives at the classifier input, we are required to find the pattern to which the observation belongs. Let us take a fixed positive integer k_n. For the observation that is being classified, let us find k_n-nearest neighbours in the learning sample for the Euclidean metric ρ and determine the pattern to which a majority of neighbours belong. This pattern will be taken to be the solution.

The k_n-nearest neighbour classification ($k_n NN$-classifiers) has been widely studied (see [1, 2]). The drawbacks that hinder its practical application are large processing time and computer memory needed for storing the learning sample and for computing and sorting the distances between observation and sampling points. Various pre-processing techniques (e.g. ordering [2, 3], structurization [4], and reduction [5] of learning samples) are available for overcoming this hurdle. Ordering of sample sets, for instance, cuts down computation volume on the average to $O(d \cdot n \cdot \log n)$ [3]. If the sample size n is large, even such a reduction is not effective. Furthermore, since sorting involves a large number of branching operations, such classifiers do not easily yield to implementation on parallel, and particularly an transputer systems [6].

We shall examine a sample pre-processing method, which consists of constructing empirical informative sets corresponding to various classes. Observations are classified by their membership of these sets. Verification of the membership of an observation to a set is an operation that is linear in computation complexity and can be easily bifurcated into parallel operations. We shall also discuss the results obtained from the solution of a model problem.

17.2 Parallel Classifier

Let (X, Y) be a pair of random variables, where X is a random vector in a d-dimensional Euclidean space $\Re^d$ with uniformly continuous density $f(x)$, and Y is a discrete random variable that takes its values in the set $\{\omega_1, \omega_2, \ldots, \omega_m\}$ with probabilities $p_1, p_2, \ldots, p_m$, $\sum p_i = 1$. The probabilities p_i are called a priori probabilities of classes, and the conditional densities $f_i(x) = f(x \mid Y = \omega_i)$ are called a priori densities of classes. In what follows we confine ourselves to a learning sample containing only two classes ($m = 2$), because generalization to a greater number of class is obvious.

Given that a vector x of observation (x, y) of the pair (X, Y) is fed at the classifier input, we are required to take a decision on the value of the unknown component y. Let $\delta(\omega_i \mid x) \in \{0, 1\}$ be the decision rule used to decide whether an observation x belongs to the class ω_i or not. In case of rejection, we assume that the observation is assigned to an extraneous "background" class ω_0, i.e. $\delta(\omega_0 \mid x) = 1$. A special procedure similar to the technique described in [5] is required to classify the observations belonging to the "background" class.

For a fixed $c > 0$, let us consider two sets

$$B_1(c) := \{x \mid p_1 f_1(x) \geq c p_2 f_2(x)\}$$

and

$$B_2(c) := \{x \mid p_2 f_2(x) \geq c p_1 f_1(x)\},$$

which we call as *theoretical informative sets* for the classes ω_1 and ω_2 [7].

Let us define a decision rule as follows:

$$\delta(\omega_i \mid x) := 1, \quad \text{if } x \in B_i(c), \ i = 1, 2, \qquad (17.1)$$
$$\delta(\omega_0 \mid x) := 1, \quad \text{otherwise.}$$

For $c \geq 1$ the decision rule (17.1) defines a Bayes $(0, 1)$-classifier with failures.

The sets $B_1(c)$ and $B_2(c)$ can be interpreted as domains of high (compared to c) "concentrations" of points from the corresponding class, and the decision rule (17.1) is used to decide whether an observation belongs to a high concentration domain or not. Let us observe that the informative sets may intersect if $c < 1$.

The simply connected components of the set $B_i(c)$ form clusters of the i-th pattern, which can be regarded as clusters of a high density compared to the level that is a function of x [8, 9].

Let a learning sample $\{(x_j, y_j)\}$, $j = 1, 2, \ldots, n$ consist of the realizations of n independent random vectors (X_j, Y_j) having the same distribution as the vector (X, Y). The vectors (X_j, Y_j) and (X, Y) are assumed to be independent. Using this sample set, we now construct a decision rule $\delta_n(\omega_i \mid x)$ for classifying a new observation x.

Let $O_{k_n}(x_j)$ be the minimal hyperball with centre at the sample point x_j that contains at least k_n points from the sample set, and let $r_{k_n}(x)$ and $V_{k_n}(x)$ denote the radius and volume of this hyperball. The k_n-nearest neighbour estimate [4] for the density $f(x)$ is

$$\widehat{f}_n(x) = k_n / (n V_{k_n}(x)).$$

Let a point x in the learning sample have $k_{ni}(x)$ "neighbours" belonging to the pattern ω_i. Clearly, the relation $k_{n1}(x) + k_{n2}(x) = k_n$ is satisfied for every point x.

Using the sets

$$B_{n1}(c) := \bigcup_{j=1}^{n} \{O_{k_n}(x_j) \mid k_{n1}(x_j) \geq ck_{n2}(x_j)\}$$

and

$$B_{n2}(c) := \bigcup_{j=1}^{n} \{O_{k_n}(x_j) \mid k_{n2}(x_j) \geq ck_{n1}(x_j)\},$$

we shall construct a decision rule for the k_n-nearest neighbour parallel classifier

$$\delta_n(\omega_i \mid x) := 1, \quad \text{if } x \in B_{ni}(c), \ i = 1, 2, \qquad (17.2)$$
$$\delta_n(\omega_0 \mid x) := 1, \quad \text{otherwise.}$$

Classification by rule (17.2), unlike the usual k_n nearest neighbour classification, is an operation of linear computational complexity. To classify by this rule, there is no need to utilize the entire learning sample; it suffices to use a subset consisting of the sampling points belonging either to the set $B_{n1}(c)$ or to the set $B_{n2}(c)$. This subset can additionally be contracted by eliminating those points whose hyperballs are contained in hyperballs of larger radii or the union of a few such hyperballs.

Another property of the decision rule (17.2) is its natural parallelism. To decide whether or not an observation x belongs to the empirical informative set, it suffices to compute its distances from the sampling points in this set and compare them with radii of hyperballs computed previously. Clearly, this operation can be implemented concurrently on a few separate processors with local memory.

Consistency of classifier (17.2) is resulted from the Theorem 17.1.

Theorem 17.1 *Let*

(i) $f_1(x)$ and $f_2(x)$ be uniformly continuous densities, and $f(x) = p_1 f_1(x) + p_2 f_2(x)$, where $p_1, p_2 > 0$;

(ii) $x_1, x_2, \ldots, x_n$ be observations of identically distributed independent random vectors in $\Re^d$ with density $f(x)$;

(iii) k_n be a sequence of positive integers such that k_n/n tends to zero and $k_n/\log n$ tends to infinity as n tends to infinity; and

(iv) for a given $c > 0$ the informative sets $B_1(c)$ and $B_2(c)$ have nonzero measure, and the measure of their boundaries is zero, i.e.

$$\mu\{x \mid p_1 f_1(x) = cp_2 f_2(x)\} = 0,$$

$$\mu\{x \mid p_2 f_2(x) = c p_1 f_1(x)\} = 0.$$

Then for $i = 1, 2$,

$$\lim_{n \to \infty} P\Big(x \in \{B_i(c) \triangle B_{ni}(c)\}\Big) = 0,$$

where $\triangle$ is a symmetric set differencing operator.

Proof is based on the strong uniform consistency of kernel density estimates [4] and k_n nearest neighbour density estimates [11] and given in [12].

17.3 Signal Recognition Against a Noise Background: A Numerical Example

We shall examine the recognition of signals (class ω_1) arriving at a noise background (class ω_2). Both signals and noise are represented as d-dimensional vectors in a Euclidean space. We assume that the arrival of a signal is a relatively rare event (i.e. $p_1 \ll p_2$) and there is an "expensive" (say, as regards computer resources) classifier that recognizes these signals with the required accuracy. But the predominant noise at the input renders its application ineffective. The problem now is, given a learning sample, is it possible to conduct an "inexpensive" (fast) preliminary classification that would filter off most of the input noise for a given signal transmission probability? Such kinds of problems are encountered, for instance, in testing computer chips.

We shall solve this preliminary selection problem with a parallel classifier (17.2), focusing our interest on verifying the membership of observations to the informative set $B_{n1}(c)$. The parameter c determines the upper bound of the signal transmission probability and takes values in the interval $(0, 1)$.

Let us now examine a model example in which observations are distributed with a density $f(x)$, which is a mixture of two densities of the classes

$$f(x) = p_1 f_1(x) + p_2 f_2(x),$$

where $p_1 = 0.05$, $p_2 = 0.95$, $f_1(\cdot) \sim 0.5 N(\mu_{11}, \Sigma_1) + 0.5 N(\mu_{12}, \Sigma_1)$, and $f_2(\cdot) \sim N(\mu_2, \Sigma_2)$ with the parameters

$$\mu_{11} = (-3, 0), \quad \mu_{12} = (3, 0), \quad \mu_2 = (0, 0),$$

$$\Sigma_1 = \begin{pmatrix} 1 & 0 \\ 0 & 1 \end{pmatrix}, \quad \Sigma_2 = \begin{pmatrix} 4 & 0 \\ 0 & 4 \end{pmatrix}.$$

Let $c = 0.2$. The informative set $B_1(c)$ in this case consists of two simply connected components. The probability for an observation to belong to the informative set is $P(x \in B_1(c)) \approx 0.13$.

Using a PC, we conducted 25 experiments. In each run we generated a learning sample of size $n = 750$ and made a test classification of 300 observations with the optimal classifier (17.1), parallel classificator (17.2) for $k_n = 18$, and the usual $k_n NN$-classificator described in the introduction.

In all these experiments the usual $k_n NN$-classifier took about one hour of machine time, while the parallel classifier finished in about two minutes. This clearly demonstrates, though indirectly, the superior computational merits of the parallel classificator.

The mean frequency 12.6% we found for an observation to fall into the informative set was close to the probability value. Parallel classifier yielded the following results: 11.8% of the points from the sampling set were properly assigned to the informative set, 0.8% points were not placed in the informative set, and 9.8% were erroneously included in the informative set. For the usual $k_n NN$-classifier these figures were 7.%8, 4.8%, and 2.2%, respectively.

On average, the parallel classifier took erroneous decisions more often than the $k_n NN$-classifier. But, while including more "improper" points in the informative set, it rarely failed to include the points properly belonging to the set.

Most probably, the parallel classifier exhibits such an ability (which can be regarded either as a merit or a demerit, depending on the formulation of the problem) because the informative set is almost wholly covered by hyperballs which at the same time encroach superfluous domains of the space as well. This presumption is corroborated by our experiments which demonstrate that with the increasing size of the learning sample set the number of proper points not included in the informative set rapidly tends to zero, while the number of "improper" included in the set slowly decreases. The behaviour of the last is apparently attributable to the slow rate at which the radii of the covering hyperballs tend to zero.

To evaluate how the density estimation accuracy affects the classification quality, we compared the results yielded by a parallel classifier and its model in which the radii $r_{k_n}(x_j)$ of the covering hyperballs were determined not from the density estimate but from the true density value, *i.e.* by a formula, which for $d = 2$ is

$$r_{k_n}(x_j) = \left\{ k_n / (\pi n f(x_j)) \right\}^{1/2}.$$

Let us now examine the averaged results obtained from 25 experiments. In each experiment we generated a learning sample of size $n = 1500$ and then conducted a test classification of 500 observations by the optimal classificator (17.1), and parallel classificator (17.2) for for $k_n = 18$ and its model. As would be expected, the model classifier did not fail to place any proper point into the informative set, whereas the parallel classifier misplaced on the average 0.05% points of the sample set. Unexpectedly, both the classifiers imbedded almost the same number of "improper" points into the informative set (8.9% for the parallel classifier and 8.5% for its model), *i.e.* true density, when substituted for density estimate, has virtually no effect on the quality of parallel classification! Supplementary experiments conducted with different n and k_n corroborated this conclusion; so, parallel classifiers may be expected to yield reliable results even with small samples.

Acknowledgements

This work was supported by the International Association for the Promotion of Co-operation with Scientists from the Independent States of the Former Soviet Union under grant INTAS-93-0725.

References

[1] Goldenstein M. (1972) K-nearest neighbor classification. *IEEE Trans. Inf. Theory* IT-18 627–630.

[2] Patrik E.A. (1980) *Elements of Pattern Recognition Theory* (In Russian), Sovetskoe Radio, Moscow.

[3] Friedman J., Baskett F. and Shustek J. (1975) An algorithm for finding nearest neighbors. *IEEE Trans. Comput* C-24: 1000–1006, October.

[4] Devroye L. and Machell F. (1985) Data structures in kernel density estimation. *IEEE Trans. Pattern Anal. and Mach. Intell* V. PAMI-7 (3): 360–366.

[5] Niemann H. (1983) *Klassifikation von Mustern*, Springer-Verlag, Berlin.

[6] Krasnov S.A. (1990) Transputers, transputer computing system, and Occams. In: *Computing Processes and Systems* (in Russian), Marchuk G.I. (ed) 3–82, Nauka, Moscow.

[7] Grusho A.A. (1987) Conditions for consistent detection of inclusions in a sample from a uniform distribution. In: *Probability Problems of Discrete Mathematics. Inter-college Collection* (in Russian), Ivchenko G.I. (ed) Institute of Electronic Machines, Moscow, 65–73.

[8] Bock H.H. (1974) *Automatische Klassifikation. Theoretische und praktische Methoden zur Gruppierung und Strukturierung von Daten (Clusteranalyse)*, Vandenhoeck &Ruprecht, Göttingen.

[9] Hartigan J.A. (1975) *Clustering Algorithms*, John Wiley & Sons, New York.

[10] Devroye L. and Wagner T.J. (1977) The strong uniform consistency of nearest neighbor density estimates. *The Annals of Statistics*, (5) 3: 536–540.

[11] Devroye L. and Wagner T.J. (1980) The strong uniform consistency of kernel density estimates. *5th Int. Symp. on Multivariate Analysis*, 59–77, New York.

[12] A. Kovalenko (1995) A parallel classifier based on the k_n-nearest neighbor technique. *Automation and Remote Control* (56) 1: Part 2.

18

Extreme Values of Functionals Characterizing Stability of Statistical Decisions

A.V. Nagaev[1]

18.1 Introduction

The basic goal of this chapter is to discuss a number of problems related to the stability of decision rules. The problems can be embedded in the following general scheme.

Let $\mathcal{F}$ be a class of distributions. It can be both parametric or semiparametric. Consider a functional $l(F)$ defined on $\mathcal{F}$. The problem is to evaluate

$$\sup_{F \in \mathcal{F}} l(F) \text{ and } \inf_{F \in \mathcal{F}} l(F).$$

If $l(F)$ is continuous and $\mathcal{F}$ is closed[2] then apart from the extreme values it worth trying to describe the sets

$$\arg\sup_{F \in \mathcal{F}} l(F) \text{ and } \arg\inf_{F \in \mathcal{F}} l(F).$$

---

[1] Supported by INTAS-93-725 programme

[2] We do not care what "continuity" and "closeness" mean.

If we simultaneously consider a number of functionals, say, $l_1(F), \ldots, l_k(F)$, then the problem is to find the boundary of the set

$$(x : x = (x_1, \ldots, x_k),\ x_i = l_i(F),\ i = 1, \ldots, k)$$

or, perhaps, the convex hull of this set.

Within the context of mathematical statistics $\mathcal{F}$ contains all the prior information about the underlying distribution. Moreover, $\mathcal{F}$ can be based on a sample, that is, be random.

It is obvious that the stability of a statistical decision rule can always be expressed in terms of a certain functional of the underlying distribution. The broader $\mathcal{F}$ the more uncertainty and, hence, the less stability of a decision rule.

It worth noting that it is not very easy to solve the extreme problems which arise within the context. Nevertheless the, already established, results look very promising.

The chapter is organized as follows. In Section 18.2 we state a number of problems which arise in belief networks. They are concerned with the stability of decisions within the framework of a semiparametric model of elliptically contoured distributions. We do not refer to the networks as Bayesian because they are minimax rather than Bayesian. The results presented here are stated as propositions because they aim only to outline the conditions which lead to a reasonable setting of extreme problems.

In Section 18.3 we assume that the underlying distribution is so close to the normal one that it cannot be discriminated by a given goodness of fit test. Here, as an objective functional, we choose a quantile. Formally, the theorems stated in Section 18.3 are new, though the omitted proofs do not differ too much from those given in [2].

18.2 Stability within a Belief Network

18.2.1 Properties of Elliptically Contoured Distributions

Let a random vector $x = (x_1, \ldots, x_d)$ have the absolutely continuous distribution with the density

$$p(x) = p_{f,V} = (detV)^{-1/2} f(\sqrt{(x-\mu)^T V^{-1}(x-\mu)}) \qquad (18.1)$$

where V is a positive definite matrix and

$$f(t) \in \mathcal{F} = (f : f(t) \geq 0,\ 0 \leq t < \infty,\ \int f(|x|)dx = \int x_1^2 f(|x|)dx = 1). \qquad (18.2)$$

When

$$f(t) = f_0(t) = (2\pi)^{-d/2} \exp\left(-\frac{t^2}{2}\right),$$

that is, x has the Gaussian distribution, V contains all the information about dependence between various subvectors of x. In particular, it uniquely determines the so-called graphical model or belief network (see, for example, [4] or [1]). It should be emphasized that, as to the graphic itself, it does not determine the underlying distribution uniquely.

It is easy to see that the covariance matrix or, simply, the variance of x equals V. Without loss of generality, we assume that $\mu = 0$.

Let x_a and x_b be subvectors of x corresponding to $a = \{i_1, \ldots, i_k\} \subset \{1, 2, \ldots, d\}$ and $b = \{j_1, \ldots, j_m\} \subset \{1, 2, \ldots, d\}$, $a \cap b = \emptyset$. We adhere to the notation from Whittaker [4]. Denote by V_{aa} the covariance matrix of x_a or the variance of x_a. Then

$$V_{a \cup b} = \begin{pmatrix} V_{aa} & V_{ab} \\ V_{ba} & V_{bb} \end{pmatrix}$$

where $V_{ab} = V_{ba}$ is called the covariance of (x_a, x_b). Consider the well-known formulæ

$$x_{a \cup b}^T V_{a \cup b, a \cup b}^{-1} x_{a \cup b} = x_a^T V_{aa}^{-1} x_a + (x_b - \hat{x}_b)^T S_{ba}^{-1}(x_b - \hat{x}_b) \qquad (18.3)$$

and

$$det V_{a \cup b, a \cup b} = det V_{aa} det S_{ba} \qquad (18.4)$$

where

$$\hat{x}_b = \hat{x}_b(x_a) = V_{ba} V_{aa}^{-1} x_a \qquad (18.5)$$

and

$$S_{ba} = V_{bb} - V_{ba} V_{aa}^{-1} V_{ab}. \qquad (18.6)$$

From (18.1) and (18.3) it follows that the marginal density of x_a has the form

$$p_{f_{d-k}, V_{aa}}(u) =$$
$$(det V_{aa})^{-1/2} \int_{R^{d-k}} f\left(\sqrt{|w|^2 + u^T V_{aa}^{-1} u}\right) dw =$$
$$(det V_{aa})^{-1/2} f_{d-k}(\sqrt{u^T V_{aa}^{-1} u}), \; u \in R^k, \qquad (18.7)$$

where

$$f_{d-k}(t) = c_{d-k} \int_0^\infty r^{d-k-1} f(\sqrt{r^2 + t^2}) dr$$

with

$$c_k = \frac{2\pi^{k/2}}{\Gamma(k/2)}. \qquad (18.8)$$

According to (18.3)–(18.7) the conditional density of x_b given x_a has the form

$$p_{mk}(x_b|x_a) =$$
$$(det S_{ba})^{-\frac{1}{2}} \int_{R^{d-m-k}} f(\sqrt{|w|^2 + x_a^T V_{aa}^{-1} x_a + (x_b - \hat{x}_b)^T S_{ba}^{-1}(x_b - \hat{x}_b)}) dw \times$$

$$\left(\int_{R^{d-k}} f(\sqrt{|w|^2 + x_a^T V_{aa}^{-1} x_a}) dw\right)^{-1}. \tag{18.9}$$

Hence, the conditional expectation of x_b given x_a equals $\hat{x}_b$. Thus, all the predictions do not depend upon $f(\cdot)$. As to the corresponding conditional variance, in contrast with the Gaussian model, it, strictly speaking, depends upon x_a and $f(\cdot)$. It easily follows from (18.10) that the conditional variance equals $\sigma_f^2(x_a) S_{ba}$ with

$$\sigma_f^2(u) = \frac{\int_{R^{d-k}} w_1^2 f(\sqrt{|w|^2 + u^T V_{aa}^{-1} u}) dw}{\int_{R^{d-k}} f(\sqrt{|w|^2 + u^T V_{aa}^{-1} u}) dw}.$$

Here w_1 is the first component of $w \in R^{d-k}$. The last formula can be rewritten as follows:

$$\sigma_f^2(u) = (d-k)^{-1} \frac{\int_0^\infty r^{d-k+1} f(\sqrt{r^2 + u^T V_{aa}^{-1} u}) dr}{\int_0^\infty r^{d-k-1} f(\sqrt{r^2 + u^T V_{aa}^{-1} u}) dr}. \tag{18.10}$$

It is evident that $E\sigma_f^2(x_a) \equiv 1$. Consider another characteristic of the prediction accuracy

$$\pi_f^{(z)}(u) = P((x_b - \hat{x}_b)^T S_{ba}^{-1} (x_b - \hat{x}_b) \geq z^2 | x_a = u).$$

It is obvious that

$$\pi_f^{(z)}(u) = \frac{\int_{|v|>z} \int_{R^{d-m-k}} f(\sqrt{|w|^2 + |v|^2 + u^T V_{aa}^{-1} u}) dv dw}{\int_{R^{d-k}} f(\sqrt{|w|^2 + u^T V_{aa}^{-1} u}) dw} \tag{18.11}$$

and

$$E\pi_f^{(z)}(x_a) = \int_{|v|>z} \int_{R^{d-m}} f(\sqrt{|w|^2 + |v|^2}) dv dw. \tag{18.12}$$

18.2.2 Extreme Problems for the Variation Distance

Let $\mathcal{B}^d$ be the σ-algebra of the Borel subsets of R^d. Denote by P_f the measure on $\mathcal{B}^d$ generated by $p_{f,V}$. Consider the functional

$$l(f) = \sup_{A \in \mathcal{B}^d} |P_f(A) - P_{f_0}(A)|.$$

It is clear that $l(f)$ characterizes the stability of all the decision rules based on the assumption of normality.

The question arises: what can we say about $\sup_{f \in \mathcal{F}} l(f)$? If not to narrow $\mathcal{F}$ then the answer is rather trivial.

EXTREME VALUES OF STABILITY

Proposition 1. *If $\mathcal{F}$ is defined as (18.2) then*

$$\sup_{f \in \mathcal{F}} l(f) = 1.$$

Proof. It is convenient to represent $l(f)$ as follows:

$$l(f) = \frac{1}{2} \int_0^\infty |p_f(t) - p_{f_0}(t)| dt$$

where (see (18.8))

$$p_f(t) = c_d t^{d-1} f(t).$$

Consider the family of functions

$$f_{\rho r}(t) = \begin{cases} \lambda, & \text{if } 0 \leq t < \rho \\ \mu, & \text{if } \rho \leq t < r \\ 0, & \text{if } t > r \end{cases} \qquad (18.13)$$

with $0 \leq \mu \leq \lambda$, $0 \leq \rho \leq r$. If $f_{\rho r} \in \mathcal{F}$ then

$$\lambda \rho^d + \mu(r^d - \rho^d) = \frac{d}{c_d}$$

and

$$\lambda = \rho^{d+2} + \mu(r^{d+2} - \rho^{d+2}) = \frac{d+2}{c_d}$$

whence

$$\lambda = \frac{d}{c_d \rho^d r^d} \frac{r^d(r^2 - d - 2) - \rho^d(\rho^2 - d - 2)}{r^2 - \rho^2}$$

and

$$\mu = \frac{d}{c_d \rho^d r^d} \frac{\rho^d(d + 2 - \rho^2)}{r^2 - \rho^2}.$$

Since $\mu \geq 0$ we have $\rho \leq \sqrt{d+2}$ and, therefore, $r \geq \sqrt{d+2}$. Let us fix $\rho \in (0, \sqrt{d+2})$ and let $r \to \infty$. Then

$$\lambda \to \frac{d}{c_d \rho^d}, \quad \mu = O(r^{-d-2}). \qquad (18.14)$$

Further,

$$l(f_{\rho r}) \to \frac{1}{2} \int_0^\rho |p_{f_{\rho\infty}}(t) - p_{f_0}(t)| dt + \frac{1}{2} \int_\rho^\infty p_{f_0}(t) dt = 1 - \int_0^\rho p_{f_0}(t) dt.$$

It remains to recall that ρ can be arbitrarily small. The proposition is proved.

The class $\mathcal{F}$ proved to be too wide. It contains the functions leading to the distributions abnormally concentrated near the origin. Let us narrow $\mathcal{F}$ as follows:

$$\mathcal{F} = \mathcal{F}_K = (f : f \; non-negative, \; non-increasing \; in \; (0, \infty),$$

$$f(0) \leq K, \; \int f(|x|) dx = \int x_1^2 f(|x|) dx = 1). \qquad (18.15)$$

Proposition 2. *If $\mathcal{F}_K$ is defined as (18.15) then*
$$\sup_{f \in \mathcal{F}} l(f) < 1.$$

Proof. For $f \in \mathcal{F}_K$ we have
$$p_f(t) \leq K_d t^{d-1}, \quad K_d = Kc_d.$$

Since $f(t)$ is non-increasing
$$1 \geq c_d \int_0^t s^{d-1} f(s) ds \geq c_d d^{-1} f(t) t^d$$

and
$$d \geq c_d \int_0^t s^{d+1} f(s) ds \geq c_d(d+2)^{-1} f(t) t^{d+2}.$$

Therefore,
$$p_f(t) \leq g(t) = \min(K_d t^{d-1}, \ dt^{-1}, \ d(d+2)t^{-3}).$$

Note that
$$\max g(t) = d^{1-1/d} K_d^{1/d}$$

and
$$t_0 = \arg \max g(t) = (d/K_d)^{1/d}.$$

Moreover,
$$\int_0^{t_0} g(t) dt = 1$$

whence
$$\int_0^\infty g(t) dt > 1.$$

Let Q and Q_0 be the measures on $\mathcal{B}^1$ generated by p_f and p_{f_0}, respectively. It is evident that
$$\sup_{A \in \mathcal{B}^1} |Q(A) - Q_0(A)| = \sup_{A \in (A: A \in \mathcal{B}^1, \ Q(A) > Q_0(A))} (Q(A) - Q_0(A)).$$

If $Q(A) > Q_0(A)$ and $\int_A g(t) dt > 1$ then
$$Q(A) - Q_0(A) \leq 1 - Q_0(A')$$

where $A' \subset A$ and $\int_{A'} g(t) dt = 1$. But if $Q(A) > Q_0(A)$ and $\int_A g(t) dt < 1$ then
$$Q(A) - Q_0(A) \leq \int_A (g(t) - p_{f_0}(t)) dt <$$
$$1 - \int_{A''} p_{f_0}(t) dt = 1 - Q_0(A'')$$

where $A \subset A''$ and $\int_{A''} g(t)dt = 1$. Thus,

$$\sup_{A \in \mathcal{B}^1} |Q(A) - Q_0(A)| \leq 1 - \inf_{A \in \mathcal{A}} Q_0(A)$$

where

$$\mathcal{A} = \left(A : \int_A g(t) dt = 1 \right).$$

It remains to prove that

$$\inf_{A \in \mathcal{A}} Q_0(A) > 0.$$

Assume that

$$\inf_{A \in \mathcal{A}} Q_0(A) = 0.$$

Then for any $\epsilon > 0$ there exists A' such that

$$\int_{A'} g(t) dt = 1 \text{ and } Q_0(A') < \epsilon.$$

Then

$$mes(A' \cap (T^{-1}, T)) \leq \epsilon / \min(p_{f_0}(T^{-1}), p_{f_0}(T)).$$

Let us choose T so that

$$\int_0^{T^{-1}} g(t) dt + \int_T^\infty g(t) dt \leq 1/4.$$

Then

$$1 = \int_0^\infty g(t) dt \leq \frac{\epsilon \max_t g(t)}{\min(p_{f_0}(T^{-1}), p_{f_0}(T))} + 1/4.$$

Since ϵ is arbitrary we come to contradiction. The proposition is proved. Thus, within $\mathcal{F}_K$ the problem becomes non-trivial. Of course, one can suggest various restrictions to be imposed on f. For example, $\mathcal{F}$ can be taken as $\mathcal{F}_K \cap \mathcal{F}^*$ where $\mathcal{F}^*$ is a confidence region for an unknown $f^* \in \mathcal{F}_K$ based on a non-parametric estimator of f^*.

18.2.3 On the Stability of Predictions

Consider the following functionals which characterize the accuracy of prediction:

$$l_1(f) = \max_t \frac{\int_0^\infty s^{d-k+1} f(\sqrt{s^2 + t^2}) ds}{\int_0^\infty s^{d-k-1} f(\sqrt{s^2 + t^2}) ds},$$

$$l_2(f) = \max_t \frac{\int_{q>z} \int_0^\infty q^{m-1} s^{d-m-k-1} f(\sqrt{s^2 + q^2 + t^2}) dq ds}{\int_0^\infty s^{d-k-1} f(\sqrt{s^2 + t^2}) ds}$$

and

$$l_3(f) = \int_{q>z} \int_0^\infty q^{m-1} s^{d-m-1} f(\sqrt{s^2 + q^2}) dq ds.$$

It is evident that these functionals, up to constant factors, coincide with $\max_{u \in R^k} \sigma_f(u), \max_{u \in R^k} \pi_f^{(z)}(u)$ and $E\pi_f^{(z)}(x_a)$ (see (18.10)–(18.12)).

Proposition 3. *If $k \geq 2$ then*
$$\inf_{f \in \mathcal{F}_K} l_1(f) > 0.$$

Proof. It is obvious that
$$l_1(f) \geq \frac{c_d \int_0^\infty s^{d-k+1} f(s) ds}{\int_0^\infty s^{-k} g(s) ds}$$

where $g(\cdot)$ is as in the proof of Proposition 2.

Further,
$$\int_0^\infty s^{d-k+1} f(s) ds \geq \int_0^1 s^{d-1} f(s) ds$$

while
$$c_d \int_S^\infty s^{d-1} f(s) ds \leq dS^{-2}.$$

Since $f(s)$ is non-increasing
$$c_d \int_0^1 s^{d-1} f(s) ds \geq S^{-1}(1 - dS^{-2}) > 0$$

provided $S > d^{1/2}$. The proposition is proved.

Proposition 4. *If $z > \sqrt{d+2}$ and $K > dc_d^{-1}(d+2)^{-d/2}$ then there exists $f \in \mathcal{F}_K$ such that $l_i(f) = 0$, $i = 2, 3$.*

If $t_0 \leq z < \sqrt{d+2}$ and $K > dc_d^{-1}(d+2)^{-d/2}$ then
$$\inf_{f \in \mathcal{F}_K} l_i(f) = 0, \quad i = 2, 3.$$

But if $z < t_0$ then
$$\inf_{f \in \mathcal{F}_K} l_i(f) > 0, \quad i = 2, 3.$$

Here t_0 is as in the proof of Proposition 2.

Proof. It is sufficient to deal only with $i = 2$. Let $z > \sqrt{d+2}$. Set in (18.13) $\rho = 0$, $\mu = \sqrt{d+2}$. It is evident that $f_{0,\sqrt{d+2}} \in \mathcal{F}_K$ and $l_2(f_{0,\sqrt{d+2}}) = 0$.

If $z \in [t_0, \sqrt{d+2})$ we first fix $\rho \in (z, \sqrt{d+2})$ and then let $r \to \infty$. In view of (18.14) we obtain

$$\int_{q>z} \int_0^\infty q^{m-1} s^{d-m-k-1} f_{\rho r}(\sqrt{s^2 + q^2}) dq ds \to$$

$$dc_d^{-1} \rho^{-d} \int_{q>z} \int_{s^2+q^2 < \rho^2} q^{m-1} s^{d-m-k-1} dq ds = 0,$$

that is,
$$l_2(f_{\rho r}) \to 0.$$

If $z < t_0$ then
$$l_2(f) \geq \frac{c_d}{\int_0^\infty s^{-k} g(s) ds} \int_{q>z} \int_0^\infty q^{m-1} s^{d-m-k-1} f(\sqrt{s^2+q^2}) dq ds.$$

Let $t_1 \in (z, t_0)$. It is easy to see that
$$\int_{q>z} \int_0^\infty q^{m-1} s^{d-m-k-1} f(\sqrt{s^2+q^2}) dq ds =$$
$$\int_0^{\pi/2} (\cos\phi)^{m-1} (\sin\phi)^{d-m-k-1} d\phi \int_{z/\cos\phi}^\infty s^{d-k-1} f(s) ds \geq$$
$$\int_0^{\phi_1} (\cos\phi)^{m-1} (\sin\phi)^{d-m-k-1} d\phi \int_{t_1}^\infty s^{d-k-1} f(s) ds$$

where $\cos\phi_1 = z/t_1$. Further,
$$\int_{t_1}^\infty s^{d-k-1} f(s) ds \geq t^{-k} \int_{t_1}^t s^{d-1} f(s) ds.$$

It is obvious that
$$\int_{t_1}^t s^{d-1} f(s) ds = c_d^{-1} - \int_0^{t_1} s^{d-1} f(s) ds - \int_t^\infty s^{d-1} f(s) ds$$

or
$$\int_{t_1}^t s^{d-1} f(s) ds \geq 2^{-1} c_d^{-1} (1 - d^{-1} K_d t_1^d) > 0$$

provided t is sufficiently large.

Thus,
$$l_2(f) \geq c > 0$$

for all $f \in \mathcal{F}_K$. The proposition is proved.

From Propositions 3 and 4 it follows that evaluation of $\inf_{f \in \mathcal{F}_K} l_i(f)$, $i = 1, 2, 3$, is not trivial.

It should be noted also that
$$l_2(f) \leq \min(1, z^{-2}).$$

It implies that it is also worth trying to evaluate $\sup_{f \in \mathcal{F}_K} l_i(f)$, $i = 2, 3$.

18.3 Extreme Problems for Quantiles

Let $x = (x_1, \ldots, x_n)$ be a sample drawn from an unknown distribution. Suppose that a goodness-of-fit test does not reject the hypothesis of normality.

It means that the true distribution is either normal or, what is much more plausible, rather close to a normal one.

If we test normality with the help of the sample skewness and kurtosis coefficients it is natural to assume that the underlying distribution belongs to the semiparametric class

$$F(\frac{u-a}{b}), \quad -\infty < a < \infty, \ b > 0, \ F \in \mathcal{F} \qquad (18.16)$$

where

$$\mathcal{F} = (F : F \text{ is absolutely continuous, symmetrical, unimodal,}$$
$$\int u^2 dF(u) = 1, \ t_- \leq \int u^4 dF(u) \leq t_+, \ t_- \leq 3 \leq t_+).$$

If we use Shapiro-Wilk's test, then it is reasonable to assume that

$$\mathcal{F} = (F : F \text{ is absolutely continuous, symmetrical, unimodal,}$$
$$\int u^2 dF(u) = 1, \ t_- \leq \sum_{j=1}^{n} c_j E_F x_{(j)} \leq t_+, \ t_- \leq 1 \leq t_+).$$

where $x_{(j)}$, $j = 1, \ldots, n$, are the order statistics corresponding to x, c_j are the coefficients of the best, under normality, unbiased linear estimator of b (see [3]) and E_F corresponds to "parameter" $(0, 1, F)$ in (18.16).

Consider a functional $l(F)$, $F \in \mathcal{F}$. The problem is to find the extreme values of $l(F)$ when f runs $\mathcal{F}$. In what follows $l(F)$ is a quantile of F.

18.3.1 Lower Extreme Values for Quantiles of the Greater Levels

If $f \in \mathcal{F}$ then the equation

$$p = F(z)$$

has the unique root $z_p = z_p(F)$ for all $p \in (0, 1)$. First, we consider the case in which $9/(50t_+) \leq p \leq 1/2$. Here we come to exceedingly simple formulæ.

Theorem 18.1 *Let $p \in [9/(50t_+), 1/2)$. Then*

$$-\inf_{F \in \mathcal{F}} z_p = \begin{cases} 3^{-5/2}(5t_+)^{3/2}(9/(5t_+) - 2p), & \text{if } 9/(50t_+) \leq p \leq 3/(10t_+) \\ (2^{1/2}/3)p^{-1/2}, & \text{if } 3/(10t_+) \leq p \leq 1/6 \\ 3^{1/2}(1 - 2p), & \text{if } 1/6 \leq p \leq 1/2. \end{cases}$$

Furthermore, the function $F_0(u)$ which provides $\inf_{F \in \mathcal{F}} z_p$ has one of the following representations depending on the value of p:

$$F_0(u) = \begin{cases} (1 - 9/(5t_+))E(u) + (9/(5t_+))R(\frac{u}{\sigma}), & \text{if } 9/(50t_+) \leq p \leq 3/(10t_+) \\ (1 - 6p)E(u) + 6pR\left(\frac{u}{\sqrt{2/p}}\right), & \text{if } 3/(10t_+) \leq p \leq 1/6 \\ R(\frac{u}{2\sqrt{3}}), & \text{if } 1/6 \leq p \leq 1/2. \end{cases}$$

Here
$$\sigma = 2\sqrt{5t_+/3},$$

$$E(u) = \begin{cases} 0, & \text{if } u \leq 0 \\ 1, & \text{if } u > 0 \end{cases}$$

while
$$R(u) = \begin{cases} 0, & \text{if } u \leq -1/2 \\ u + 1/2, & \text{if } |u| < 1/2 \\ 1, & \text{if } u \geq 1/2. \end{cases}$$

It should be noted that the extremal distributions do not belong to $\mathcal{F}$. They are limiting points of the set. It is of interest that for $p \geq 3/(10t_+)$ the extreme value does not depend upon t_+ (cf Theorem 1 in [2]).

18.3.2 Lower Extreme Values for Quantiles of the Smallest Levels

The case $p \in (0, 3/50]$ leads to more complicated formulae.

First, consider the system of the linear equations

$$\begin{cases} k^2\alpha^3(r-k) + (r - (r-k)\alpha)^3 = 3r \\ k^4\alpha^5(r-k) + (r - (r-k)\alpha)^5 = 5t \end{cases} \quad (18.17)$$

where $t > 9/5$. Repeating the arguments utilized in Nagaev [2] (see his Lemmas 1-4) we can verify that for $9/(5t) \leq \alpha < 1$, $0 \leq k \leq r$ the system has the unique solution $(k(\alpha), r(\alpha))$. In particular,

$$k(9/(5t)) = 0, \ r(9/(5t)) = 3^{-5/2}(5t)^{3/2},$$

$$k(1) = 3^{1/2}, \ r(\alpha) \sim (5t-9)^{1/4}(1-\alpha)^{-5/4}$$

as $\alpha \to 1$. Let us set

$$w(\alpha) = (r^+(\alpha)/k^+(\alpha))(1/\alpha - 1)$$

where superscript '+' indicates that the system (18.17) are solved with $t = t_+$.

For $0 < p \leq 9/(50t_+)$ the equation

$$1 - 2p = \alpha + (1-\alpha)\frac{4w^2(\alpha) + 7w(\alpha) + 2}{5w^2(\alpha) + 10w(\alpha) + 4}$$

has the unique root α_p (see Lemma 10 *ibidem*).

Theorem 18.2 *If $p \in (0, 9/(50t_+))$ then*

$$-\inf_{F \in \mathcal{F}} z_p = k^+(\alpha_p)\alpha_p + r^+(\alpha_p)(1 - 2p - \alpha_p)$$

Table 18.1: The range of the absolute value of p-quantile for small p

p	min	norm	max
.060	1.293148	1.555	1.788854
.055	1.319541	1.599	1.827421
.050	1.346336	1.645	1.868981
.045	1.373615	1.696	1.914291
.040	1.401484	1.751	1.964409
.035	1.430082	1.812	2.020876
.030	1.459604	1.881	2.086044
.025	1.490326	1.960	2.163738
.020	1.522678	2.054	2.260729
.015	1.557378	2.171	2.390592
.010	1.595833	2.327	2.586564
.005	1.641665	2.576	2.969275

while (cf. Theorem 18.1)

$$F_0'(u) = \begin{cases} \frac{1}{2k^+(\alpha_p)}, & if \ |u| \leq k^+(\alpha_p)\alpha_p \\ \frac{1}{2r^+(\alpha_p)}, & if \ k^+(\alpha_p)\alpha_p \leq |u| \leq k^+(\alpha_p)\alpha_p + r^+(\alpha_p)(1-\alpha_p) \\ 0 & otherwise. \end{cases}$$

It is worth emphasizing that for small p the left extreme value of z_p depends upon the right extreme value of the kurtosis. The extreme values for a number of quantiles are given in Table 18.1 for $t = 3$.

18.3.3 Upper Extreme Values

Let us turn to $\sup_{F \in \mathcal{F}} z_p$. Let $k^-(\alpha)$ and $r^-(\alpha)$ correspond to $t = t_-$ in (18.17). We remind that $t > 9/5$.

Theorem 18.3 *Let* $p \in (0, 1/2)$. *Then*

$$-\sup_{F \in \mathcal{F}} z_p = \begin{cases} (1-2p)k^-(1-2p), & if \ 0 < p \leq 9/(10t_-) \\ 0, & if \ 9/(10t_-) \leq p \leq 1/2. \end{cases}$$

Furthermore, for $p \leq 9/(10t_-)$ *(cf. Theorems 18.1 and 18.2)*

$$F_0'(u) = \begin{cases} \frac{1}{2k^-(1-2p)}, & if \ |u| \leq (1-2p)k^- \\ \frac{1}{2r^-(1-2p)}, & if \ (1-2p)k^- \leq |u| \leq (1-2p)k^- + 2pr^- \\ 0 & otherwise, \end{cases}$$

while for $9/(10t_-) \leq p < 1/2$ (see Theorem 18.1)

$$F_0(u) = (1 - 9/(5t_+))E(u) + (9/(5t_+))R\left(\frac{u}{\sigma}\right).$$

Here, for brevity, $k^- = k^-(1-2p)$, $r^- = r^-(1-2p)$.

Thus, to evaluate the right extreme value for $p < 9/(10t_-)$ we have to solve (18.17) with $t = t_-$. Table 18.1 contains the results of computation for small p and $t = 3$. The column 'norm' contains the absolute values of the quantiles of the standard normal distribution.

It is worth noting that, as $p \to 0$, we have $-x_p \sim cp^{-1/4}$ with $c = 8 \cdot 6^{1/4}/5$.

References

[1] Buntine W.L. (1994) Operations for learning with graphical models. *Artificial Intelligence Research* 2: 159–225.

[2] Nagaev A.V. (1994) Extreme problems and stability of statistical decisions. *Appl. Math. Statist*, (1). Russian Surveys in Applied and Industrial Mathematics, TPA, Moscow (in Russian).

[3] Sarhan A.E. and Greenberg B.G. (1962) *Contribution to Order Statistics.* John Wiley, New York.

[4] Whittaker J. (1990) *Graphical Models in Applied Multivariate Statistics.* John Wiley, New York.

Index

abduction, 77
action, 125
action calculus, 132
action/observation exchange, 133
active set, 147
agrees, 138
ALARM network, 222
Alarm problem, 155
alternating-position crossover operator, 215
analogical reasoning, 81
Annealed VC-entropy, 10
approximate algorithms, 217
automated knowledge acquistion, 145
axiom, 132
axioms, 259

back-door criterion, 142
back-propagation method, 22
basic variables, 147
Bayesian, 50
Bayesian belief network, 170–172
Bayesian Confidence Propagation Neural Network, 202
Bayesian Networks, 211
BBN, 173, 176, 183
big121 problem, 158

c-k-influence diagram, 244
calculus of
cancellation set, 151
cancellation variable, 151
CATCHEM project, 170
causal effects, 128
causal model, 146
causal models, 145

causal relationships, 237
causal structure, 201
causal theory, 128
child node, 172
classification, 199, 205
classification error, 5
common knowledge, 244
complete calculus, 134
complex columns, 203
composition, 133
comprehensive causal theory, 129
comprehensive structural model, 128
compression, 69
computational learning, xiii
computing as compression, 67, 69
conditional independence, 238
conditioning, 133
context-free phrase-structure grammars, 71
correlation, 147
crossover operator, 213
cycle crossover operator, 214

d-separation theorem, 240
data mining, 70
decision strategy, 130
decision variable, 128
dependency graph, 201
description length, 152
desirable variables, 128
diphtheria in children, 191–194
Directed Acyclic Graph, 211
directed model, 146
directed models, 154
discounting of future events, 191
discrepancy, 4

displacement mutation operator, 216
distributional linguistics, 72
disturbed marginal, 148
dynamic programming, 259

empirical risk, 6
Empirical Risk Minimization induction principle, 6
entropy of the set of functions, 8
equilibral rationality, 248
estimation, 46, 53
exact algorithms, 217
exchange mutation operator, 216
execution of functions, 82
expert systems, 145
explanation, 48

fault diagnosis, 205
Fisher information, 50
formulas, 132
Fourier transform, 146
fusions of BNs, 225

G-decomposable, 138
generalization, 70, 80
generalized process, 186, 187
genetic algorithms, 212
geometric analogies, 81
geometric analogy problem, 69
grammar, 45
graphical models, 146
graphical parsimony, 246
Growth function, 10

Hadamard transform, 146
hierarchical model, 148
hierarchical sequence of graphs (HSG), 188
higher order network, 199

identifiability of causal effects, 130
implication, 75
improved reverse algorithm, 151, 152
independence properties, 148
indicator functions, 4

inductive inference, xiii, 34
inductive logic programming (ILP), 67
inductive principles, xiii
inductive reasoning, 78, 81
inference rules, 132
inference trees, 134
influence diagram, 238
information, 34
information compression, 67
information retrieval, 82
information theory, 69
insertion mutation operator, 216
insertion/deletion of actions, 133
insertion/deletion of observations, 133
interaction, 147
Interaction Model, 146
interaction model, 147
interaction structure, 145
interaction variables, 147
inversion mutation operator, 216
Iterative Proportional Scaling algorithm, 149

K2 algorithm, 223
knowledge acquistion, 145
Kolmogorov complexity, 34

language learning, 70
latent variable, 128
leading graph, 189
learning, 67, 70–74, 82
local process, 187, 188, 191, 193
log-linear models, 146
loss, 4

marginal, 147
marginalization, 133
Markov random fields, 146
martingale, 143
material implication, 75
maximum likelihood (ML) estimate, 149
maximum likelihood estimation, 146

MDL principle, 152
medical ethics, 127
minimizing the risk functional, 6
Minimum Description Length, 35
Minimum Length Encoding, 70
Mitten's framework, 270
MLE, 70, 81
model complexity, 40
model of learning from examples, 3
model selection, 152
modus ponens, 75, 76
modus tollens, 76
multi-armed bandit problem, 127
multidimensional binary sample, 145
multiple alignment, 67–78, 80, 81
mutation operator, 213
mutual information, 204

negation, 75, 76
negation as failure, 76
non-serial dynamic programming, 260

optimal vertex elimination sequence, 217
order crossover operator, 214
order-based crossover operator, 215
Orthogonal Interaction Model, 146
over fitting, 26

P-true, 132
parameter coding, 50
parameter estimation, 149
parent node, 172
parsing, 82
partially-mapped crossover operator, 213
pattern, 68
pattern matching, 67, 69
pattern recognition, 82
planning, 82
position-based crossover operator, 215
posterior, 59
posterior probability, 51

potencials, 146
predictive code length, 39
prequential, 143
prequential principle, 39
PRESS, 171, 173–177, 180, 183
prior probability, 49
probabilistic expert systems, 145
probabilistic reasoning, xiii
probabilities, 69
probability, 74, 80
probability estimation, 199
problem solving, 82
Prolog, 76
propagation of evidence, 216

qualitative rationality, 239
qualitative statements, 237
quantification, 76

random variables, 128
rational equilibrium reduction, 249
reasoning, 67, 74–78, 80, 81
recognition, 74
regression function, 5
reverse algorithm, 150, 151
risk functional, 4
rule induction, 70
rule of cut, 137

S-true, 131
scale of patient state severity, 187
scramble mutation operator, 216
search, 67, 69–71, 74, 81
Shannon's theory, 48
simple-inversion mutation operator, 216
sound calculus, 133
SP, 67
specific case analysis, 169
statistical inference, 46
stochastic complexity, 35
stochastic mechanisms, 125
strong law of large numbers, 137
structural linguistics, 72
structural model, 128

structure, 146
structure learning, 219
sufficiency theorem, 244

telephone exchange computers, 205
theorems, 133
theory selection, 53
time structure, 188, 191
Travelling Salesman Problem, 212
triangulation, 217

unification, 67, 69–71, 73, 75, 81
universal model, 36
universal quantifier, 76

valid (chance) influence diagram, 240
valuation network, 259
variable, 76
VC-dimension of a set of indicator functions, 12
VC-dimension of the set of real functions, 13
VC-entropy, 9
voting recombination crossover operator, 215

Walsh functions, 146